25.326

MONOGRAPHS ON
STATISTICS AND APPLIED PROBABILITY

General Editors

**D.R. Cox, D.V. Hinkley, N. Keiding, N. Reid,
D.B. Rubin and B.W. Silverman**

1 Stochastic Population Models in Ecology and Epidemiology
M.S. Bartlett (1960)
2 Queues *D.R. Cox and W.L. Smith* (1961)
3 Monte Carlo Methods *J.M. Hammersley and D.C. Handscomb* (1964)
4 The Statistical Analysis of Series of Events *D.R. Cox and
P.A.W. Lewis* (1966)
5 Population Genetics *W.J. Ewens* (1969)
6 Probability, Statistics and Time *M.S. Bartlett* (1975)
7 Statistical Inference *S.D. Silvey* (1975)
8 The Analysis of Contingency Tables *B.S. Everitt* (1977)
9 Multivariate Analysis in Behavioural Research *A.E. Maxwell* (1977)
10 Stochastic Abundance Models *S. Engen* (1978)
11 Some Basic Theory for Statistical Inference *E.J.G. Pitman* (1979)
12 Point Processes *D.R. Cox and V. Isham* (1980)
13 Identification of Outliers *D.M. Hawkins* (1980)
14 Optimal Design *S.D. Silvey* (1980)
15 Finite Mixture Distributions *B.S. Everitt and D.J. Hand* (1981)
16 Classification *A.D. Gordon* (1981)
17 Distribution-free Statistical Methods, 2nd edition *J.S. Maritz* (1995)
18 Residuals and Influence in Regression *R.D. Cook
and S. Weisberg* (1982)
19 Applications of Queueing Theory, 2nd edition *G.F. Newell* (1982)
20 Risk Theory, 3rd edition *R.E. Beard, T. Pentikainen and
E. Pesonen* (1984)
21 Analysis of Survival Data *D.R. Cox and D. Oakes* (1984)
22 An Introduction to Latent Variable Models *B.S. Everitt* (1984)
23 Bandit Problems *D.A. Berry and B. Fristedt* (1985)
24 Stochastic Modelling and Control *M.H.A. Davis and R. Vinter* (1985)
25 The Statistical Analysis of Compositional Data *J. Aitchison* (1986)
26 Density Estimation for Statistics and Data Analysis
B.W. Silverman (1986)

27 Regression Analysis with Applications *G.B. Wetherill* (1986)
28 Sequential Methods in Statistics, 3rd edition
   G.B. Wetherill and K.D. Glazebrook (1986)
29 Tensor Methods in Statistics *P. McCullagh* (1987)
30 Transformation and Weighting in Regression *R.J. Carroll and
   D. Ruppert* (1988)
31 Asymptotic Techniques for Use in Statistics *O.E. Barndorff-Nielsen
   and D.R. Cox* (1989)
32 Analysis of Binary Data, 2nd edition *D.R. Cox and E.J. Snell* (1989)
33 Analysis of Infectious Disease Data *N.G. Becker* (1989)
34 Design and Analysis of Cross-Over Trials *B. Jones and
   M.G. Kenward* (1989)
35 Empirical Bayes Methods, 2nd edition *J.S. Maritz and T. Lwin* (1989)
36 Symmetric Multivariate and Related Distributions *K.-T. Fang
   S. Kotz and K.W. Ng* (1990)
37 Generalized Linear Models, 2nd edition *P. McCullagh and
   J.A. Nelder* (1989)
38 Cyclic and Computer Generated Designs, 2nd edition
   J.A. John and E.R. Williams (1995)
39 Analog Estimation Methods in Econometrics *C.F. Manski* (1988)
40 Subset Selection in Regression *A.J. Miller* (1990)
41 Analysis of Repeated Measures *M.J. Crowder and D.J. Hand* (1990)
42 Statistical Reasoning with Imprecise Probabilities *P. Walley* (1991)
43 Generalized Additive Models *T.J. Hastie and R.J. Tibshirani* (1990)
44 Inspection Errors for Attributes in Quality Control
   N.L. Johnson, S. Kotz and X. Wu (1991)
45 The Analysis of Contingency Tables, 2nd edition *B.S. Everitt* (1992)
46 The Analysis of Quantal Response Data *B.J.T. Morgan* (1993)
47 Longitudinal Data with Serial Correlation: A State-space Approach
   R.H. Jones (1993)
48 Differential Geometry and Statistics *M.K. Murray
   and J.W. Rice* (1993)
49 Markov Models and Optimization *M.H.A. Davis* (1993)
50 Networks and Chaos – Statistical and Probabilistic Aspects
   *O.E. Barndorff-Nielsen,
   J.L. Jensen and W.S. Kendall* (1993)
51 Number-theoretic Methods in Statistics *K.-T. Fang
   and Y. Wang* (1994)
52 Inference and Asymptotics *O.E. Barndorff-Nielsen
   and D.R. Cox* (1994)

53 Practical Risk Theory for Actuaries *C.D. Daykin, T. Pentikäinen and M. Pesonen* (1994)
54 Biplots *J.C. Gower and D.J. Hand* (1996)
55 Predictive Inference: An Introduction *S. Geisser* (1993)
56 Model-Free Curve Estimation *M.E. Tarter and M.D. Lock* (1993)
57 An Introduction to the Bootstrap *B. Efron and R.J. Tibshirani* (1993)
58 Nonparametric Regression and Generalized Linear Models *P.J. Green and B.W. Silverman* (1994)
59 Multidimensional Scaling *T.F. Cox and M.A.A. Cox* (1994)
60 Kernel Smoothing *M.P. Wand and M.C. Jones* (1995)
61 Statistics for Long Memory Processes *J. Beran* (1995)
62 Nonlinear Models for Repeated Measurement Data *M. Davidian and D.M. Giltinan* (1995)
63 Measurement Error in Nonlinear Models *R.J. Carroll, D. Ruppert and L.A. Stefanski* (1995)
64 Analyzing and Modeling Rank Data *J.I. Marden* (1995)
65 Time Series Models – In econometrics, finance and other fields *D.R Cox, D.V. Hinkley and O.E. Barndorff-Nielsen* (1996)
66 Local Polynomial Modelling and Its Applications *J. Fan and I. Gijbels* (1996)

(Full details concerning this series are available from the Publishers).

To our teachers

To Angel, Mary and Yonghua

To Ludo

Local Polynomial Modelling and Its Applications

J. Fan

*Department of Statistics
University of North Carolina
Chapel Hill, USA*

and

I. Gijbels

*Institute of Statistics
Catholic University of Louvain
Louvain-la-Neuve, Belgium*

CHAPMAN & HALL

London · Glasgow · Weinheim · New York · Tokyo · Melbourne · Madras

Published by Chapman & Hall, 2-6 Boundary Row, London SE1 8HN, UK

Chapman & Hall, 2-6 Boundary Row, London SE1 8HN, UK

Blackie Academic & Professional, Wester Cleddens Road, Bishopbriggs, Glasgow G64 2NZ, UK

Chapman & Hall GmbH, Pappelallee 3, 69469 Weinheim, Germany

Chapman & Hall USA., 115 Fifth Avenue, New York, NY 10003, USA

Chapman & Hall Japan, ITP-Japan, Kyowa Building, 3F, 2-2-1 Hirakawacho, Chiyoda-ku, Tokyo 102, Japan

Chapman & Hall Australia, 102 Dodds Street, South Melbourne, Victoria 3205, Australia

Chapman & Hall India, R. Seshadri, 32 Second Main Road, CIT East, Madras 600 035, India

First edition 1996

© 1996 Chapman & Hall

Printed in Great Britain by St Edmundsbury Press, Bury St Edmunds, Suffolk.

ISBN 0 412 98321 4

Apart from any fair dealing for the purposes of research or private study, or criticism or review, as permitted under the UK Copyright Designs and Patents Act, 1988, this publication may not be reproduced, stored, or transmitted, in any form or by any means, without the prior permission in writing of the publishers, or in the case of reprographic reproduction only in accordance with the terms of the licences issued by the Copyright Licensing Agency in the UK, or in accordance with the terms of licences issued by the appropriate Reproduction Rights Organization outside the UK. Enquiries concerning reproduction outside the terms stated here should be sent to the publishers at the London address printed on this page.

The publisher makes no representation, express or implied, with regard to the accuracy of the information contained in this book and cannot accept any legal responsibility or liability for any errors or omissions that may be made.

A Catalogue record for this book is available from the British Library

∞ Printed on permanent acid-free text paper, manufactured in accordance with ANSI/NISO Z39.48 - 1992 and ANSI/NISO Z39.48 - 1984 (Permanence of Paper).

Contents

Preface   xiii

1 Introduction   1
    1.1 From linear regression to nonlinear regression   1
    1.2 Local modelling   4
    1.3 Bandwidth selection and model complexity   7
    1.4 Scope of the book   9
    1.5 Implementation of nonparametric techniques   11
    1.6 Further reading   12

2 Overview of existing methods   13
    2.1 Introduction   13
    2.2 Kernel estimators   14
        2.2.1 Nadaraya-Watson estimator   14
        2.2.2 Gasser-Müller estimator   15
        2.2.3 Limitations of a local constant fit   17
    2.3 Local polynomial fitting and derivative estimation   18
        2.3.1 Local polynomial fitting   19
        2.3.2 Derivative estimation   22
    2.4 Locally weighted scatter plot smoothing   22
        2.4.1 Robust locally weighted regression   24
        2.4.2 An example   26
    2.5 Wavelet thresholding   27
        2.5.1 Orthogonal series based methods   28
        2.5.2 Basic ingredient of multiresolution analysis   31
        2.5.3 Wavelet shrinkage estimator   34
        2.5.4 Discrete wavelet transform   35
    2.6 Spline smoothing   39
        2.6.1 Polynomial spline   40
        2.6.2 Smoothing spline   43

	2.7	Density estimation	46
		2.7.1 Kernel density estimation	46
		2.7.2 Regression view of density estimation	50
		2.7.3 Wavelet estimators	52
		2.7.4 Logspline method	54
	2.8	Bibliographic notes	55
3	**Framework for local polynomial regression**		**57**
	3.1	Introduction	57
	3.2	Advantages of local polynomial fitting	60
		3.2.1 Bias and variance	61
		3.2.2 Equivalent kernels	63
		3.2.3 Ideal choice of bandwidth	66
		3.2.4 Design adaptation property	68
		3.2.5 Automatic boundary carpentry	69
		3.2.6 Universal optimal weighting scheme	74
	3.3	Which order of polynomial fit to use?	76
		3.3.1 Increases of variability	77
		3.3.2 It is an odd world	79
		3.3.3 Variable order approximation	80
	3.4	Best linear smoothers	84
		3.4.1 Best linear smoother at interior: optimal rates and constants	84
		3.4.2 Best linear smoother at boundary	89
	3.5	Minimax efficiency of local polynomial fitting	91
		3.5.1 Modulus of continuity	92
		3.5.2 Best rates and nearly best constant	94
	3.6	Fast computing algorithms	94
		3.6.1 Binning implementation	96
		3.6.2 Updating algorithm	99
	3.7	Complements	100
	3.8	Bibliographic notes	105
4	**Automatic determination of model complexity**		**109**
	4.1	Introduction	109
	4.2	Rule of thumb for bandwidth selection	110
	4.3	Estimated bias and variance	113
	4.4	Confidence intervals	116
	4.5	Residual squares criterion	118
		4.5.1 Residual squares criterion	118
		4.5.2 Constant bandwidth selection	119
		4.5.3 Variable bandwidth selection	122

		4.5.4 Computation and related issues	122
	4.6	Refined bandwidth selection	123
		4.6.1 Improving rates of convergence	123
		4.6.2 Constant bandwidth selection	123
		4.6.3 Variable bandwidth selection	124
	4.7	Variable bandwidth and spatial adaptation	128
		4.7.1 Qualification of spatial adaptation	128
		4.7.2 Comparison with wavelets	129
	4.8	Smoothing techniques in use	132
		4.8.1 Example 1: modelling and model diagnostics	133
		4.8.2 Example 2: comparing two treatments	136
		4.8.3 Example 3: analyzing a longitudinal data set	137
	4.9	A blueprint for local modelling	141
	4.10	Other existing methods	148
		4.10.1 Normal reference method	149
		4.10.2 Cross-validation	149
		4.10.3 Nearest neighbor bandwidth	151
		4.10.4 Plug-in ideas	152
		4.10.5 Sheather and Jones' bandwidth selector	153
	4.11	Complements	154
	4.12	Bibliographic notes	157
5	**Applications of local polynomial modelling**		**159**
	5.1	Introduction	159
	5.2	Censored regression	160
		5.2.1 Preliminaries	160
		5.2.2 Censoring unbiased transformation	165
		5.2.3 Local polynomial regression	170
		5.2.4 An asymptotic result	173
	5.3	Proportional hazards model	175
		5.3.1 Partial likelihood	175
		5.3.2 Local partial likelihood	179
		5.3.3 Determining model complexity	183
		5.3.4 Complete likelihood	187
	5.4	Generalized linear models	189
		5.4.1 Exponential family models	190
		5.4.2 Quasi-likelihood and deviance residuals	193
		5.4.3 Local quasi-likelihood	194
		5.4.4 Bias and variance	196
		5.4.5 Bandwidth selection	197
	5.5	Robust regression	199
		5.5.1 Robust methods	199

	5.5.2	Quantile regression	201
	5.5.3	Simultaneous estimation of location and scale functions	207
5.6	Complements		208
5.7	Bibliographic notes		214

6 Applications in nonlinear time series 217
- 6.1 Introduction 217
- 6.2 Nonlinear prediction 218
  - 6.2.1 Mixing conditions 218
  - 6.2.2 Local polynomial fitting 220
  - 6.2.3 Estimation of conditional densities 224
- 6.3 Percentile and expectile regression 228
  - 6.3.1 Regression percentile 229
  - 6.3.2 Expectile regression 230
- 6.4 Spectral density estimation 233
  - 6.4.1 Smoothed log-periodogram 235
  - 6.4.2 Maximum local likelihood method 237
  - 6.4.3 Smoothed periodogram 242
- 6.5 Sensitivity measures and nonlinear prediction 243
  - 6.5.1 Sensitivity measures 244
  - 6.5.2 Noise amplification 247
  - 6.5.3 Nonlinear prediction error 247
- 6.6 Complements 249
- 6.7 Bibliographic notes 260

7 Local polynomial regression for multivariate data 263
- 7.1 Introduction 263
- 7.2 Generalized additive models 265
- 7.3 Generalized partially linear single-index models 272
  - 7.3.1 Partially linear models 273
  - 7.3.2 Single-index models 274
  - 7.3.3 Generalized partially linear single-index models 276
- 7.4 Modelling interactions 283
  - 7.4.1 Interactions in generalized additive models 283
  - 7.4.2 Interactions in generalized partially linear single-index models 287
  - 7.4.3 Multivariate adaptive regression splines 289
- 7.5 Sliced inverse regression 290
- 7.6 Local polynomial regression as a building block 295
- 7.7 Robustness 296
- 7.8 Local linear regression in the multivariate setting 297

	7.8.1 Multivariate local linear regression estimator	297
	7.8.2 Bias and variance	301
	7.8.3 Optimal weight function	302
	7.8.4 Efficiency	303
7.9	Bibliographic notes	304

References   307

Author index   330

Subject index   336

Preface

This book is on data-analytic approaches to regression problems arising from many scientific disciplines. These approaches are also called nonparametric regression in the literature. The aim of nonparametric methods is to relax assumptions on the form of a regression function, and to let data search for a suitable function that describes well the available data. These approaches are powerful in exploring fine structural relationships and provide very useful diagnostic tools for parametric models.

Over the last two decades, vast efforts have been devoted to nonparametric regression analyses. This book hopes to bring an up-to-date picture on the state of the art of nonparametric regression techniques. The emphasis of this book is on methodologies rather than on theory, with a particular focus on applications of nonparametric techniques to various statistical problems. These problems include least squares regression, quantile and robust regression, survival analysis, generalized linear models and nonlinear time series. Local polynomial modelling is employed in a large fraction of the book, but other key ideas of nonparametric regression are also discussed.

This book intends to provide a useful reference for research and applied statisticians and to serve as a textbook to graduate students and others who are interested in the area of nonparametric regression. While advanced mathematical arguments have provided valuable insights into nonparametric techniques, the methodological power of nonparametric regression can be introduced without advanced mathematics. This book, in particular the core of the local modelling approach, will be comprehensible to students with a statistical training on a Master's level. We hope that readers will be stimulated to choose or to develop their own methodology in the universe of nonparametric statistics. Key technical arguments are collected in the 'complements sections' at the end of Chapters

3–6. This gives interested researchers an idea of how the theoretical results are obtained. Hopefully, they will find the technical arguments useful in their future research.

Each chapter of this book is self-contained. One can skip a chapter without much effect on the understanding of the others. The only exception is Chapter 6, where we use part of the materials in Chapter 4 and Section 5.5. One can begin by reading/teaching a few sections in Chapter 2. Sections 2.2, 2.3, 2.5 and 2.6 provide a good picture of this chapter. One can then move on to Chapter 3, where Section 3.2 alone provides a good summary of the framework of local polynomial fitting. Sections 4.1–4.8 are the core of applications of local polynomial regression techniques in the least squares context. From here, one can choose a few sections in Chapters 5–7, depending on one's expertise and interest. Since all sections in these chapters are self-contained, skipping a few of them does not have an adverse effect on understanding other sections. However, it is helpful to read Section 5.4 before moving to Chapter 7.

There are many researchers whose work is reflected in this book. Indeed, this is such an active area that it seems impossible to give suitable credits to the vast literature when discussing a topic. While we have tried our best to be thorough and fair in the bibliographical notes, omissions and discrepancies are inevitable. We apologize for their occurrence.

We would like to thank our teachers Peter J. Bickel, Herman Callaert, David L. Donoho, Kai-Tai Fang and Noël Veraverbeke for their valuable inspiration on our research efforts. We are grateful to Peter J. Bickel and Peter G. Hall for encouraging us to write this book. We also acknowledge with gratitude Kjell A. Doksum, Nils L. Hjort, Bernard W. Silverman and other anonymous reviewers for their careful reading of the book and for their many useful suggestions. For their various offers of help, including kindly supplying us with data analyzed in this book, we are grateful to several people, including Barbara Ainsworth, Leo Breiman, Trevor J. Hastie, Iain Johnstone, Ying Lu, Steve J. Marron, Léopold Simar, Howell Tong, Matt P. Wand and Qiwei Yao. We also would like to thank our various co-authors for their friendly and very stimulating collaborations. Several results from joint papers are reflected in this book.

J. Fan's research was supported by the National Science Foundation and a National Science Foundation postdoctoral Fellowship. I. Gijbels' research was supported by the National Science Foundation (FNRS) and by the Programme d'Actions de Recherche

Concertées from the Belgian Government. The book was written with partial support from the Department of Statistics, University of North Carolina at Chapel Hill, and the Institute of Statistics, Catholic University of Louvain at Louvain-la-Neuve. We would like to acknowledge gratefully their generous support.

J. Fan
I. Gijbels
Chapel Hill and Louvain-la-Neuve
May 1995

CHAPTER 1

Introduction

Regression analysis is one of the most commonly used techniques in statistics. The aim of the analysis is to explore the association between dependent and independent variables, to assess the contribution of the independent variables and to identify their impact on the dependent variable. The main theme of this book is the application of *local modelling* techniques to various regression problems in different statistical contexts. The approaches are data-analytic in which regression functions are determined by data, instead of being limited to a certain functional form as in parametric analyses. Before we introduce the key ideas of local modelling, it is helpful to have a brief look at parametric regression.

1.1 From linear regression to nonlinear regression

Linear regression is one of the most classical and widely used techniques. For given pairs of data $(X_i, Y_i), i = 1, \cdots, n$, one tries to fit a line through the data. The part that cannot be explained by the line is often treated as noise. In other words, the data are regarded as realizations from the model:

$$Y = \alpha + \beta X + \text{error.} \qquad (1.1)$$

The error is often assumed to be independent identically distributed noise. The main purposes of such a regression analysis are to quantify the contribution of the covariate X to the response Y per unit value of X, to summarize the association between the two variables, to predict the mean response for a given value of X, and to extrapolate the results beyond the range of the observed covariate values.

The linear regression technique is very useful if the mean response is linear:

$$E(Y|X = x) \equiv m(x) = \alpha + \beta x.$$

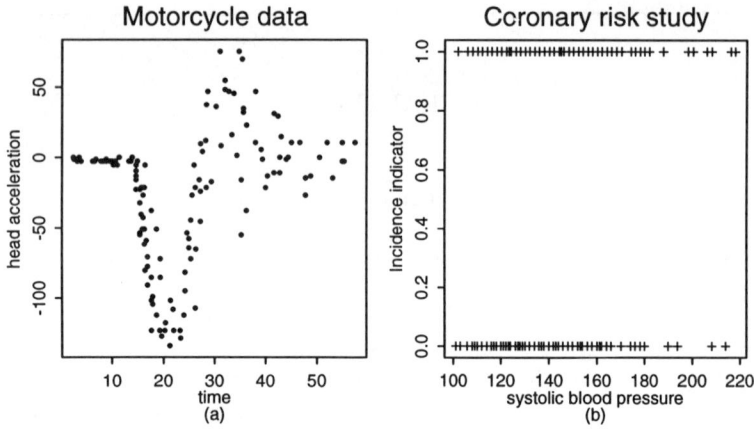

Figure 1.1. *Scatter plot diagrams for motorcycle data and coronary risk-factor study data.*

This assumption, however, is not always granted. It needs to be validated at least during the exploration stage of the study. One commonly used exploration technique is the scatter plot, in which we plot X_i against Y_i, and then examine whether the pattern appears linear or not. This relies on a vague 'smoother' built into our brains. This smoother cannot however process the data beyond the domain of visualization. To illustrate this point, Figure 1.1 gives two scatter plot diagrams. Figure 1.1 (a) concerns 133 observations of motorcycle data from Schmidt, Mattern and Schüler (1981). The time (in milliseconds) after a simulated impact on motorcycles was recorded, and serves as the covariate X. The response variable Y is the head acceleration (in g) of a test object. It is not hard to imagine the regression curve, but one does have difficulty in picturing its derivative curve. In Figure 1.1 (b), we use data from the coronary risk-factor study surveyed in rural South Africa (see Rousseauw *et al.* (1983) and Section 7.1). The incidence of Myocardial Infarction is taken as the response variable Y and systolic blood pressure as the covariate X. The underlying conditional probability curve is hard to image. Suffice to say that 'brain smoothers' are not enough even for scatter plot smoothing problems. Moreover, they cannot be automated in multidimensional regression problems, where scatter plot smoothing serves as building blocks.

What can we do if the scatter plot appears nonlinear such as in Figure 1.1? Linear regression (1.1) will create a very large mod-

FROM LINEAR REGRESSION TO NONLINEAR REGRESSION

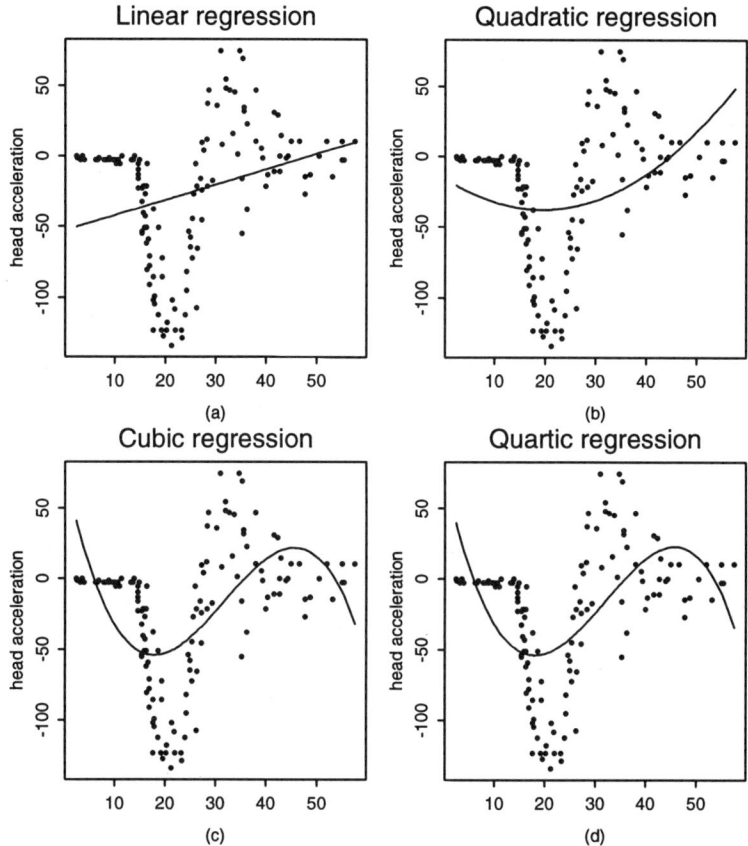

Figure 1.2. *Polynomial fits to the motorcycle data. The modelling bias is large since the family of polynomial functions is smooth everywhere.*

elling bias. A popular approach is to increase the number of parameters by using *polynomial regression*. Figure 1.2 shows such a family of polynomial fits, which have large biases. While this approach has been widely used, it suffers from a few drawbacks. One is that polynomial functions are not very flexible in modelling many problems encountered in practice since polynomial functions have all orders of derivatives everywhere. Another is that individual observations can have a large influence on remote parts of the curve. A third point is that the polynomial degree cannot be controlled continuously.

1.2 Local modelling

There are several ways to repair the drawbacks of polynomial fitting. One is to allow possible discontinuities of derivative curves. This leads to the *spline approach*. The locations of discontinuity points, called *knots*, can be selected by data via a *smoothing spline* method or a stepwise deletion method. See Section 2.6. Another possible proposal is to expand the regression function into an orthogonal series, then choose a few useful subsets of the basis functions, and use them to approximate the regression function. This approach is called the *orthogonal series method* and will be discussed in Section 2.5. A third approach is that instead of increasing the number of parameters, we apply the linear regression model locally. That is, for any given point x, model $m(\cdot)$ linearly around x and apply the linear regression technique to a fraction of the data around x. We term this approach the *local (linear) modelling approach*. The size of the local neighborhood, called the *bandwidth*, will be chosen either subjectively by data analysts or objectively by data. See Sections 2.2–2.4 for details. Thus, instead of solving a parametric problem with many parameters, we solve many linear regression problems with only two parameters.

To obtain the regression curve at a given point x, we apply the linear regression technique to a strip of data around x. Let h be the size of the local neighborhood. Then, we model the data in the strip by

$$Y_i = a(x) + b(x)X_i + \text{error}, \quad \text{for } X_i \in x \pm h, \qquad (1.2)$$

where the dependence of the parameters a and b on x is stressed. One can also incorporate a weight scheme to the local least squares problem (1.2) to down-weigh the contributions of a data point away from x. Let K be a unimodal nonnegative function. We can assign a weight $K\{(X_i-x)/h\}$ to the point (X_i, Y_i), leading to the following weighted least squares problem:

$$\sum_{i=1}^{n}\{Y_i - a(x) - b(x)X_i\}^2 K\left(\frac{X_i - x}{h}\right) I\left(\frac{|X_i - x|}{h} \leq 1\right). \qquad (1.3)$$

The above indicator can be absorbed into the kernel function if K has a support contained in $[-1, 1]$. Thus, problem (1.3) can be written in a more compact way as follows

$$\sum_{i=1}^{n}\{Y_i - a(x) - b(x)X_i\}^2 K\left(\frac{X_i - x}{h}\right). \qquad (1.4)$$

The function K, called the *kernel function*, does not have to be compactly supported, as long as it decays fast enough to eliminate the impact of a remote data point. Let $a(x)$ and $b(x)$ be the minimizers of the least squares problem (1.4). Then, the estimated regression curve at point x would be

$$\widehat{m}(x) = \widehat{a}(x) + \widehat{b}(x)x.$$

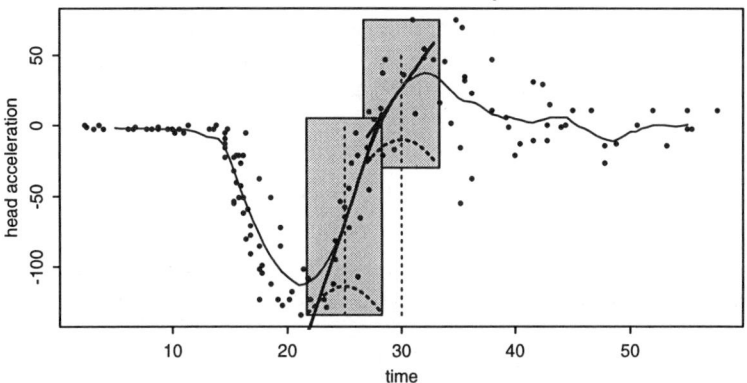

Figure 1.3. *Illustration of local modelling approach. For each given x, fit a linear model for the data points contained in the strip $x \pm 3.3$, using the weight function indicated at the bottom of the strip.*

As an illustration, we consider the motorcycle data presented in Figure 1.1. To obtain the regression curve at a given point $x = 25$, we fit a line through the data in the neighborhood 25 ± 3.3 with kernel function $K(t) = (1 - t^2)_+$, where the subscript means that the positive part should be taken, and obtain the estimated value $\widehat{m}(25) = -69.01$. Now if we want to estimate the regression function at another point $x = 30$, we fit a line using the data in the neighborhood 30 ± 3.3. The whole curve is obtained by estimating the regression function in a grid of points. For example, the curve in Figure 1.3 resulted from running 101 linear regressions, associated with 101 grid points.

The above idea can easily be generalized to other regression contexts, yielding a general local modelling principle. We can abstract the problem as follows. Suppose that we are interested in estimating a function $g(x)$ based on the observed sample $\{(X_i, Y_i), i = 1 \cdots, n\}$. For example, the function g can be the logit transform of the conditional probability function for the data given in Figure 1.1 (b) in the *generalized linear model* context, or can be a hazard regression function in the *proportional hazards model*. Various examples of g are given in Chapters 5 and 6. In a parametric linear approach, one models g linearly: $g(x) = \alpha + \beta x$ and tries to minimize or maximize an objective function:

$$L(\alpha, \beta) = \sum_{i=1}^{n} \ell_i(X_i, Y_i, \alpha + \beta X_i). \qquad (1.5)$$

Here ℓ_i is usually a discrepancy loss or the contribution of an individual log-likelihood. The local modelling approach aims simply to relax the global linear assumption to the local linear model, leading to the new objective function

$$L\{a(x), b(x)\} = \sum_{i=1}^{n} \ell_i\{X_i, Y_i, a(x) + b(x)X_i\} K\left(\frac{X_i - x}{h}\right). \qquad (1.6)$$

Let $\widehat{a}(x)$ and $\widehat{b}(x)$ solve the optimization problem. Then the function $g(x)$ can be estimated by $\widehat{g}(x) = \widehat{a}(x) + \widehat{b}(x)x$.

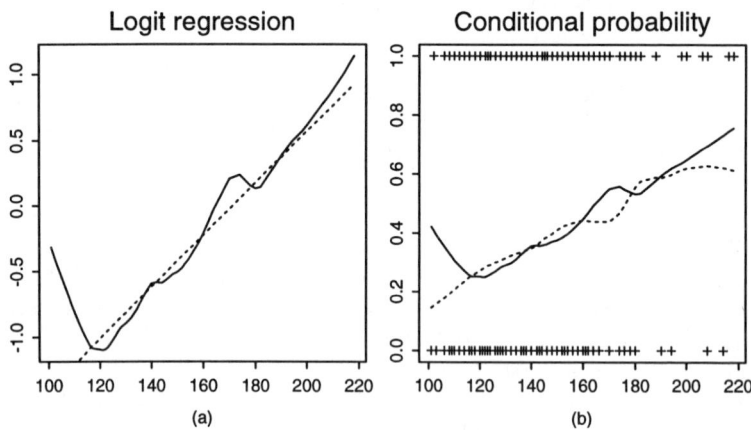

Figure 1.4. *Illustration of local modelling approach for binary data. (a) Estimated logit transform of the conditional probability function. (b) Estimated conditional probability function. Solid curve – local modelling approach using about 1/3 of the data points in each neighborhood; dashed curve – parametric logit linear model, using all data points.*

As an illustration, we consider once again the data presented in Figure 1.1 (b). We take g to be the logit of the conditional

probability function
$$p(x) = E(Y|X = x) = P(Y = 1|X = x).$$
In this case, ℓ_i is simply the log-likelihood function from the *Bernoulli model*:
$$\ell_i\{X_i, Y_i, g(X_i)\} = \log[p(X_i)^{Y_i}\{1 - p(X_i)\}^{1-Y_i}],$$
where $g(x) = \text{logit}\{p(x)\} = \log \frac{p(x)}{1-p(x)}$ or equivalently
$$p(x) = \exp\{g(x)\}/[1 + \exp\{g(x)\}].$$
Figure 1.4 (a) is obtained by maximizing (1.6) for each point x. Figure 1.4 (b) is a simple transform of Figure 1.4 (a).

The above expositions are based on the local modelling idea. Its theoretical basis is a Taylor expansion: for each y in a neighborhood of x,
$$g(y) \approx g(x) + g'(x)(y - x).$$
That is, each smooth function can locally be approximated by a linear function. This class of functions is far wider than the class of linear functions, and allows flexibility for exploring nonlinearity.

We would like to remark further that the above local linear modelling can be generalized to local low-degree polynomial modelling. One often reparametrizes the local polynomial model around the point x
$$g(y) = a_0(x) + a_1(x)y + \cdots + a_p(x)y^p$$
by
$$g(y) = \alpha_0(x) + \alpha_1(x)(y - x) + \cdots + \alpha_p(x)(y - x)^p. \qquad (1.7)$$
The local parameters $\alpha_i(x)$ have a more appealing interpretation: $\alpha_i(x) = g^{(i)}(x)/i!$.

1.3 Bandwidth selection and model complexity

A natural question is how wide the local neighborhood should be so that the local model or approximation (1.7) is valid. This is equivalent to asking how large the bandwidth parameter h should be in (1.6) or (1.4). If we take a very small h, the modelling bias (approximation error) will be small. However, since the number of data points falling in this local neighborhood is also small, the variance of the estimated local parameters $\widehat{\alpha}_i(x)$ will be large. On the other hand, if the bandwidth h is large, it can create a large modelling bias, depending on the underlying function.

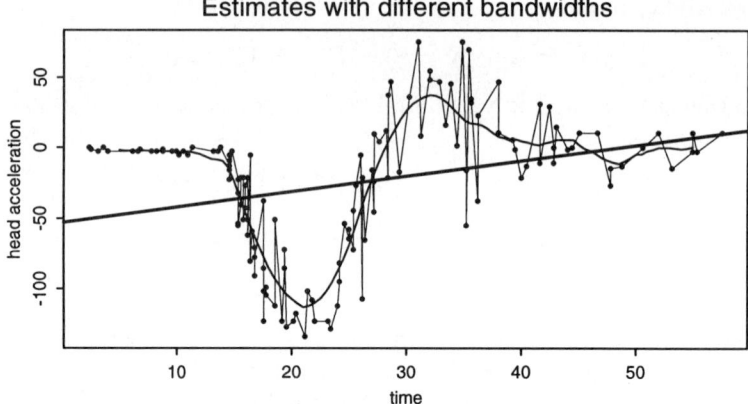

Figure 1.5. *Local linear fit with $h = 0$ (thin curve), 3.3 (solid curve) and ∞ (thick line). As h runs from 0 to ∞, the resulting estimate runs from the most complicated curve to the simplest curve.*

To illustrate this we present in Figure 1.5 the local linear estimates of the motorcycle data using $h = 0$, 3.3 and $+\infty$. As h is small, the resulting estimate essentially interpolates the data points and when h is large, it is the same as the parametric linear regression estimate. Viewing the family of models indexed by the continuous parameter h, the whole family of local linear models $\{\mathcal{M}_h : h \in (0, \infty)\}$ runs from the most complex model $\mathcal{M}_0$ – the interpolation model – to the simplest model $\mathcal{M}_\infty$ – the parametric linear model. Thus, the bandwidth parameter effectively controls the *model complexity*. Hence, the bandwidth selection problem can be regarded as choosing a model $\mathcal{M}_{h_0}$, from the family $\{\mathcal{M}_h\}$, that is most suitable for a given data set.

As illustrated above, there is a bias and variance trade-off in selecting the bandwidth h. A data-analytic approach is to let data choose a bandwidth that minimizes the risk of an estimated curve. This automates the smoothing technique, which is very useful in multivariate regression analysis. A subjective choice of the bandwidth can be viewed in a similar way, except that the trade-off now relies on our visualization.

Local modelling techniques provide parsimony to parametric modelling in the sense that when $h = \infty$, the local modelling becomes a global modelling. Therefore, the boundary between parametric and nonparametric modelling becomes moot when one takes the bandwidth parameter h into account. The key difference is that in parametric modelling, one always uses $h = \infty$ or tries different families of parametric models with or without validation, while in nonparametric modelling, one tries a few different bandwidths or lets data pick a bandwidth so that the resulting curve describes well a given data set.

The aim of nonparametric techniques is to relax the restrictive form of a regression function. It provides a useful tool for validating or suggesting a parametric form. This is particularly useful when one deals with multivariate data, in which one can hardly decide whether a regression surface is a plane or not. By using multivariate dimensionality reduction techniques, such as those described in Chapter 7, one can examine whether the contribution of each covariate is linear or not. However, nonparametric techniques have no intention of replacing parametric techniques. In fact, a combination of them can lead one to discover many interesting findings that are difficult to accomplish by any single method.

1.4 Scope of the book

Before we outline the scope of the book, it is helpful to provide a road map for the universe of nonparametric techniques. The road map in Figure 1.6 should be amplified on a much larger scale. For example, local smoothing methods alone would include the Nadaraya-Watson kernel regression, the Gasser-Müller smoother, the LOWESS smoother and so on. Spline techniques would include at least smoothing splines and polynomial splines. Orthogonal series approaches contain for example the truncated Fourier series method and the thresholded wavelet method.

It poses a challenge to any authors to exhaust all of the above $5 \times 3 \times 4 \times 4$ complete 'design' statistical problems. For example, one can try to introduce quantile regression problems, using additive modelling techniques, with smoothing splines as basic building blocks and using a cross-validation type of smoothing parameter selector. If we consider the amplified map, the number of elements in each factor would be far larger. Many of the 'design points' have not even been explored. In addition to the example just mentioned, we have not seen a published paper handling quantile regression problems in nonlinear time series, using partial linear modelling with local smoothing as building blocks and incorporating a plug-in bandwidth selector. Nevertheless, we have tried our best to give an objective picture of the literature. Even though it is not possi-

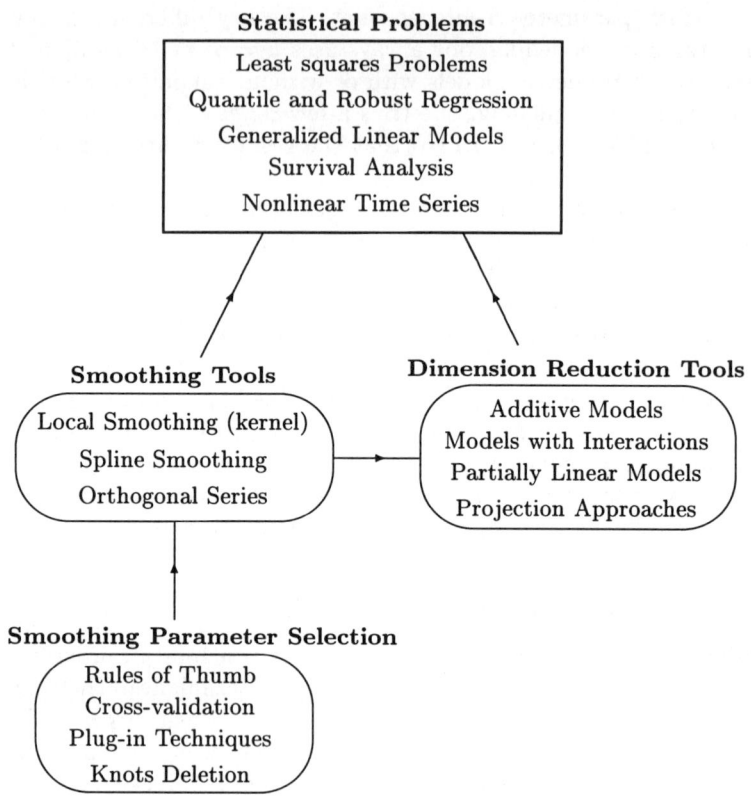

Figure 1.6. *A road map for nonparametric techniques.*

ble to cover all of the 'design points', this book provides materials detailed enough so that readers can hopefully be inspired to handle unsolved problems.

A general picture of one-dimensional smoothing tools is provided in Chapter 2. There, we explain the key ideas of the most commonly used smoothers. This basically amplifies the smoothing tool box in the above diagram.

Chapter 3 singles out a particular smoother, the local polynomial regression estimator, for detailed study. In particular, we provide some fundamental insights into this local polynomial regression estimator and outline the basic sampling and asymptotic properties of this method.

Bandwidth selection for local polynomial regression estimators is discussed in Chapter 4. We focus on the kind of plug-in procedures that we found useful in our implementations. With this method, the local polynomial regression estimator is automated. Various applications are also given so that readers can get a good picture of how nonparametric techniques are used in data analyses. Various bandwidth selection techniques are also outlined at the end of this chapter. This is intended to give readers a summary of the bandwidth selection box in Figure 1.6.

In Chapter 5, we show how nonparametric regression techniques can be applied to frequently encountered statistical problems. We focus only on applications of the local polynomial methods – other smoothers can also be employed. The aim of this chapter is to provide a picture of the statistical problems box in the above road map on nonparametrics.

Chapter 6 deals with various problems arising from nonlinear time series. In particular, we consider nonlinear prediction, quantile regression, spectral density estimation, and sensitivity measures in nonlinear stochastic systems.

Finally, in Chapter 7, we discuss various nonparametric techniques for handling multicovariate regression problems. Many useful dimensionality reduction principles are described. This chapter is intended to provide some details of the last box in the above map, the box on dimension reduction tools.

1.5 Implementation of nonparametric techniques

Most of the one-dimensional local modelling methods described in this book can easily be implemented by using existing software for parametric models by adding kernel weights. This process has to be repeated for each given x where the value of an estimated function is to be evaluated. All rule of thumb bandwidth selectors can also easily be programmed by using the existing software for parametric procedures. Part of the computation and graphics in this book was carried out using the software package S-Plus, an extension of the statistical language S (see Becker, Chambers and Wilks (1988)). Local polynomial fitting with application of specialized procedures such as the adaptive-order procedure described in Chapter 3 and the bandwidth selection rules described in Chapter 4, was implemented by using our own C-codes.

It is our hope that readers will be stimulated to use the methods described in this book for their own applications and research. We have provided information detailed enough so that readers can produce their own implementations. This would be a valuable exercise for students and readers that are new to the area.

1.6 Further reading

In this book we have not attempted to exhaust all nonparametric techniques. Other recent books and monographs provide more detailed treatments in some specific areas of the map in Figure 1.6. These books include Eubank (1988), Müller (1988), Györfi, Härdle, Sarda and Vieu (1989), Nadaraya (1989), Härdle (1990), Hastie and Tibshirani (1990), Wahba (1990), Rosenblatt (1991), Green and Silverman (1994) and Wand and Jones (1995) among others.

CHAPTER 2

Overview of existing methods

2.1 Introduction

In this chapter, a brief summary is given of the key methods available for univariate nonparametric regression function estimation. Some of the methods, in particular local polynomial fitting, will be discussed in greater detail in later chapters, but it is helpful to have an overview of the subject before examining any particular method in detail. Different techniques are like different hardware tools, having different strengths in their domains of applications. A large literature has been devoted to scatter plot smoothing, where the domain of the data is the same as that of human visualization. But there are also many important cases where the domain of smoothing is different from human visualization. In these cases, nonparametric smoothing techniques are particularly useful for assisting statistical modelling. Examples of these include estimation of derivatives, of conditional probabilities for binary data, and of certain functionals of regression such as variance explained by regression. This issue will be made clearer in this and the following chapters.

We first focus on nonparametric regression function estimation. Two data sets will be used to illustrate some of the methods. The first consists of 133 observations of motorcycle data presented in Figure 1.1. The second data set concerns 435 adults (between ages 17 and 85) suffering from burn injuries. The data were collected at the University of Southern California General Hospital Burn Center (courtesy of Professor Leo Breiman), and contain information on several variables. The binary response variable is taken to be 1 for those victims who survived their burn injuries and zero otherwise, and log(area of third degree burn + 1) is taken as a covariate. Many values of this covariate are overlapping, which results in exact overplotting. To improve the graphical view of the plotted covariate we *jittered* the data. This is accomplished by adding a very small amount of random noise to the covariate data, which prevents them from overlapping. The scatter plots for these two data sets are given in Figure 2.1 below. For Figure 2.1 (a) we cannot easily imagine the regression curve by only examining the scatter plot, while for Figure 2.1 (b) the scatter plot and the estimated curve are both in a domain similar to that of human visualization.

To put the smoothing problem into a statistical framework, it is convenient to think of each data set as a realization of a random sample from a certain population. Except where otherwise stated, it will be assumed that we have independently identically distributed observations $(X_1, Y_1), \cdots, (X_n, Y_n)$. Let (X, Y) denote a generic member of the sample, whose conditional mean and conditional variance are denoted respectively by

$$m(x) = E(Y|X = x) \quad \text{and} \quad \sigma^2(x) = \text{Var}(Y|X = x). \quad (2.1)$$

Many important applications involve estimation of the regression function $m(x)$ or its νth-derivative $m^{(\nu)}(x)$.

The performance of an estimator $\widehat{m}_\nu(x)$ of $m^{(\nu)}(x)$ is conveniently assessed via its *Mean Squared Error* (MSE) or its *Mean Integrated Squared Error* (MISE) defined respectively by

$$\text{MSE}(x) = E\big[\{\widehat{m}_\nu(x) - m^{(\nu)}(x)\}^2 \big| \mathbf{X}\big], \quad (2.2)$$

where $\mathbf{X} = (X_1, \cdots, X_n)$, and MISE $= \int \text{MSE}(x) w(x) dx$ with $w \geq 0$ a weight function. When the MSE-criterion is used, it is understood that the main objective is to estimate the function $m^{(\nu)}(\cdot)$ at the point x, and when the MISE-criterion is used, the main goal is to recover the whole curve. It can be easily seen that the MSE has the following bias-variance decomposition:

$$\text{MSE}(x) = [E\{\widehat{m}_\nu(x)|\mathbf{X}\} - m^{(\nu)}(x)]^2 + \text{Var}\{\widehat{m}_\nu(x)|\mathbf{X}\}. \quad (2.3)$$

One often refers to the term $E\{\widehat{m}_\nu(x)|\mathbf{X}\} - m^{(\nu)}(x)$ as the *bias* and to the second term in (2.3) as the *variance*.

2.2 Kernel estimators

2.2.1 Nadaraya-Watson estimator

Without assuming a specific form of the regression function m, a datum point remote from x carries little information about the value of $m(x)$. Thus, an intuitive estimator for the conditional mean function $m(x)$ is the running local average. An improved version of this is the locally weighted average. Let K be a real-valued function assigning weights. The function K is usually a symmetric probability density (which is also assumed throughout this chapter) and is called a *kernel function*. Let h be a *bandwidth* (also called a *smoothing parameter*), which is a nonnegative number controlling the size of the local neighborhood. Denote $K_h(\cdot) = K(\cdot/h)/h$. The Nadaraya-Watson kernel regression estimator is given by

$$\widehat{m}_h(x) = \frac{\sum_{i=1}^n K_h(X_i - x) Y_i}{\sum_{i=1}^n K_h(X_i - x)}. \tag{2.4}$$

See Nadaraya (1964) and Watson (1964). The asymptotic properties of the Nadaraya-Watson estimator are depicted in Table 2.1 below. See Härdle (1990) for more discussion on this estimation method.

Commonly used kernel functions include the Gaussian kernel $K(t) = (\sqrt{2\pi})^{-1} \exp(-t^2/2)$, and the 'symmetric Beta family'

$$K(t) = \frac{1}{\text{Beta}(1/2, \gamma + 1)} (1 - t^2)_+^\gamma, \quad \gamma = 0, 1, \cdots, \tag{2.5}$$

where the subscript + denotes the positive part, which is assumed to be taken *before* the exponentiation. Note that the constant factor in (2.5) is not actually used in $\widehat{m}_h$ – it is merely a normalization constant so that K is a density function. The choices $\gamma = 0, 1, 2$, and 3 lead to respectively the *uniform*, the *Epanechnikov*, the *biweight* and the *triweight* kernel function. As noted in Marron and Nolan (1988) this family includes the most widely used kernel functions, also the Gaussian kernel in the limit as $\gamma \to +\infty$. Figure 2.1 (a) shows how the Nadaraya-Watson estimator, with the Epanechnikov kernel and bandwidth $h = 1$, assigns the weights to each datum point.

2.2.2 Gasser-Müller estimator

The random denominator in (2.4) is inconvenient, when taking derivatives of the estimator and when deriving its asymptotic properties. Assume that the data have already been sorted according to the X-variable. Gasser and Müller (1979) proposed the following estimator:

$$\widehat{m}_h(x) = \sum_{i=1}^n \int_{s_{i-1}}^{s_i} K_h(u - x) du \, Y_i, \tag{2.6}$$

with $s_i = (X_i + X_{i+1})/2$, $X_0 = -\infty$ and $X_{n+1} = +\infty$. See also Gasser and Müller (1984). Note that the sum of the weights in (2.6) is one and hence no denominator is needed. This estimator

OVERVIEW OF EXISTING METHODS

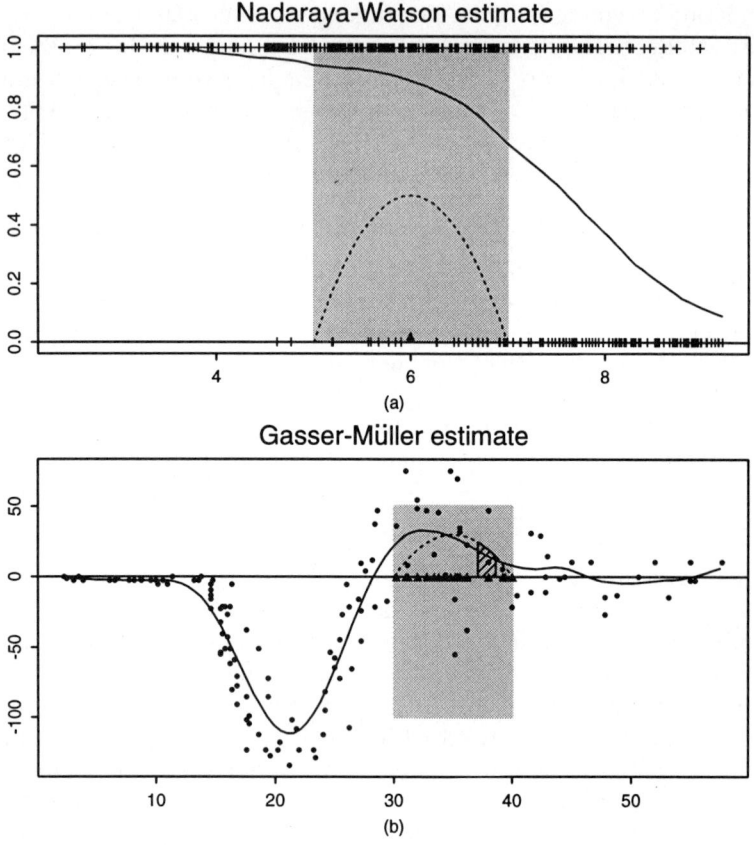

Figure 2.1. *A kernel function assigns different weights to each datum point. The weights depend on the bandwidth and the estimator that are used. The Nadaraya-Watson estimator assigns weights proportional to the heights of the rescaled kernel, while the Gasser-Müller estimator uses the areas (see the marked area). The kernel used here is the Epanechnikov kernel. (a) Burn data set and its Nadaraya-Watson estimate. (b) Motorcycle data set and its Gasser-Müller estimate. The shaded area indicates the local neighborhood around the point x.*

was originally proposed for *equispaced designs*, but can also be used for *non-equispaced designs*. The Gasser-Müller estimator is a modification of an earlier version of Priestley and Chao (1972), and is similar to that of Cheng and Lin (1981). Figure 2.1 (b) indicates how the Gasser-Müller estimator assigns weights to points in the

Table 2.1. *Pointwise asymptotic bias and variance of kernel regression smoothers. Taken from Fan (1992).*

Method	Bias	Variance
Nadaraya-Watson	$(m''(x) + \frac{2m'(x)f'(x)}{f(x)})b_n$	V_n
Gasser-Müller	$m''(x)b_n$	$1.5V_n$
Local linear	$m''(x)b_n$	V_n

Here, $b_n = \frac{1}{2}\int_{-\infty}^{\infty} u^2 K(u)du\, h^2$ and $V_n = \frac{\sigma^2(x)}{f(x)nh}\int_{-\infty}^{\infty} K^2(u)du$.

local neighborhood.

Müller (1988) gives a detailed discussion of the estimator (2.6). Its basic asymptotic properties, for a random design, are presented in Table 2.1, and originate from the work of Mack and Müller (1989) and Chu and Marron (1991).

2.2.3 Limitations of a local constant fit

From a function approximation point of view, the Nadaraya-Watson and the Gasser-Müller estimator both use local constant approximations. Suppose that $m(\cdot)$ is approximated locally by a constant θ. Then, one runs a local least squares regression and obtains the estimate

$$\widehat{\theta} = \operatorname{argmin}_\theta \sum_{i=1}^{n}(Y_i - \theta)^2 w_i = \sum_{i=1}^{n} w_i Y_i / \sum_{i=1}^{n} w_i.$$

The Nadaraya-Watson and Gasser-Müller estimator are of this form with weights $w_i = K_h(X_i - x)$ and $w_i = \int_{s_{i-1}}^{s_i} K_h(u - x)du$, respectively.

As noted by the numerical analyst G. Strang (1993) 'Through all of scientific computing runs this common theme: Increase the accuracy at least to second order. What this means is: *Get the linear term right* '. In other words, a local constant approximation is unsatisfactory, and a local linear fit is desirable. Let f denote the *design density*, i.e. the density of the random variable X. Table 2.1 summarizes the first order asymptotic performance of the Nadaraya-Watson estimator, the Gasser-Müller estimator, and the local linear regression estimator (to be introduced in the next section) at an interior point of the support of the design density.

In comparison with the local linear fit, the Nadaraya-Watson estimator locally uses one parameter less without reducing the asymptotic variance. This extra parameter enables the local linear fit to reduce the bias. (A similar phenomenon holds for more general local polynomial fits and will be discussed in Section 3.3.) The Nadaraya-Watson estimator suffers from large bias particularly at the region where the derivative of the regression function or of the design density is large. It can have a large bias even when the true regression curve is linear. It does not adapt to nonuniform designs: the bias can be large when $f'(x)/f(x)$ is large, as will be demonstrated in Figure 2.2. Moreover, it was shown that the Nadaraya-Watson estimator has zero minimax efficiency. See Section 3.4 and Fan (1992, 1993a) for a more detailed discussion of these issues. The key arguments will be summarized in Sections 3.2.1 and 3.4.1. The Gasser-Müller estimator on the other hand corrects the bias of the Nadaraya-Watson estimator but at the expense of increasing its variability. When the design is fixed however, the variability will not increase. This can intuitively be explained, as in Figure 2.2, by the fact that the Gasser-Müller estimator assigns fluctuating weights. Further, both the Nadaraya-Watson and the Gasser-Müller estimator have a large order of bias when estimating a curve at a boundary region. Many proposals such as *reflection methods* and *boundary kernel methods* are proposed to handle this problem, but they are less efficient than the automatic boundary correction of the local linear fit. Indeed, it is shown by Cheng, Fan, and Marron (1993) that the local linear fit is efficient in correcting boundary bias, in an asymptotic minimax sense. See Sections 3.2.6 and 3.4.2 for more detailed discussions on bias correction at the boundary.

Comparisons between local linear and local constant fit were discussed in detail by Chu and Marron (1991), Fan (1992) and Hastie and Loader (1993), and will be outlined in Sections 2.3 and 3.3.

2.3 Local polynomial fitting and derivative estimation

The idea of local polynomial regression has been around for a long time. It was systematically studied by Stone (1977, 1980, 1982) and Cleveland (1979). Recent work on local polynomial fitting includes Fan (1992, 1993a), Fan and Gijbels (1992) and Ruppert and Wand (1994). These papers give a detailed picture of the advantages of local polynomial fitting. This section briefly introduces the basic ingredient of local polynomial fitting. A thorough study of this

LOCAL POLYNOMIAL FITTING AND DERIVATIVE ESTIMATION   19

topic can be found in Chapters 3 and 4.

2.3.1 Local polynomial fitting

Suppose that locally the regression function m can be approximated by

$$m(z) \approx \sum_{j=0}^{p} \frac{m^{(j)}(x)}{j!}(z-x)^j \equiv \sum_{j=0}^{p} \beta_j(z-x)^j, \qquad (2.7)$$

for z in a neighborhood of x, by using Taylor's expansion. From a statistical modelling viewpoint, (2.7) models $m(z)$ *locally* by a simple polynomial model. This suggests using a locally weighted polynomial regression

$$\sum_{i=1}^{n}\{Y_i - \sum_{j=0}^{p}\beta_j(X_i-x)^j\}^2 K_h(X_i-x), \qquad (2.8)$$

where $K(\cdot)$ denotes a kernel function and h is a bandwidth. Denote by $\widehat{\beta}_j$ $(j=0,\cdots,p)$ the minimizer of (2.8). The above exposition suggests that an estimator for $m^{(\nu)}(x)$ is

$$\widehat{m}_\nu(x) = \nu!\,\widehat{\beta}_\nu. \qquad (2.9)$$

The whole curve $\widehat{m}_\nu(\cdot)$ is obtained by running the above local polynomial regression with x varying in an appropriate estimation domain.

At first glance, the above regression approach looks similar to the traditional parametric approach in which the function is usually *globally* modelled by a polynomial. In order to have a satisfactory modelling bias, the degree p of the polynomial often has to be large. For example, for a simple function such as $m(x) = \sin 3x + \cos x$, $x \in [0, 2\pi]$, parametric polynomial modelling requires a large p in order to have a reasonable bias. As in Figure 1.2, the motorcycle data also require a large p to reduce the modelling bias. But this large degree p introduces an over-parametrization, resulting in a large variability of the estimated parameters. As a consequence the estimated regression function is numerically unstable. In marked contrast to this parametric approach, our technique is *local*, and hence requires a small degree of the local polynomial, typically of order $p = \nu + 1$ or occasionally $p = \nu + 3$, as will be explained in detail in Section 3.3. For example, for estimating a regression function one often uses a running line regression ($p = 1$). It is clear that $h = 0$ results in an estimate which essentially interpolates the data, while $h = +\infty$ is equivalent to a linear model $m(x) = \alpha + \beta x$. As h ranges from 0 to $+\infty$, $\hat{m}_0$ ranges from the most complex model (interpolation) to the simplest model (linear model). Thus, the *model complexity* is effectively controlled by the bandwidth h. See also Figure 1.5.

A nice feature of the local polynomial fit is that it reveals simple expressions for local bias and variance as will be demonstrated in Section 3.2.1. None of the other approaches explained in this chapter share this genuine local character. The size of the local neighborhood is explicitly governed by h. As a consequence, local polynomial fitting possesses the flexibility of using a variable amount of smoothing to enhance, when needed, *spatial adaptation* in the sense that it can adapt readily to different degrees of smoothness and can neatly capture short and sharp aberrations of signals. Another feature is that it can easily be incorporated with a global parametric fit to correct the bias of the modelling. See the recent interesting paper by Hjort and Glad (1995) for details.

With $p = 1$, the estimator $\hat{m}_0(x)$ is termed a *local linear regression smoother* or a *local linear fit*. This estimator can be explicitly expressed as

$$\hat{m}_0(x) = \sum_1^n w_i Y_i / \sum_1^n w_i, \quad w_i = K_h(X_i - x)\{S_{n,2} - (X_i - x)S_{n,1}\},$$

where $S_{n,j} = \sum_1^n K_h(X_i - x)(X_i - x)^j$. Figure 2.2 shows how the Nadaraya-Watson estimator, the Gasser-Müller estimator, and the local linear regression estimator assign their weights at interior points and boundary points, using the Epanechnikov kernel.

For the interior point $x = 0.60$, the Nadaraya-Watson estimator assigns symmetric weights to both sides of $x = 0.60$. For such a non-equispaced design, this estimator overweighs the points on the left-hand side and hence creates a large bias. This explains the theoretical result that the Nadaraya-Watson estimator is not design-adaptive. If we make the design more asymmetric by deleting a few data points with $X_i > 0.6$, the bias of this estimator is more severe. The Gasser-Müller estimator corrects this by assigning seemly random weights but at the expense of increasing its variance. A local linear fit, on the other hand, adapts automatically to this random design by assigning an asymmetric weighting scheme, while maintaining the same kind of smooth weighting scheme as the Nadaraya-Watson estimator. In other words, the local linear fit takes advantage of both the Nadaraya-Watson and the Gasser-

LOCAL POLYNOMIAL FITTING AND DERIVATIVE ESTIMATION 21

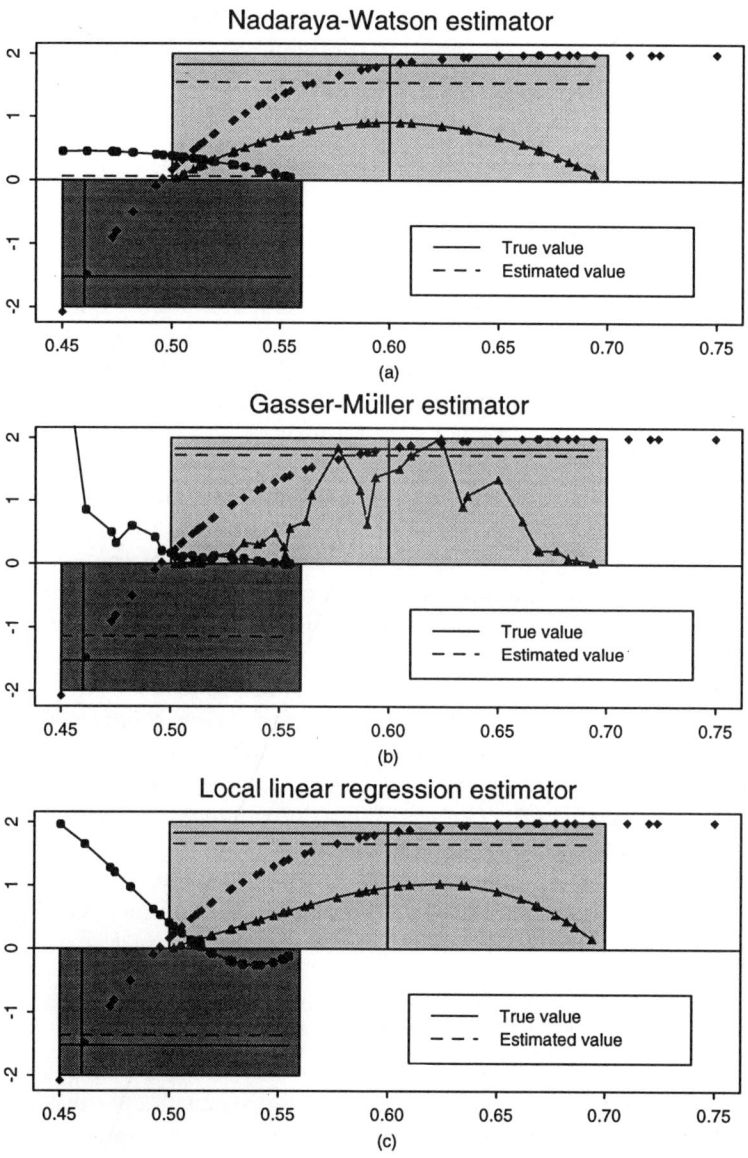

Figure 2.2. *Effective weights assigned to local data points at an interior point $x = 0.60$ (weights denoted by ▲) and a boundary point $x = 0.46$ (weights denoted by ■). (a) the Nadaraya-Watson estimator; (b) the Gasser-Müller estimator; (c) the local linear fit. For clarity, the data (♦) contain no noise. This idea is adapted from Hastie and Loader (1993).*

Müller estimate, as was already suggested by Table 2.1. A similar phenomenon can be observed at the boundary point $x = 0.46$ – the biases of the Nadaraya-Watson and the Gasser-Müller estimator are larger. In other words, the local linear fit adapts readily to boundary points, while the other two estimators do not.

2.3.2 Derivative estimation

The derivative estimators were proposed in (2.9). For estimating the first derivative $m'(\cdot)$, one takes the slope of a local quadratic regression; a local cubic fit can be used for estimating the second derivative function, and so on. In general, local polynomial fitting has certain advantages over the Nadaraya-Watson and the Gasser-Müller estimator not only for estimating regression curves, but also for derivative estimation. This is proved theoretically in Chapter 3. Here we demonstrate this fact via a graphical device in Figure 2.3. Consider three possible estimators: the derivative of the Nadaraya-Watson estimator, the derivative of the Gasser-Müller estimator and the linear coefficient of the local quadratic fit. All are linear estimators in the sense that they are weighted averages of the response variable. Clearly, the Nadaraya-Watson estimator has larger bias, exhibited by the zero crossing of the weights being too far to the left, while the Gasser-Müller estimator has larger variance, indicated by the wildly fluctuating heights. The local quadratic fit overcomes both problems. The estimated slopes obtained via the Gasser-Müller derivative estimator and the local quadratic derivative estimator are very similar in Figure 2.3. Note that when the design points are more asymmetric, namely fewer data points on the right-hand side of $x = 0.6$, the bias of the Nadaraya-Watson derivative estimator will be far larger.

2.4 Locally weighted scatter plot smoothing

The local averaging smoothers described in Sections 2.2 and 2.3 can be highly influenced by extreme observations, i.e. by *outliers*, in the response variables. This influence can result in a quite different fitted curve. In these situations it is preferable to have an estimation method which is more resistant for extreme observations. How can one develop such a *robust method*? An evaluation of the residuals of a fit can be helpful in detecting outliers in the response variables, with large residuals indicating possible outliers. In this section we discuss a particular robust fitting procedure pro-

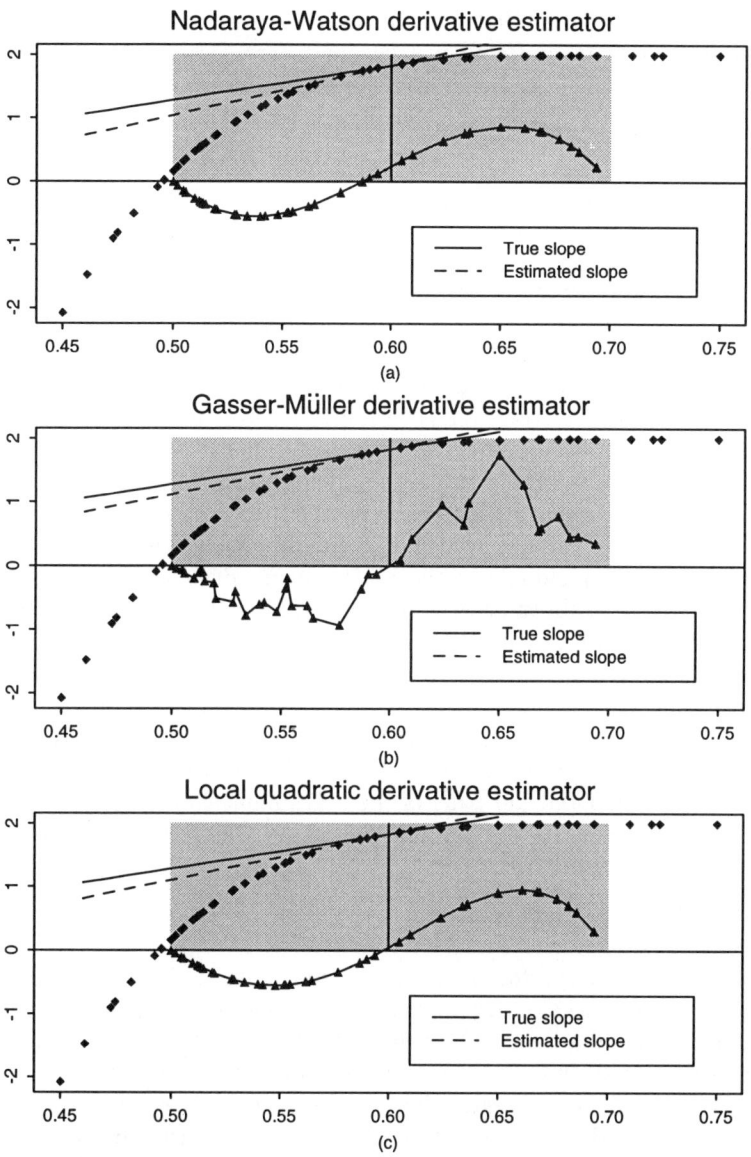

Figure 2.3. *Effective weights assigned to local data points at an interior point $x = 0.60$ by (a) the Nadaraya-Watson derivative estimator; (b) the Gasser-Müller derivative estimator; (c) the slope of the local quadratic fit. The data set (denoted by ◆) is the same as that of Figure 2.2 and the picture was adapted from Fan and Marron (1994).*

posed by Cleveland (1979). Its multivariate generalization *loess* and applications can be found in Cleveland, Grosse and Shyu (1992). See also Section 7.7. Other robust estimation methods and their applications are discussed in Section 5.5.

2.4.1 Robust locally weighted regression

LOcally WEighted Scatter plot Smoothing (LOWESS), introduced by Cleveland (1979), robustifies the locally weighted least squares method discussed in Section 2.3. The basic idea is as follows. Fit a polynomial of degree p locally via (2.8) and obtain the fitted values. Calculate the residuals and assign weights to each residual: large (respectively small) residuals receive small (respectively large) weights. Then carry through again a local polynomial fit of order p but now with assigning to each observation a new weight that is the product of the weight at the initial fit and the weight assigned to its residual from that initial fit. Hence the observations showing large residuals in the initial fit are *downweighted* in the second fit. The above process is repeated a number of times resulting in the estimator LOWESS. This robust estimation procedure has been implemented in the S-Plus package and is widely used.

We now describe in more detail this robust estimation procedure. Local polynomial fitting is the basic ingredient of the LOWESS method, and hence questions of how to choose the bandwidth, the kernel function and the order of the polynomial should get some attention. In his LOWESS method Cleveland (1979) uses the *tricube* kernel

$$K(t) = \frac{70}{81}(1 - |t|^3)^3 I_{[-1,1]}(t).$$

One of the reasons for choosing this particular weight function is that it descends smoothly to 0. The local neighborhoods are further determined by a nearest neighbor bandwidth. Let $0 < d \leq 1$, and consider nd, a quantity referring to a portion of the data. Round nd to the nearest integer, denoted by r. For given observations $X_1, \cdots, X_n$, the neighborhood of a particular observation X_k is determined by its associated bandwidth h_k, defined as the distance from X_k to its rth nearest neighbor. More precisely, h_k is the rth smallest number among $|X_k - X_j|$, for $j = 1, \cdots, n$. In Section 4.10 we discuss a related nearest neighbor type of bandwidth. After specifying the kernel and the bandwidth, we are now ready to find the fitted value at the design point X_k. To each observation X_i we

LOCALLY WEIGHTED SCATTER PLOT SMOOTHING 25

assign the weight
$$K_i(X_k) = K\left\{h_k^{-1}(X_i - X_k)\right\}. \tag{2.10}$$
These weights are now utilized in the initial locally weighted polynomial regression
$$\sum_{i=1}^{n}\{Y_i - \sum_{j=0}^{p}\beta_j(X_i - X_k)^j\}^2 h_k^{-1} K_i(X_k), \tag{2.11}$$
leading to the estimates $\widehat{\beta}_j$, $j = 0, \cdots, p$. See also Section 2.3. The fitted value for Y_k, denoted by $\widehat{Y}_k$, is given by $\widehat{Y}_k = \widehat{\beta}_0 = \widehat{\beta}_0(X_k)$. Since local polynomial regression estimators are linear smoothers (see Section 2.3 and Chapter 3), we can write
$$\widehat{Y}_k = \sum_{i=1}^{n} w_i(X_k) Y_i,$$
where the coefficients $w_i(X_k)$ clearly depend on X_k as well as the design points $\{X_i\}$.

The above step is done for each observation X_k, i.e. a local polynomial fit is carried through for each point X_k. This results in the initial fitted values $\widehat{Y}_1, \cdots, \widehat{Y}_n$.

The residuals from this initial fit are
$$r_k = Y_k - \widehat{Y}_k, \qquad k = 1, \cdots, n. \tag{2.12}$$
Let M be the median of the sequence $|r_1|, \cdots, |r_n|$. Then the *robustness weights* are defined as
$$\delta_i = B(r_i/(6M)), \qquad i = 1, \cdots, n, \tag{2.13}$$
where
$$B(t) = (1 - |t|^2)^2 I_{[-1,1]}(t),$$
is the biweight kernel as defined in (2.5) (with $\gamma = 2$). This finishes the initial step.

Now compute a new fitted value $\widehat{Y}_k$ by fitting locally a pth order polynomial, using now the weight $\delta_i K_i(X_k)$ for the observation (X_i, Y_i), $i = 1, \cdots, n$. After a number of iterations, say N, the final fitted values $\widehat{Y}_k$, $k = 1, \cdots, n$, yield the robust locally weighted regression estimator. Values of the estimated curve in points x_0 different from the design points X_k, $k = 1, \cdots, n$, can be obtained via, for example, interpolation.

To summarize, the algorithm for computing LOWESS reads as follows:

1. For each k do a locally weighted regression (2.11) with weights as in (2.10). Compute the fitted value $\widehat{Y}_k$, and the residual r_k via (2.12);
2. Calculate the robustness weights δ_i, $i = 1, \cdots, n$, as defined in (2.13);
3. For each k compute a new fitted value $\widehat{Y}_k$ by carrying through a locally weighted regression (2.11) with weights $\delta_i K_i(X_k)$ for (X_i, Y_i);
4. Repeat steps 2 and 3 a total of N times. The final fitted values yield the estimated curve.

Cleveland (1979) recommends using $p = 1$ and $N = 3$. The bandwidth depends on d and for this quantity a choice between 0.2 and 0.8 seems to be reasonable. For more details see Cleveland (1979).

Note that when $N = 0$, LOWESS corresponds to local polynomial fitting using a nearest neighbor bandwidth. When $N \to +\infty$, if the algorithm converges, the LOWESS estimator, with M kept fixed, tends to the estimator (5.58) in Section 5.5 with $x_0 = X_k$, $h = h_k$, and $\ell'(t) = tB\{t/(6M)\}$ – *Huber's bisquare function*.

2.4.2 An example

We now illustrate the usefulness of the particular robust estimation procedure LOWESS via a data set, presented by Carroll and Ruppert (1982) and discussed further in Carroll and Ruppert (1988, page 48). The data set involves 115 assays for the concentration of the enzyme esterase. The response Y is the number of bindings counted by a radio-immunoassay (RIA) procedure and the covariate X is the observed esterase concentration. The variability of the counts gets larger as esterase concentration increases. More detailed analysis of this data set will be given in Section 5.5. Figure 2.4 (a) gives the scatter plot of the esterase data set, along with the LOWESS fits using $d = 2/3, N = 3$ and $d = 2/3, N = 0$. As noted above, $N = 0$ corresponds to the least squares version, which is not resistant to outliers. To see this effect more clearly, we add 5 outliers to the data set in Figure 2.4 (b). Clearly, from comparing Figures 2.4 (a) and (b), the LOWESS fit with $N = 3$ is robust against these outliers. However, the least squares version is not. It responds to the outliers by raising its local mean curve. The effect is more pronounced with a smaller bandwidth (compare the shortest dashed curve with the longest dashed curve).

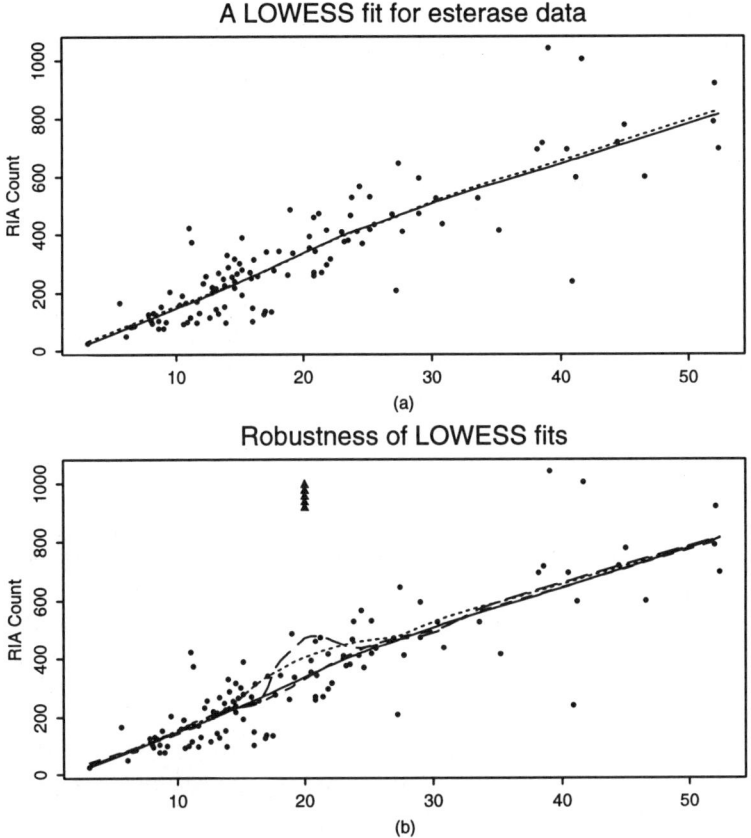

Figure 2.4. *Robustness of LOWESS smoother. (a) LOWESS fit using $d = 2/3$ and $N = 3$ (solid curve) and $d = 1/3$ and $N = 0$ (dashed curve). (b) Robustness of LOWESS fits. Five outliers (indicated by ▲) were added to the data set. Presented curves are LOWESS fits with different parameters: $d = 2/3, N = 3$ (solid curve); $d = 2/3, N = 0$; $d = 1/3, N = 3$; $d = 1/3, N = 0$ (from shortest dash to longest dash).*

2.5 Wavelet thresholding

A local polynomial fit models the unknown regression function m locally by a few parameters and uses the bandwidth to control the complexity of the modelling. It is also possible however to model the function globally with a large number of unknown parameters. Wavelet and spline expansions, for example, can be used for such a parametrization.

Wavelet transforms are a device for representing functions in a way that is local in both time and frequency domains. They have recently received a great deal of attention in applied mathematics, image analysis, signal compression, and many other fields of engineering. Good introductory references to this subject include Meyer (1990), Chui (1992), Daubechies (1992) and Strang (1989, 1993).

Wavelet-based methods, demonstrated by Donoho (1995) and Donoho and Johnstone (1994, 1995a, b), have many exciting statistical properties. They possess the spatial adaptation property in the sense that they can adapt readily to spatially inhomogeneous curves. This fact is convincingly demonstrated in Donoho and Johnstone (1994, 1995a, b) and Fan, Hall, Martin and Patil (1993). Wavelet-based methods are nearly minimax (in rate) for a large class of functions with unknown degrees of smoothness (Donoho and Johnstone (1995a, 1996)). An excellent overview on wavelet statistical estimation can be found in Donoho, Johnstone, Kerkyacharian and Picard (1995). The relationship between wavelet thresholding and local kernel smoothing methods is illuminatingly described by Hall and Patil (1995a, b), and an interesting comparison in terms of efficiency is given in Fan, Hall, Martin and Patil (1993, 1996).

This section only covers some basics of wavelet methods. Unless otherwise stated, we assume that we have n data points $(x_i, Y_i), i = 1, \cdots, n$, from the *canonical regression model*

$$Y_i = m(x_i) + \varepsilon_i, \tag{2.14}$$

where $x_i = i/n$ and $\{\varepsilon_i\}$ is a sequence of i.i.d. Gaussian noise $\varepsilon \sim N(0, \sigma^2)$. The assumptions on the equispaced design and the Gaussian white noise are important for the development of the methodology, although the basic idea should in principle be applicable to non-canonical regression models. Before introducing wavelet-based methods, we first resume the general principle of statistical estimation based on orthogonal series expansions.

2.5.1 Orthogonal series based methods

Assume that the function $m \in L^2[0,1]$, the function space of square integrable functions on $[0,1]$, and that the series $\{q_i(x), i =$

$1, 2, \cdots\}$ is a complete orthonormal basis:

$$\int_0^1 q_i(x)q_j(x)dx = \delta_{ij},$$

where δ_{ij} is the Kronecker delta function. Therefore, the function m admits the orthogonal expansion: $m(x) = \sum_{j=1}^{\infty} \theta_j q_j(x)$ with $\theta_j = \int q_j(x)m(x)dx$. A natural estimator of the parameter θ_j is

$$\widehat{\theta}_j = n^{-1} \sum_{i=1}^{n} q_j(x_i)Y_i. \tag{2.15}$$

A classical reconstruction of m is based on the truncated subseries:

$$\widehat{m}(x) = \sum_{j=1}^{N} \widehat{\theta}_j q_j(x), \tag{2.16}$$

where N, depending on n and tending to ∞, is a smoothing parameter. From a statistical modelling point of view, the estimator (2.16) can be regarded approximately as a least squares estimator:

$$\mathrm{argmin}_\theta \sum_{i=1}^{n} \{Y_i - \sum_{j=1}^{N} \theta_j q_j(x_i)\}^2,$$

based on the model assumption $m(x) = \sum_{j=1}^{N} \theta_j q_j(x)$. Indeed, the design matrix $\left(q_j(x_i)\right)$ is nearly orthonormal as evidenced by

$$n^{-1} \sum_{i=1}^{n} q_j(x_i)q_k(x_i) \approx \int_0^1 q_j(x)q_k(x)dx = \delta_{jk},$$

and hence the least squares estimate of θ_j is approximately $\widehat{\theta}_j$. For the Fourier basis, $q_1(x) = 1$, $q_{2k}(x) = \sqrt{2}\cos(2\pi kx)$, $q_{2k+1}(x) = \sqrt{2}\sin(2\pi kx)$, $k = 1, 2, \cdots$, the design matrix is exactly orthonormal and hence the estimator (2.16) is precisely a least squares estimate. Often shrinkage ideas, introduced by Stein (1956), can be applied to improve the efficiency. See for example the seminal work by Efromovich and Pinsker (1982), Efromovich (1985, 1996) and Nussbaum (1985) for further discussions.

In the case of the Fourier basis, one would expect that the energy at high frequencies is small (i.e. $|\theta_j|$ is small when j is large) – the information is compressed into low frequencies. Therefore use of the first N basis functions would be reasonable for modelling $m(\cdot)$. However, if such *primary information* is not granted, an ideal choice of the subset of basis functions consists of those basis functions having large absolute signals $|\theta_j|$. Such an ideal choice depends of course on the unknown function m. Donoho and Johnstone (1994) referred to the estimator based on such an ideal choice of basis functions as the *Oracle estimator*. In practice, one has to mimic the Oracle estimator. An intuitive way of finding such an ideal subset of size N is to compare the residual sums of squares of all possible least squares regression problems with the model $\{m(x) = \sum_{j \in S} \theta_j q_j(x), |S| = N\}$. However, this is not only an expensive procedure but also one which accumulates stochastic errors in the subset selection. A simple practical rule for selecting the subset without such expensive *data mining* is to choose basis functions with large empirical coefficient $\widehat{\theta}_j$. This idea leads to the

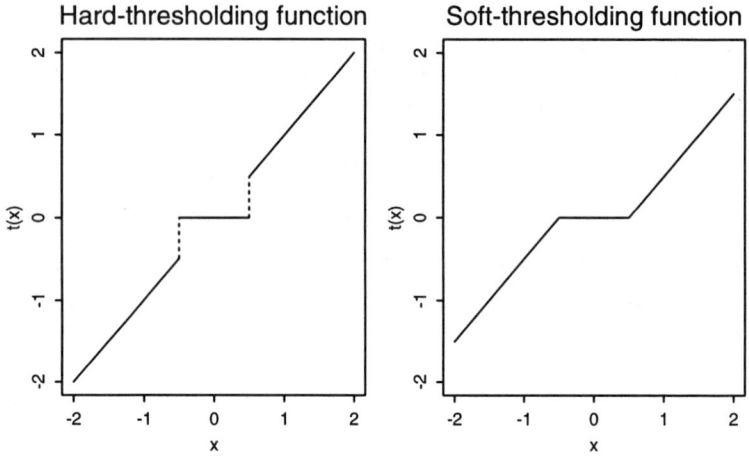

Figure 2.5. *Hard- and soft-thresholding functions with $\delta = 0.5$, i.e. $t(x) = x1\{|x| > \delta\}$ and $t(x) = sgn(x)(|x| - \delta)_+$ respectively.*

following *hard-thresholding estimator*:

$$\widehat{m}_H(x) = \sum_{j=1}^{\infty} \widehat{\theta}_j 1\{|\widehat{\theta}_j| > \delta\} q_j(x), \qquad (2.17)$$

where $\delta > 0$ is a *thresholding parameter* and $1\{|\widehat{\theta}_j| > \delta\}$ is an indicator function, giving values 1 when $|\widehat{\theta}_j| > \delta$ and 0 otherwise. Another simple choice of the subset, involving *shrinkage* of the coefficients by an amount δ, is the *soft-thresholding estimator*

$$\widehat{m}_S(x) = \sum_{j=1}^{\infty} \operatorname{sgn}(\widehat{\theta}_j)(|\widehat{\theta}_j| - \delta)_+ q_j(x), \qquad (2.18)$$

where sgn($\cdot$) denotes the sign function. This thresholding was introduced by Bickel (1983) and Donoho and Johnstone (1994). Figure 2.5 presents the hard- and soft-thresholding functions with $\delta = 0.5$, i.e. $t(x) = x\mathbf{1}\{|x| > \delta\}$ and $t(x) = \operatorname{sgn}(x)(|x| - \delta)_+$ respectively.

In summary, the truncated subseries estimator (2.16) is applied to situations where most of the information is compressed into the first N dimensions, whereas the thresholding estimators (2.13) or (2.18) are used when data information is concentrated on a few unknown dimensions, or more generally when there is no *primary information* available.

2.5.2 Basic ingredient of multiresolution analysis

The key ingredients of wavelet analysis are discussed in detail by Strang (1989, 1993), Meyer (1990) and Daubechies (1992). We here only highlight some key features of the multiresolution analysis of Meyer (1990). See also Section 5.1 of Daubechies (1992). The outlines presented here are detailed enough for many statistical applications, both theoretically or empirically.

Suppose that we decompose the function space $L^2(\mathbb{R})$ into successive closed subspaces $\cdots \subset V_{-1} \subset V_0 \subset V_1 \subset \cdots$ such that

$$\overline{\cup_{k \in \mathbb{Z}} V_k} = L^2(\mathbb{R}), \quad \cap_{k \in \mathbb{Z}} V_k = \{0\}. \qquad (2.19)$$

This *sieve method* approximates an L^2-function by a member in V_k with increasing accuracy, as k increases. The *multiresolution analysis* requires further that

$$f \in V_k \iff f(2^{-k} \cdot) \in V_0, \qquad (2.20)$$

i.e. a finer space is just a rescaling of a coarser space. Suppose further that there exists a *scale function* or *father wavelet* $\phi \in V_0$ such that its translations $\{\phi(x - \ell) : \ell \in \mathbb{Z}\}$ form an orthonormal basis in V_0. This requirement together with (2.20) implies

Requirements:

1. The subspace V_k is spanned by $\{2^{k/2}\phi(2^k x - \ell), \ell \in \mathbb{Z}\}$;

2. The sequence $\{2^{k/2}\phi(2^k x - \ell), \ell \in \mathbb{Z}\}$ is an orthonormal family in $L^2(\mathbb{R})$.

Since $\phi \in V_0 \subset V_1$, a necessary condition for the above requirements is that ϕ satisfies the so-called *scaling equation*,

$$\phi(x) = \sum_\ell c_\ell \phi(2x - \ell), \qquad (2.21)$$

where the constants c_ℓ have the property

$$\sum_\ell c_\ell c_{\ell-2m} = 2\delta_{0m}, \quad \forall m \in \mathbb{Z}. \qquad (2.22)$$

It can easily be checked that conditions (2.21) and (2.22) correspond respectively to Requirements 1 and 2 (see Strang (1989)). The well known Haar basis, defined by $\phi(x) = 1\{x \in (0,1]\}$, satisfies (2.21) and (2.22) with $c_0 = 1, c_1 = 1$ and the remaining coefficients c_ℓ zero.

The scaling coefficients $\{c_j\}$ uniquely determine the function ϕ under appropriate regularity conditions. Indeed, by (2.21), the Fourier transform $\widehat{\phi}$ of ϕ satisfies

$$\widehat{\phi}(t) = m_0(t/2)\widehat{\phi}(t/2),$$

where $m_0(t) = \sum_\ell c_\ell \exp(i\ell t)/2$. Using this formula recursively, one can easily see that $\widehat{\phi}(t) = \prod_{k=1}^\infty m_0(t/2^k)\widehat{\phi}(0)$. Consequently, $\widehat{\phi}$ and hence ϕ is determined by the coefficients $\{c_\ell\}$ to within a constant factor. It is easily seen that $\|\phi\|$, the L_2-norm of ϕ, equals 1 and hence ϕ is uniquely determined by the coefficients $\{c_\ell\}$. Further, if $\{c_\ell\}$ has bounded support $[0, N]$, so does ϕ. In the following, it will become clear that it is precisely the coefficients $\{c_\ell\}$ which are directly involved in computing wavelet transforms. Certain properties of a wavelet function, such as smoothness and moment conditions, can be tracked back to some simple conditions on the coefficients $\{c_\ell\}$.

Daubechies (1992) provides two families of wavelet constructions: the *extremal phase family* and the *least asymmetric family*, both indexed by N – the number of vanishing moments. Since $\{c_\ell\}$ uniquely determines the scaling function ϕ, tabulation of the coefficients $\{c_\ell\}$ is enough. These coefficients can be found in Tables 6.1 and 6.3 of Daubechies (1992). See also Chapter 7, in particular, pages 226 and 232 in Daubechies (1992), for the regularity of the wavelet functions in the extremal phase and the least asymmetric family. The coefficients $\{c_\ell\}$ in both families have a bounded

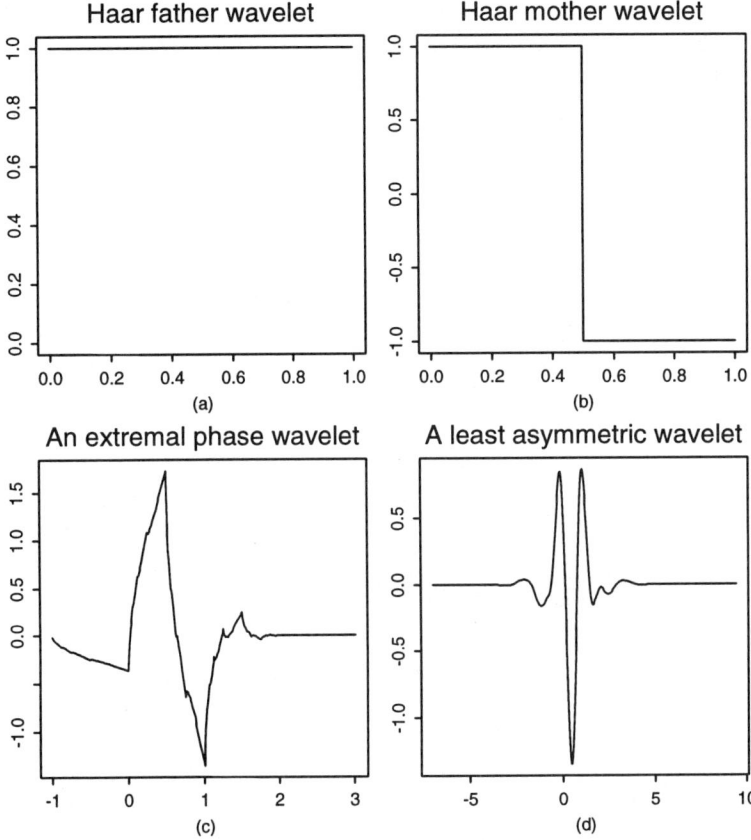

Figure 2.6. *Wavelet functions. (a) Haar father wavelet ; (b) Haar mother wavelet; (c) Daubechies' extremal phase (mother) wavelet with one vanishing moment; (d) Daubechies' least asymmetric (mother) wavelet with 7 vanishing moments.*

support and satisfy

$$\sum_\ell c_\ell = 2, \quad \sum_\ell (-1)^\ell \ell^j c_\ell = 0 \ (0 < j \le N-1), \quad \sum_\ell \ell^{2N} c_\ell^2 < \infty, \tag{2.23}$$

which in turn implies that the scaling function has a bounded support with $N-1$ vanishing moments:

$$\int x^j \phi(x) dx = \delta_{0j} \text{ for } j = 0, \cdots, N-1 \text{ and } \int |x^N \phi(x)| dx < \infty.$$

The function ϕ does not admit a closed form, but can be computed rapidly (see pages 204 and 205 in Daubechies (1992)). Figure 2.6 shows some of these functions.

Under (2.20) and Requirements 1 and 2, there exists a function ψ, called the *mother wavelet*, given by

$$\psi(x) = \sum_{\ell}(-1)^{\ell} c_{1-\ell}\phi(2x - \ell), \qquad (2.24)$$

such that

1. $\{2^{k/2}\psi(2^k x - \ell), \ell \in \mathbb{Z}\}$ is an orthonormal basis of W_k, where W_k is the orthogonal complement such that $V_{k+1} = V_k \oplus W_k$;
2. $\{2^{k/2}\psi(2^k x - \ell), \ell \in \mathbb{Z}, k \in \mathbb{Z}\}$ is an orthonormal basis of $L^2(\mathbb{R})$;

Further, if the coefficients $\{c_\ell\}$ satisfy (2.23), then

3. the zeroth and first $N-1$ moments of ψ vanish:

$$\int x^j \psi(x) dx = 0 \text{ for } j = 0, \cdots, N-1 \text{ and } \int |x^N \psi(x)| dx < \infty.$$

Since

$$L^2(\mathbb{R}) = \overline{\cup_{k \in \mathbb{Z}} V_k} = V_0 \oplus W_0 \oplus W_1 \oplus \cdots,$$

the basis functions

$$\{\phi(x - \ell), 2^{k/2}\psi(2^k x - \ell), \ell \in \mathbb{Z}, k \in \mathbb{Z}_+\}$$

form an orthonormal basis of $L^2(\mathbb{R})$.

Let $p > 0$ denote the *level of primary resolution*, and define $p_k = p2^k$. Put

$$\phi_\ell(x) = p^{1/2}\phi(px - \ell) \quad \text{and} \quad \psi_{k\ell}(x) = p_k^{1/2}\psi(p_k x - \ell)$$

for an integer $\ell \in \mathbb{Z}$. Then, as noted in the previous paragraph, the basis functions

$$\{\phi_\ell(x), \psi_{k\ell}(x), \ell \in \mathbb{Z}, k \in \mathbb{Z}_+\} \qquad (2.25)$$

are also an orthonormal basis of $L^2(\mathbb{R})$.

2.5.3 Wavelet shrinkage estimator

We are now in a position to apply the general principle described in Section 2.5.1 to construct estimators based on wavelet transforms.

Suppose that the regression function $m \in L^2(\mathbb{R})$. Then, by (2.25), it has the following representation:

$$m(x) = \sum_{\ell} \theta_\ell \phi_\ell(x) + \sum_{k=0}^{\infty} \sum_{\ell} \theta_{k\ell} \psi_{k\ell}(x),$$

with *wavelet coefficients*

$$\theta_\ell = \int m(x)\phi_\ell(x)dx, \quad \text{and} \quad \theta_{k\ell} = \int m(x)\psi_{k\ell}(x)dx. \quad (2.26)$$

Unbiased estimates of the wavelet coefficients are given by

$$\widehat{\theta}_\ell = n^{-1}\sum_{i=1}^n \phi_\ell(x_i)Y_i, \quad \widehat{\theta}_{k\ell} = n^{-1}\sum_{i=1}^n \psi_{k\ell}(x_i)Y_i.$$

For high resolution (i.e. large p_k), the estimate $\widehat{\theta}_{k\ell}$ will basically be noise, since $\psi_{k\ell}$ is supported only in a small neighborhood around ℓ/p_k and hence very few data points are used to calculate $\widehat{\theta}_{k\ell}$. Therefore, we terminate the analysis at a certain high resolution level q. For the wavelet basis, we have no prior information on where the large coefficients $|\theta_{ij}|$ are. Following (2.17) and (2.18), the hard- and soft-thresholding estimators are given respectively by

$$\widehat{m}_H(x) = \sum_\ell \widehat{\theta}_\ell \phi_\ell(x) + \sum_{k=0}^q \sum_\ell \widehat{\theta}_{k\ell} 1\{|\widehat{\theta}_{k\ell}| > \delta\}\psi_{k\ell}(x) \quad (2.27)$$

and

$$\widehat{m}_S(x) = \sum_\ell \widehat{\theta}_\ell \phi_\ell(x) + \sum_{k=0}^q \sum_\ell \text{sgn}(\widehat{\theta}_{k\ell})(|\widehat{\theta}_{k\ell}| - \delta)_+ \psi_{k\ell}(x). \quad (2.28)$$

Donoho and Johnstone (1994) proposed using $\delta = \sqrt{2\log n}\sigma$, which suppresses all noise contaminated in the estimated wavelet coefficients. They refer to such a thresholding as *visual shrinkage* since it often produces the best visual pictures. The choice of this particular thresholding parameter is based on extreme value theory for the Gaussian white noise:

$$(\sqrt{2\log n}\sigma)^{-1} \max_{1\leq i\leq n} \varepsilon_i \to 1 \quad \text{a.s.}$$

This choice is also asymptotically optimal as described in Donoho and Johnstone (1994).

2.5.4 Discrete wavelet transform

Direct calculation of the hard- and soft-thresholding estimators in (2.27) and (2.28) would require very intensive computations. In practical implementations they are calculated approximately via *discrete wavelet transforms* using a *pyramid algorithm* of $O(n)$-operations. See for example Mallat (1989). The discrete wavelet transform consists of an orthogonal transform of the original data by successively applying low-pass and high-pass linear filters. These two filters are defined as follows. For a given sequence $\{z_j\}$, the *low-pass filter* is given by

$$L(z)_j = \frac{1}{\sqrt{2}} \sum_\ell c_{\ell-j} z_\ell = \frac{1}{\sqrt{2}} \sum_\ell c_\ell z_{\ell+j}, \qquad (2.29)$$

where $\{c_\ell\}$ are the coefficients given in the scale equation (2.21) and the *high-pass filter* is defined by

$$H(z)_j = \frac{1}{\sqrt{2}} \sum_\ell d_{\ell-j} z_\ell = \frac{1}{\sqrt{2}} \sum_\ell d_\ell z_{\ell+j}, \qquad (2.30)$$

with $d_\ell = (-1)^\ell c_{1-\ell}$, the coefficients of equation (2.24).

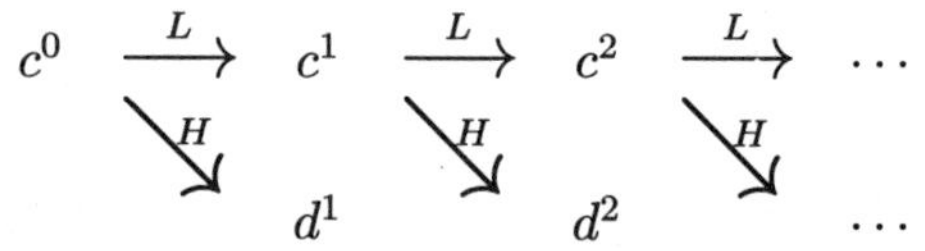

$$c^0 \xrightarrow{L} c^1 \xrightarrow{L} c^2 \xrightarrow{L} \cdots$$
$$\searrow^H \qquad \searrow^H \qquad \searrow^H$$
$$\qquad d^1 \qquad \quad d^2 \qquad \quad \cdots$$

Figure 2.7. *Schematic representation of wavelet decomposition.*

Suppose we have $n = 2^m$ observations of Y from model (2.14). To be consistent with the notation of Daubechies (1992), we denote this vector of observations by c^0 indexed from 0 to $n-1$. We first apply the low-pass filter to obtain a vector of length n and then keep only the even-indexed sequence (since the odd-indexed sequence will not be used, it is not actually computed). Denote this sequence of length 2^{m-1} by $c^1 = L(c^0)$. This step basically performs a smoothing of the original sequence. Next, apply the high-pass filter to get a vector of length n and keep only the even-indexed sequence (again, the odd-indexed sequence is not computed). Denote by $d^1 = H(c^0)$ the resulting vector of length 2^{m-1}. This vector corresponds to the detail removed after smoothing from c^0 to c^1. More precisely, the vector d^1 contains the discrete wavelet coefficients corresponding to the highest resolution level in the wavelet expansion. From a linear transform point of view, we apply an orthonormal matrix to the vector $\{c^0\}$ to obtain the new vectors of half-length $\{c^1\}$ and $\{d^1\}$. We now further decompose vector $\{c^1\}$ into the smoothed vector $\{c^2 = L(c^1)\}$ and the discrete wavelet coefficients $\{d^2 = H(c^1)\}$. Continuing this process m times, we obtain $d^1, d^2, \cdots, d^m$, which corresponds to the discrete wavelet coefficients at resolution $m-1, m-2, \cdots, 0$. The systematic scheme of the above decomposition can be expressed as in Figure 2.7. The discrete coefficients $d^1, d^2, \cdots, d^m$ are usually stacked in a pyramid form as shown in Figure 2.8 (b) – the coefficients d^j are drawn at the $(m-j)^{th}$ resolution level. They are just obtained by an orthogonal transform (denoted by Γ) of the original vector $\{y = c^0\}$, namely $d \equiv (d^1, \cdots, d^m) = \Gamma(c^0)$.

The simplest example to understand the above process is the Haar basis, for which $c_0 = 1, c_1 = 1$ and the rest of the coefficients c_ℓ are zero. In this case, the low-pass filter corresponds to the average of two neighboring data points, and the high-pass filter corresponds to the difference of these two points. Starting from a vector of length 2^m, the difference (divided by $\sqrt{2}$) of y_{2j-1} and y_{2j} forms the discrete wavelet coefficient at the highest resolution, and the average (multiplied by $\sqrt{2}$) of y_{2j-1} and y_{2j} forms the new vector to be further decomposed.

The boundary handling of linear filters can be delicate. The simplest convention is the periodic one, which regards data points as an infinite series by circularly padding the original series. We omit detailed discussions here.

We use the following simple example to illustrate the decomposition scheme and its relation with the orthogonal transform. Let $c^0 = (2, 3, 0, 1, 1, 0, 4, 3)^T$ be the vector to be transformed. Consider Daubechies' extremal phase wavelet with one vanishing moment, presented in Figure 2.6 (c). Then, from Table 6.1 of Daubechies (1992), we have the coefficients

$$\frac{1}{\sqrt{2}}(c_0, c_1, c_2, c_3)^T = (0.483, 0.837, 0.224, -0.129)^T.$$

Hence, we have the coefficients

$$\frac{1}{\sqrt{2}}(d_{-2}, d_{-1}, d_0, d_1)^T = (-0.129, -0.224, 0.837, -0.483)^T.$$

Applying the high-pass and the low-pass filter to c^0, we obtain

$$\begin{aligned} d^1 &= (-0.963, -1.413, 0.613, 1.77)^T, \\ c^1 &= (3.348, 1.061, 0.992, 4.504)^T. \end{aligned}$$

The orthogonal matrix Γ_1 that transforms the vector c_0 to $\begin{pmatrix} d^1 \\ c^1 \end{pmatrix}$

38 OVERVIEW OF EXISTING METHODS

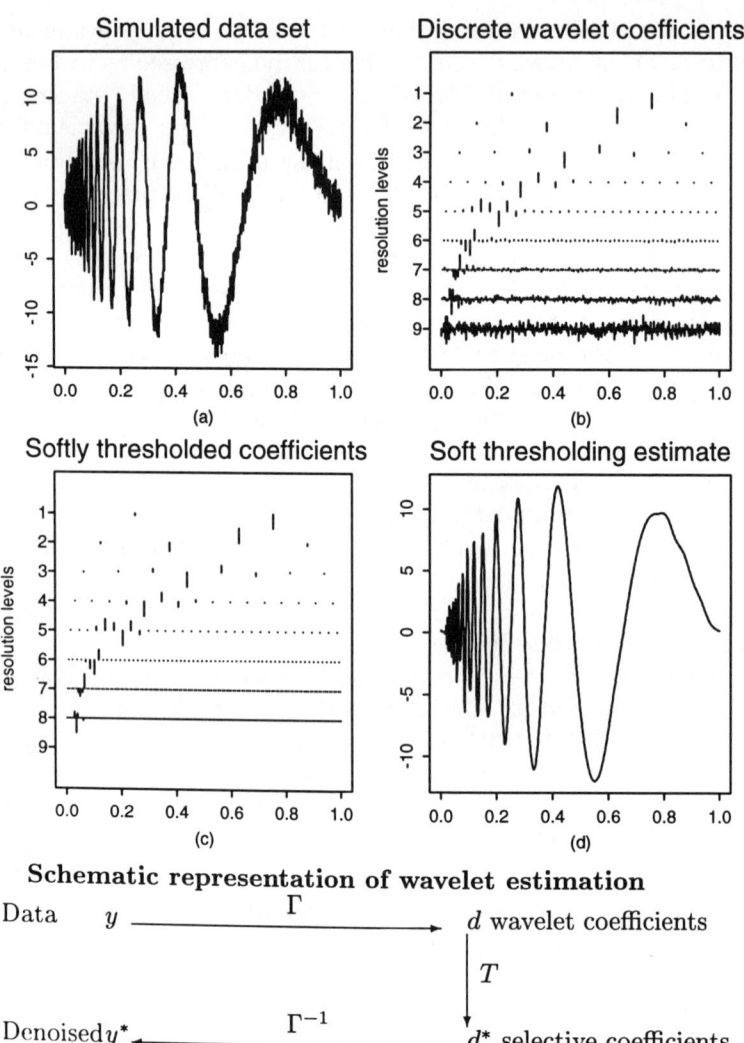

Figure 2.8. *Wavelet decomposition and reconstruction by using the least asymmetric Daubechies wavelet with $N = 8$. Soft-thresholding applies to all wavelet coefficients in (b) except for the first 5 resolution levels. This idea is adapted from Donoho and Johnstone (1994).*

can easily be obtained as:

$$\begin{pmatrix} 0.837 & -0.483 & & & & & -0.129 & -0.224 \\ -0.129 & -0.224 & 0.837 & -0.483 & & & & \\ & & -0.129 & -0.224 & 0.837 & -0.483 & & \\ & & & & -0.129 & -0.224 & 0.837 & -0.483 \\ 0.483 & 0.837 & 0.224 & -0.129 & & & & \\ & & 0.483 & 0.837 & 0.224 & -0.129 & & \\ & & & & 0.483 & 0.837 & 0.224 & -0.129 \\ 0.224 & -0.129 & & & & & 0.483 & 0.837 \end{pmatrix}.$$

Note that this matrix is formed by stacking the coefficients in the high-pass and the low-pass filters. As in the first row, circular padding is used when needed. One can further decompose the vector c^1 in the same fashion. Therefore, the wavelet coefficients are obtained by consecutively applying a few orthogonal transforms.

Once the discrete wavelet coefficients $d = \Gamma(y)$ are obtained, one can apply the hard- or soft-thresholding to obtain a new sequence of the empirical wavelet coefficients $d^* = T(d)$, as is illustrated in Figure 2.8 (c). Finally, apply the inverse transform Γ^{-1} to construct the smoothed curve $y^* = \Gamma^{-1}(d^*)$, which is depicted in Figure 2.8 (d). Figure 2.8 demonstrates how a wavelet estimator is constructed from a sample of $n = 2^{10} = 2048$ observations.

The *universal thresholding* suggested by Donoho and Johnstone (1994) is $\delta = \sqrt{2 \log n} \hat{\sigma}$, where $\hat{\sigma}$ is the median absolute deviation of the discrete wavelet coefficients at the highest resolution, divided by 0.6745. This is based on robust estimation of the standard deviation under the normal model. However, in practical implementation, one may also want to try other values of the thresholding parameter.

The software developed by Nason and Silverman (1994), and Bruce, Gao and Ragozin (1993) in S-Plus makes implementation of the above algorithm very easy.

2.6 Spline smoothing

A polynomial function, possessing all derivatives at all locations, is not very flexible for approximating curves with different degrees of smoothness at different locations. Examples of such curves include those presented in Figures 2.1 (b) and 2.8 (d). See also Figure 1.2. One way to overcome this drawback is to use *locally* a polynomial approximation of low degree, such as running lines introduced in Section 2.3. Another way to enhance the flexibility is to allow the derivatives of the approximating function to have discontinuities at certain locations. This can be accomplished by fitting piecewise polynomials or *splines*, resulting in the *spline method*. The points where the derivatives of the approximating function could have discontinuities are called *knots*. Useful reference books on spline approximations and their statistical applications include Eubank (1988) and Wahba (1990). Recently, Green and Silverman (1994) gave a comprehensive account on applications of spline techniques to various fields of statistics.

Consider the homoscedastic model

$$Y_i = m(X_i) + \varepsilon_i, \qquad i = 1, \cdots, n, \qquad (2.31)$$

where ε_i are i.i.d. with zero mean and common variance σ^2. The homoscedasticity of the model (2.31) is important to the development of spline smoothing techniques, although those techniques can also be applied to heteroscedastic models. In the following, we first introduce the spline method with fixed knots, and then offer two procedures for automatic selection of knots. The first procedure is based on the *knot deletion* idea of Smith (1982) and Breiman, Friedman, Olshen and Stone (1993), and the second method relies on the *smoothing spline* of Wittaker (1923) and Wahba (1975).

2.6.1 Polynomial spline

Suppose that we want to approximate the unknown regression function m by a spline function. Frequently, a cubic spline, i.e. a piecewise polynomial with continuous first two derivatives, is used for such an approximation. For concreteness, we focus only on the cubic spline approximation, but the idea can also be applied to splines of other order. Let $t_1, \cdots, t_J$ be a fixed knot sequence such that $-\infty < t_1 < \cdots < t_J < +\infty$. Then, the cubic spline functions are twice continuously differentiable functions s such that the restriction of s to each of the intervals $(-\infty, t_1], [t_1, t_2], \cdots, [t_{J-1}, t_J], [t_J, +\infty)$ is a cubic polynomial. The collection of all cubic spline functions forms a $(J+4)$-dimensional linear space. There are two popular cubic spline bases for this linear space:

- *power basis*: $(x - t_j)_+^3, (j = 1, \cdots, J), 1, x, x^2, x^3$;

- *B-spline basis* (see page 108 of de Boor (1978) for the definition).

Usually, the B-spline basis is numerically more stable, because the multiple correlation among the basis functions is smaller, but the power spline basis has the advantage that deleting the basis function $(x - t_j)_+^3$ is the same as deleting the knot t_j. Figure 2.9 displays these two bases.

SPLINE SMOOTHING

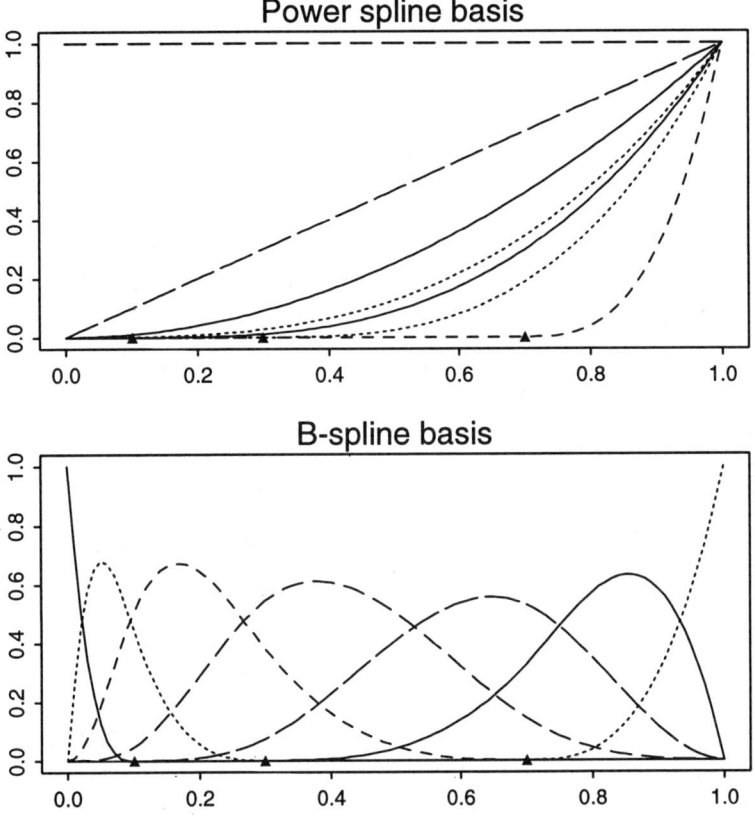

Figure 2.9. *The power spline basis and B-spline basis for given knots 0.1, 0.3 and 0.7.*

Let $B_1, \cdots, B_{J+4}$ be a cubic spline basis. Then, a cubic spline function can be expressed as

$$s(x) = \sum_{j=1}^{J+4} \theta_j B_j(x).$$

For the given knots the spline method consists of finding the best spline approximation via the following least squares regression:

$$\min_{\theta} \sum_{i=1}^{n} \{Y_i - \sum_{j=1}^{J+4} \theta_j B_j(X_i)\}^2. \qquad (2.32)$$

Let $\hat{\theta}_j$ $(j = 1, \cdots, J+4)$ be the least squares estimate. Then, the spline approach estimates m by the *spline function* $\hat{m}(x) = \sum_{j=1}^{J+4} \hat{\theta}_j B_j(x)$.

The spline method described above is sensitive to the choice of the number of knots J as well as to the location of the knots. The knots are often placed at locations where curvatures have a reasonably large change. Trial-and-error procedures on placing knots can be very cumbersome and time consuming. An automatic procedure for selecting knots is clearly desirable.

We first introduce the *knot deletion* idea. Let $t_1, \cdots, t_J$ be the initial knots that might be deleted in the knot selection process. These knots are often placed at the order statistics of the X-variable and the number of knots J is often taken to be $J = [n/2]$ or $J = [n/3]$. An example of the initial knots are $t_j = X_{(2j)}, j = 1, \cdots, [n/2]$. Let the corresponding power basis be

$$\begin{cases} B_j(x) = (x-t_j)_+^3, (j=1,\cdots,J), \\ B_{J+1}(x) = 1, B_{J+2}(x) = x, B_{J+3}(x) = x^2, B_{J+4}(x) = x^3. \end{cases}$$

Denote by $\hat{\theta}_j$ the least squares estimate resulting from (2.32) and let $SE(\hat{\theta}_j)$ be its estimated standard error. Then, delete the j_0^{th} knot $(1 \leq j_0 \leq J)$ having the smallest absolute t-value: $|\hat{\theta}_j|/SE(\hat{\theta}_j)$ $(1 \leq j \leq J)$. Repeat the above deleting process (delete one knot at each step). We arrive at a sequence of models indexed by j $(0 \leq j \leq J)$: the jth model contains $J+4-j$ free parameters with *residual sum of squares* RSS_j. Choose the model $\hat{j}$ that minimizes the modified *Mallows' C_p criterion* (see Mallows (1973)):

$$C_j = RSS_j + \alpha(J+4-j)\hat{\sigma}^2, \qquad (2.33)$$

where $\hat{\sigma}$ is the estimated standard deviation at the 0th model (full model), and α denotes a smoothing parameter penalizing for too complex models. This completes the knot selection process. The knots involved in the $\hat{j}^{th}$ model are used for the final spline modelling. This idea is often employed by Stone and his collaborators. See for example Kooperberg and Stone (1991) and Kooperberg, Stone and Truong (1995a, b).

Under the Gaussian assumption $\varepsilon_j \sim N(0, \sigma^2)$, the criterion (2.33) is the same as *Akaike's information criterion* (see Akaike (1970, 1974)) with penalty parameter α for the jth model. Kooperberg and Stone (1991) recommend using $\alpha = 3$ instead of the more traditional value $\alpha = 2$, applied in the context of *logspline* density estimation. Our experiences tend to support their recommenda-

SPLINE SMOOTHING 43

tion. Schwarz (1978) recommends using $\alpha = \log n$ in a different context. For more detailed studies on the choice of the parameter α, see Shibata (1976), Woodroofe (1982) and Zhang (1992).

2.6.2 Smoothing spline

An alternative approach to automatic selection of knots is the smoothing spline. To motivate this procedure, let us first consider a naive least squares problem: find a function m that minimizes

$$\sum_{i=1}^{n} \{Y_i - m(X_i)\}^2.$$

The solution to this naive least squares problem is any function m which interpolates the data. Such a solution is undesirable for most statistical applications: it does not at all analyze the data and produces a model as complex as the original data. From a statistical modelling point of view, it over-parametrizes the model, resulting in large variability of the estimated parameters. The residuals $\widehat{\varepsilon}_i = Y_i - \widehat{m}(X_i)$ from this naive approach are zero. This strongly contradicts our model (2.31) – one cannot expect that the realizations of the uncorrelated random noise ε_i are all zero. This is another reason why the naive least squares approach is not desirable.

What is lacking in this naive least squares approach? The defect is that we did not impose a penalty for over-parametrization. A convenient way for introducing such a penalty is via the roughness, popularly measured by $\int \{m''(x)\}^2 dx$. This leads to the following penalized least squares regression: find $\widehat{m}_\lambda$ that minimizes

$$\sum_{i=1}^{n} \{Y_i - m(X_i)\}^2 + \lambda \int \{m''(x)\}^2 dx, \qquad (2.34)$$

for a nonnegative real number $\lambda > 0$, called a smoothing parameter. Expression (2.34) consists of two parts. The first part penalizes the lack of fit, which is in some sense the modelling bias. The second part puts a penalty on the roughness, which relates to the over-parametrization. It is clear that $\lambda = 0$ corresponds to interpolation, whereas $\lambda = +\infty$ results in a linear regression $m(x) = \alpha + \beta x$. As λ ranges from zero to infinity, the estimate ranges from the most complex model (interpolation) to the simplest model (linear model). Thus, the model complexity of the smoothing spline approach is effectively controlled by the smoothing parameter λ. The estimator $\widehat{m}_\lambda$ is called the *smoothing spline estimator*.

Nychka (1995) studies the local properties of such estimators. An extension of (2.34), using a variable smoothing parameter, can be found in Abramovich and Steinberg (1996).

Expression (2.34) has also a Bayesian interpretation. See for example Good and Gaskins (1971) for the density estimation setup and Wahba (1978) for the regression setup.

It is well known that a solution to the minimization of (2.34) is a cubic spline on the interval $[X_{(1)}, X_{(n)}]$. The solution is also unique in this data range. Moreover, it can easily be argued that $\widehat{m}_\lambda$ is linear in the responses:

$$\widehat{m}_\lambda(x) = n^{-1} \sum_{i=1}^{n} W_i(x, \lambda; X_1, \cdots, X_n) Y_i, \qquad (2.35)$$

where the weight W_i does not depend on the response $\{Y_i\}$. See for example pages 59-60 of Härdle (1990). The connections between kernel regression and smoothing splines have been established theoretically by Silverman (1984, 1985), and related approximations to a Green's function have been studied by Speckman (1981) and Cox (1984). In particular, Silverman (1984) pointed out that the smoothing spline is basically a local kernel average with a variable bandwidth. For X_i away from the boundary, and for n large and λ relatively small,

$$W_i(x, \lambda; X_1, \cdots, X_n) \approx f(X_i)^{-1} h(X_i)^{-1} K_s\{(X_i - x)/h(X_i)\}, \qquad (2.36)$$

where $h(X_i) = [\lambda/\{nf(X_i)\}]^{1/4}$ and

$$K_s(t) = 0.5 \exp(-|t|/\sqrt{2}) \sin(|t|/\sqrt{2} + \pi/4).$$

This approximation is also valid for calculating the mean and variance of the smoothing spline estimator $\widehat{m}_\lambda(x)$, as evidenced by Messer (1991). The approximation (2.36) provides intuition to help in understanding how the smoothing spline assigns weights to the local neighborhood around a point x.

The smoothing parameter λ can be chosen objectively by the data. One approach to select λ is via the minimization of the *cross-validation* criterion

$$\text{CV}(\lambda) = n^{-1} \sum_{i=1}^{n} \{Y_i - \widehat{m}_{\lambda,i}(X_i)\}^2, \qquad (2.37)$$

where $\widehat{m}_{\lambda,i}$ is the estimator arising from (2.34) without using the ith observation. This cross-validation method uses ideas of Allen

(1974) and Stone (1974). The cross-validation criterion (2.37) is computationally intensive. An improved version of this is the *Generalized Cross-Validation* (GCV), proposed by Wahba (1977) and Craven and Wahba (1979). This criterion can be described as follows. By (2.35) the fitted values can be expressed as

$$(\widehat{m}_\lambda(X_1), \cdots, \widehat{m}_\lambda(X_n))^T = H(\lambda)Y,$$

where $H(\lambda)$ is the $n \times n$ hat matrix, depending only on the X-variate, and $Y = (Y_1, \cdots, Y_n)^T$, where the superscript T denotes the transpose. Then the GCV approach selects λ that minimizes

$$\text{GCV}(\lambda) = [n^{-1}\text{tr}\{I - H(\lambda)\}]^{-2} \text{MASE}(\lambda) \tag{2.38}$$

where $\text{MASE}(\lambda) = n^{-1}\sum_{i=1}^n\{Y_i - \widehat{m}_\lambda(X_i)\}^2$. Both quantities $\text{CV}(\lambda)$ and $\text{GCV}(\lambda)$ are consistent estimates of the MISE of $\widehat{m}_\lambda$. There are also other methods for selecting λ. See Wahba and Wang (1990) for a description of all these methods.

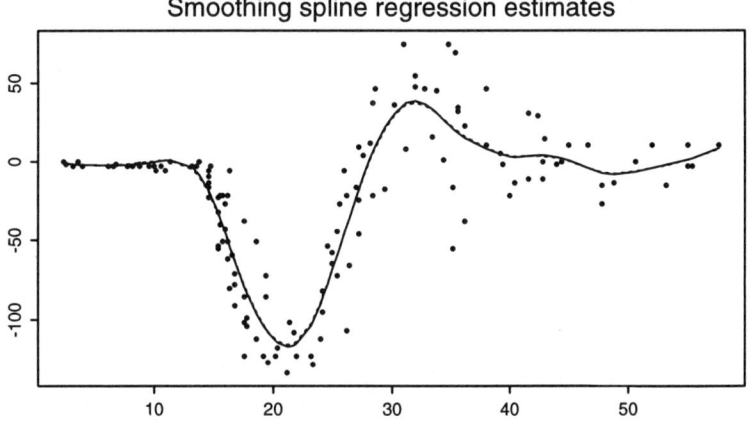

Figure 2.10. *Smoothing spline estimate of the regression function for the motorcycle data, using respectively CV and GCV to select the smoothing parameter. The resulting two estimated curves are indistinguishable for this data set.*

Fast computation issues of smoothing splines can be found in Section 5.3 of Eubank (1988) and Chapter 11 of Wahba (1990). Figure 2.10 displays the smoothing spline regression estimator with the smoothing parameter selected by respectively CV and GCV, defined by respectively (2.37) and (2.38). The two estimates are almost indistinguishable for this example.

2.7 Density estimation

A technique closely related to nonparametric regression is *density estimation*. It is used to examine the overall pattern of a data set. This includes number and locations of peaks and valleys as well as the symmetry of the density. A comprehensive account of this subject and its applications is given in detail in Silverman (1986) and Scott (1992). Here we only discuss some important aspects of the technique.

2.7.1 Kernel density estimation

Suppose that we are given n observations $X_1, \cdots, X_n$. The empirical distribution function $\widehat{F}_n(x)$ of this data set is obtained by putting mass $1/n$ at each datum point. However, the data structure can hardly be examined via the plot of the function $\widehat{F}_n(x)$. A better visualization device is to attempt to plot its density function $\widehat{F}'_n(x)$, but this function does not exist. An improved idea over the empirical distribution function is to smoothly redistribute the mass $1/n$ at each datum point to its vicinity. See Figure 2.11. This leads to the following kernel density estimate:

$$\widehat{f}_h(x) = n^{-1} \sum_{i=1}^{n} K_h(X_i - x). \qquad (2.39)$$

If the data $X_1, \cdots, X_n$ can be thought of as a random sample from a population with a density f, then $\widehat{f}_h(x)$ estimates this unknown density function f. The MSE of the kernel density estimate can be expressed as

$$\text{MSE}(x) \equiv E\{\widehat{f}_h(x) - f(x)\}^2$$
$$\approx \frac{1}{4}\{\int_{-\infty}^{\infty} u^2 K(u) du\}^2 \{f''(x)\}^2 h^4 + \int_{-\infty}^{\infty} K^2(u) du \frac{f(x)}{nh}. \qquad (2.40)$$

An important aspect of the kernel density estimator (2.39) is the selection of the bandwidth. A too large bandwidth results in an oversmoothed estimate which obscures the fine structure of the data, while a too small bandwidth makes an undersmoothed estimate, producing a wiggly curve and artificial modes. See Figure 2.12. An ideal bandwidth is the one that minimizes the asymptotic

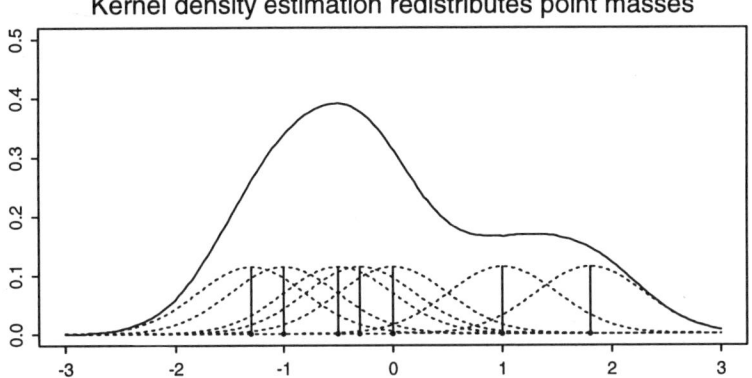

Figure 2.11. *Kernel density estimation redistributes the point mass at each datum point and adds the redistributed masses together to get the final estimate.*

MISE (with weight function $w \equiv 1$), resulting in

$$h_{opt} = \alpha(K) \left[\int \{f''(x)\}^2 dx \right]^{-1/5} n^{-1/5}, \qquad (2.41)$$

where $\alpha(K) = \{\int u^2 K(u) du\}^{-2/5} \{\int K^2(u) du\}^{1/5}$ is a known constant. However, this bandwidth depends on the unknown parameter $\int f''^2$. When f is a Gaussian density with standard deviation σ, one can easily deduce from (2.41) that

$$h_{opt,n} = (8\sqrt{\pi}/3)^{1/5} \alpha(K) \sigma n^{-1/5}. \qquad (2.42)$$

The *normal reference bandwidth selector* (see for example Bickel and Doksum (1977) and Silverman (1986)) is the one obtained by replacing the unknown parameter σ by the sample standard deviation s_n. When K is the Gaussian kernel, the normal reference bandwidth selector is $\widehat{h}_{opt,n} = 1.06 s_n n^{-1/5}$; for the Epanechnikov kernel, $\widehat{h}_{opt,n} = 2.34 s_n n^{-1/5}$. See Section 3.4 of Silverman (1986) for more discussions. An improved rule can be obtained by writing an Edgeworth expansion for f around the Gaussian density. Such a rule is provided in Hjort and Jones (1996b) and is given by

$$\widehat{h}^*_{opt,n} = \widehat{h}_{opt,n} \left(1 + \frac{35}{48} \widehat{\gamma}_4 + \frac{35}{32} \widehat{\gamma}_3^2 + \frac{385}{1024} \widehat{\gamma}_4^2 \right)^{-1/5},$$

where $\widehat{h}_{opt,n}$ is obtained from (2.42), and $\widehat{\gamma}_3$ and $\widehat{\gamma}_4$ are respectively

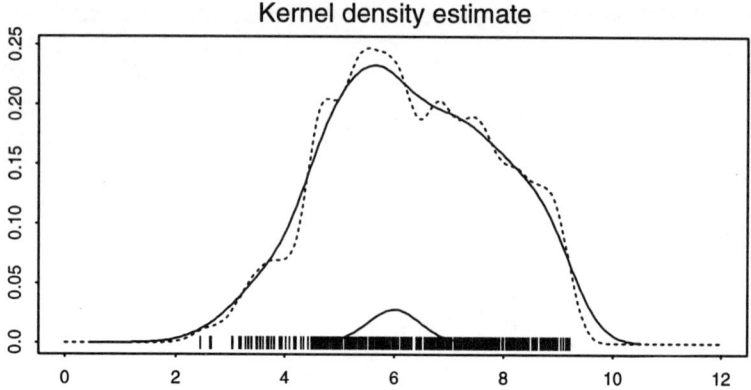
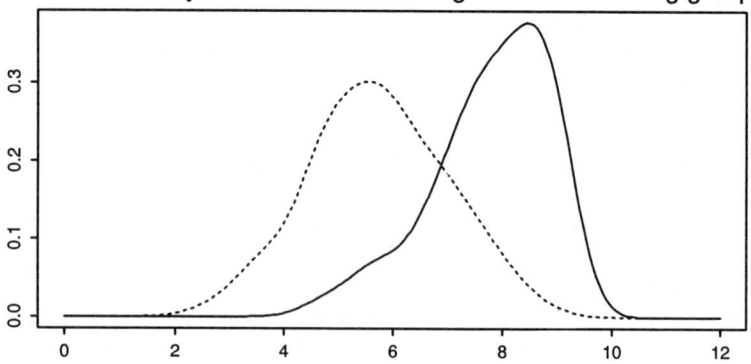

Figure 2.12. *Kernel density estimation for burn data using the Gaussian kernel with the normal reference bandwidth selection rule. The data here are the log(area of third degree burn + 1). Top panel: the data are indicated at the bottom of the plot. The Gaussian density at the bottom of the plot indicates the amount of smoothing. The solid curve uses bandwidth (2.42), while the dashed curve uses half of the bandwidth (2.42). Bottom panel: separate kernel density estimates for the non-surviving group (solid line) and surviving group (dashed line).*

the sample skewness and kurtosis.

The normal reference bandwidth selector is a good selector when the data are nearly Gaussian distributed, and is often reasonable in many applications. However, it can oversmooth the estimate when the data are not nearly Gaussian distributed. In that case, one can

DENSITY ESTIMATION

either subjectively tune the bandwidth, or objectively select the bandwidth by more sophisticated bandwidth selectors.

The kernel density estimate provides a powerful tool for the comparison of two experiments, namely the study of a *two-sample problem*. Figure 2.12 illustrates kernel density estimation on the burn data set. Clearly, the area of third degree burn injuries is lower in the surviving group with a large standard deviation.

Kernel density estimation is also a very useful device for analyzing longitudinal data. See Example 3 of Section 4.8 for a detailed analysis.

There are quite a few important techniques for selecting the bandwidth such as smoothing cross-validation (CV) and plug-in bandwidth selectors. A conceptually simple technique, with good theoretical and empirical performance, is the *plug-in technique*. This technique relies on finding an estimate of the functional $\int f''^2$ in (2.41). A good implementation of this approach is proposed by Sheather and Jones (1991). We will discuss this procedure and other related techniques in Section 4.10.

The derivative $f^{(\nu)}(x)$ can be estimated by taking the νth derivative of the kernel density estimator (2.39):

$$\widehat{f}_h^{(\nu)}(x) = (nh^\nu)^{-1} \sum_{i=1}^n K_h^{(\nu)}(X_i - x), \qquad (2.43)$$

where $K_h^{(\nu)}(t) = K^{(\nu)}(t/h)/h$. The MSE of this estimator is

$$E\{\widehat{f}_h^{(\nu)}(x) - f^{(\nu)}(x)\}^2 \approx \frac{1}{4}\{\int_{-\infty}^\infty u^2 K(u)du\}^2 \{f^{(\nu+2)}(x)\}^2 h^4$$

$$+ \int_{-\infty}^\infty \{K^{(\nu)}(u)\}^2 du \frac{f(x)}{nh^{2\nu+1}}. \qquad (2.44)$$

Minimizing the asymptotic MISE (with weight function $w \equiv 1$) of $\widehat{f}_h^{(\nu)}$ leads to the asymptotically optimal bandwidth

$$h_{\nu,\text{opt}} = (2\nu+1)^{\frac{1}{2\nu+5}} \alpha_\nu(K) [\int \{f^{(\nu+2)}(x)\}^2 dx]^{-\frac{1}{2\nu+5}} n^{-\frac{1}{2\nu+5}}, \qquad (2.45)$$

where $\alpha_\nu(K) = \{\int u^2 K(u)du\}^{-2/(2\nu+5)}[\int \{K^{(\nu)}(u)\}^2 du]^{1/(2\nu+5)}$. The normal referencing technique can easily be applied to (2.45) to obtain a data-driven bandwidth selector. For example, for estimating the derivative function f',

$$\widehat{h}_{1,\text{opt}} = (16\sqrt{\pi}/5)^{1/7} \alpha_1(K) \widehat{s}_n n^{-1/7}, \qquad (2.46)$$

where $\widehat{s}_n$ is the sample standard deviation. In particular, when the Gaussian kernel is employed, we have

$$\widehat{h}_{1,\text{opt}} = (4/5)^{1/7}\widehat{s}_n n^{-1/7} = 0.9686 \widehat{s}_n n^{-1/7}. \qquad (2.47)$$

2.7.2 Regression view of density estimation

The connection between density estimation and nonparametric regression can be made clear via a binning technique. Suppose that we are interested in estimating the density function $f(x)$ or its derivatives on an interval $[a, b]$. Partition the interval $[a, b]$ into N subintervals $\{I_k, k = 1, \ldots, N\}$ of equal length $\Delta = (b - a)/N$. Let x_k be the center of I_k and y_k be the proportion of the data $\{X_i, i = 1, \ldots, n\}$ falling in the interval I_k, divided by the bin length Δ. Then it is clear that the bin counts $n\Delta y_k$ have a binomial distribution:

$$n\Delta y_k \sim \text{Binomial}(n, p_k) \quad \text{with} \quad p_k = \int_{I_k} f(x)dx. \qquad (2.48)$$

When the partition is fine enough, i.e. when $N \to \infty$, it can easily be calculated from (2.48) that

$$Ey_k \approx f(x_k), \quad \text{Var}(y_k) \approx \frac{f(x_k)}{n\Delta}. \qquad (2.49)$$

Thus, we could regard the density estimation problem as a heteroscedastic nonparametric regression problem with $m(x) = f(x)$ and $\sigma(x) = \frac{f(x)}{n\Delta}$, based on the data $\{(x_k, y_k) : k = 1, \cdots, N\}$, which are approximately independent. See Theorem 4.1 of Fan (1996) for an exact formulation and justification. Hence, the nonparametric regression techniques introduced in this chapter are in principle applicable to the density estimation setup.

Consider for example local polynomial fitting of order p. Let $\widehat{\beta}_0, \ldots, \widehat{\beta}_p$ be the solution to the local polynomial regression problem:

$$\sum_{k=1}^{N} \{y_k - \sum_{j=0}^{p} \beta_j (x_k - x)^j\}^2 K_h(x_k - x).$$

Then, the estimator for the νth derivative is

$$\widehat{f}_\nu(x) = \nu! \widehat{\beta}_\nu.$$

Under the assumption that $Nh \to \infty$, $h \to 0$ and $\Delta/h \to 0$, the asymptotic properties for $\widehat{f}_\nu(x)$ are the same as those obtained by

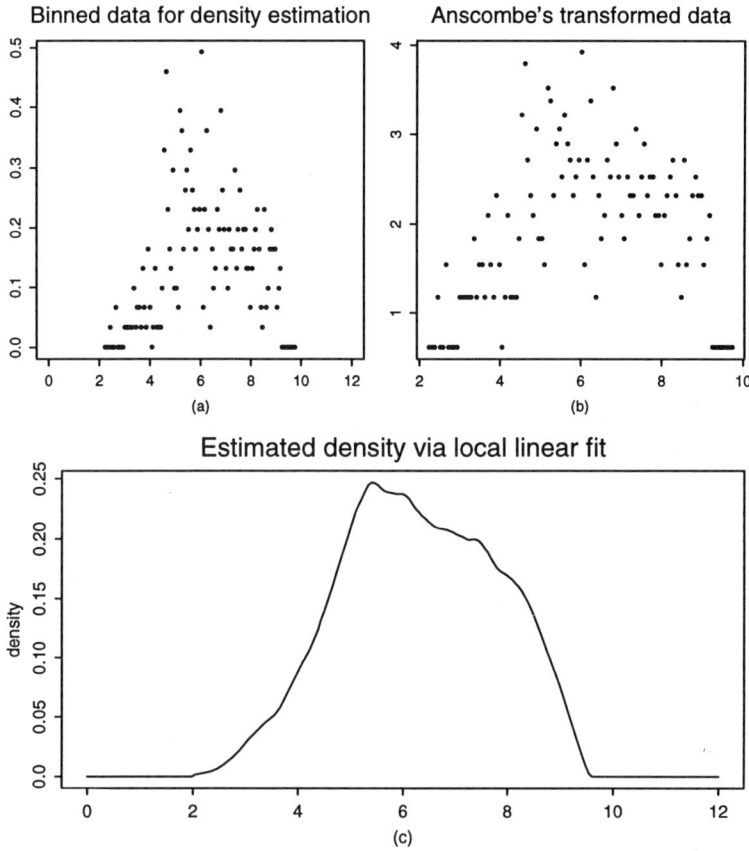

Figure 2.13. *Regression view of density estimation for burn data using the local linear fit with the Epanechnikov kernel and bandwidth 1.0. The data here are the log(area of third degree burn + 1) and the number of bins is $N = 108$.*

the kernel density estimator (2.43) (see Cheng, Fan, and Marron (1993)).

As indicated in (2.49), the induced data $\{(x_k, y_k), k = 1, \cdots, N\}$ are heteroscedastic. Often, homoscedastic data are preferable. This is particularly true when spline and wavelet methods are used. The homoscedasticity can be accomplished via Anscombe's (1948) *variance-stabilizing transformation* to the bin count $c_k = n\Delta y_k$, i.e.

$$y_k^* = 2\sqrt{c_k + \frac{3}{8}}.$$

The density estimator is then obtained by applying a nonparametric smoother to the transformed data set $\{(x_k, y_k^*), k = 1, \cdots, N\}$, and taking the inverse of Anscombe's transformation. More precisely, let $\widehat{m}^*(x)$ be a regression smoother based on the transformed data. Then the density estimator is defined by

$$\widehat{f}(x) = C[\frac{\{\widehat{m}^*(x)\}^2}{4} - \frac{3}{8}]_+,$$

where the constant C is a normalization constant such that $\widehat{f}(x)$ integrates to 1. Figure 2.13 shows how local linear regression can be used to estimate the density.

One can also estimate the density function by using the local polynomial fit from a local likelihood prospective. See Loader (1995) and Hjort and Jones (1996a).

2.7.3 Wavelet estimators

Wavelet techniques can also be applied to density estimation to invoke variable degrees of smoothing. This is convincingly demonstrated in Kerkyacharian and Picard (1993a, b) and Donoho, Johnstone, Kerkyacharian and Picard (1993). It is shown by Donoho and Johnstone (1996) that wavelet transforms with a single thresholding attain, within a logarithm factor, the optimal minimax risk rates for Hölder classes of densities with unspecified degree of smoothness. This kind of spatial adaptation property is also shown by Fan, Hall, Martin, and Patil (1993) using a completely different formulation.

Let f be square integrable. By (2.25), it admits the following wavelet expansion:

$$f(x) = \sum_\ell \theta_\ell \phi_\ell(x) + \sum_{k=0}^\infty \sum_\ell \theta_{k\ell} \psi_{k\ell}(x), \qquad (2.50)$$

with wavelet coefficients

$$\theta_\ell = \int f(x) \phi_\ell(x) dx, \quad \theta_{k\ell} = \int f(x) \psi_{k\ell}(x) dx.$$

DENSITY ESTIMATION

This suggests unbiased estimates of the wavelet coefficients:

$$\widehat{\theta}_\ell = n^{-1} \sum_{i=1}^n \phi_\ell(X_i), \quad \widehat{\theta}_{k\ell} = n^{-1} \sum_{i=1}^n \psi_{k\ell}(X_i).$$

Following Donoho and Johnstone (1994), we select useful estimated coefficients $\widehat{\theta}_{k\ell}$ by *thresholding*, leading to the hard- and soft-thresholding estimators:

$$\widehat{f}_H(x) = \sum_\ell \widehat{\theta}_\ell \phi_\ell(x) + \sum_{k=0}^q \sum_\ell \widehat{\theta}_{k\ell} 1\{|\widehat{\theta}_{k\ell}| \geq \delta\} \psi_{k\ell}(x), \quad (2.51)$$

and

$$\widehat{f}_S(x) = \sum_\ell \widehat{\theta}_\ell \phi_\ell(x) + \sum_{k=0}^q \sum_\ell \text{sgn}(\widehat{\theta}_{k\ell})(|\widehat{\theta}_{k\ell}| - \delta)_+ \psi_{k\ell}(x), \quad (2.52)$$

where δ is a thresholding parameter. In order not to downgrade the noise cumulation in the selection process of the empirical wavelet coefficients, δ is taken to be of the form

$$\delta = c(\|f\|_\infty \log n/n)^{1/2} \quad (2.53)$$

with $c \geq \sqrt{6}$. See, for example, Fan, Hall, Martin and Patil (1993).

Density estimation is a heteroscedastic problem. Indeed, it can easily be seen that when p_k is large $\text{Var}(\widehat{\theta}_{k\ell}) \approx f(\ell/p_k)$, which depends on the unknown density at a dyadic location ℓ/p_k. A good thresholding policy should be based on the random variable $\widehat{\theta}_{k\ell}/\sqrt{\text{Var}(\widehat{\theta}_{k\ell})}$. For the hard-thresholding estimator, this policy is of the form $1\{\widehat{\theta}_{k\ell}/\sqrt{\text{Var}(\widehat{\theta}_{k\ell})} \geq \delta^*\}$ for some constant δ^*, yielding a *variable thresholding* policy. However, the thresholding parameter δ in (2.53) simply replaces the unknown density by its maximum. This could filter out many important wavelet coefficients in a low density region. For this heteroscedastic model, more appealing wavelet shrinkage estimators can be defined as follows:

$$\widehat{f}_H(x) = \sum_\ell \widehat{\theta}_\ell \phi_\ell(x) + \sum_{k=0}^q \sum_\ell \widehat{\theta}_{k\ell} 1\{|\widehat{\theta}_{k\ell}| \geq \delta^* \sqrt{\check{f}(\ell/p_k)}\} \psi_{k\ell}(x),$$

$$\widehat{f}_S(x) = \sum_\ell \widehat{\theta}_\ell \phi_\ell(x) + \sum_{k=0}^q \sum_\ell \text{sgn}(\widehat{\theta}_{k\ell})\{|\widehat{\theta}_{k\ell}|/\sqrt{\check{f}(\ell/p_k)} - \delta^*\}_+$$
$$\times \psi_{k\ell}(x),$$

where $\delta^* = c(\log n/n)^{1/2}$ with $c \geq \sqrt{6}$ and $\check{f}$ is a pilot estimate. These variable thresholding estimators share the same spatial adaptation as those defined by (2.51) and (2.52), but are more appealing in practical applications. A simple pilot estimator can be constructed by using the kernel density estimator (2.39) with the normal reference bandwidth rule (2.42).

2.7.4 Logspline method

Logspline density estimation, proposed by Stone and Koo (1986) and Kooperberg and Stone (1991), captures nicely the tail of a density. See also Breiman, Stone and Kooperberg (1990) for its applications to estimating extreme upper tail quantiles. The basic idea is to model the logarithm of a density by a spline function, and then to use the maximum likelihood approach to estimate the parameters in the model. The logarithm transform has a few advantages. First of all, it penalizes the error in tail estimation more severely than the ordinary spline modelling. Secondly, it preserves the positivity constraint of a density function. Finally, it makes the log-likelihood function (2.54) strictly concave, an attractive feature for computation. This concavity makes computation fast and guarantees the uniqueness of the maximizer of (2.54) if this exists. So no special care is needed for handling local maximizers.

To reasonably model the tails of a density, Kooperberg and Stone (1991) require further that the logarithm of a density is linear at both tails. This requirement is based on the consideration that fewer data points are available at the tails and hence one can only estimate fewer parameters at these regions. Another important reason is that the linear tail constraint makes it easy to construct a *logspline function* which is integrable – a basic requirement for a logspline function to be a density. Now, we are in a position to construct such a logspline function. Let $t_1 < \cdots < t_J$ be the knots of a spline function. Then, a cubic spline corresponds to a function which is piecewise cubic on the intervals $(-\infty, t_1], [t_1, t_2], \cdots,$ $[t_{J-1}, t_J]$ and $[t_J, +\infty)$. Here, we add the further restriction that the function is linear on $(-\infty, t_1]$ and $[t_J, +\infty)$. The space of all restricted cubic spline functions has dimension J. Let $1, B_1, \cdots,$ B_{J-1} be the basis of these restricted cubic spline functions. We can choose $B_1, \cdots, B_{J-1}$ such that B_1 is a linear function with positive slope on $(-\infty, t_1]$ and is constant on $[t_J, +\infty)$, $B_2, \cdots, B_{J-2}$ are constants on $(-\infty, t_1]$ and $[t_J, +\infty)$, and B_{J-1} is a linear function with negative slope on $[t_J, +\infty)$ and constant on $(-\infty, t_1]$.

Next, the logarithm of the density function f is modelled by a restricted spline function

$$\log f(x;\theta) = c(\theta) + \theta_1 B_1(x) + \cdots + \theta_{J-1} B_{J-1}(x)$$

where $\theta = (\theta_1, \cdots, \theta_{J-1})^T$ such that $\theta_1 > 0$ and $\theta_{J-1} > 0$. Then, the fact that $\int f(x;\theta)dx = 1$ gives

$$c(\theta) = -\log[\int \exp\{\theta_1 B_1(x) + \cdots + \theta_{J-1} B_{J-1}(x)\}dx].$$

The log-likelihood function corresponding to the logspline family

$$\ell(\theta) = \sum_{i=1}^{n} \log f(X_i;\theta) \qquad (2.54)$$

is strictly concave. The parameters θ can be obtained via maximizing (2.54).

The knot deletion idea introduced in Section 2.6.1 can also be used here for automatic determination of knots. We omit the details. See Kooperberg and Stone (1991) for the performance of the logspline density estimation and the implementation of the knot deletion idea.

2.8 Bibliographic notes

There is a vast literature on smoothing techniques. Several references were already given in the various sections of this chapter. Here we only provide some additional references.

The early published papers dealing explicitly with density estimation are Rosenblatt (1956a) and Parzen (1962). The optimal rates of convergence were obtained by Farrell (1972), Hasminskii (1978) and Stone (1980). The asymptotic distributions of the kernel density estimator under the L_2 and the L_∞-norm have been studied by Bickel and Rosenblatt (1973). Eddy (1980) discussed kernel estimation of modes. Theoretical contributions on consistency properties were established by Devroye and Wagner (1980) and Devroye (1981). Developments on the theory of kernel methods are comprehensively summarized in Prakasa Rao (1983). The L_1 view on kernel density estimation was treated thoroughly in Devroye and Györfi (1985). Some references on bandwidth selection will be given in Chapter 4.

A list of references on local polynomial fitting is provided in Chapter 3.

The orthogonal series method was introduced by Čencov (1962). Rutkowski (1982) used this method to estimate a regression function nonparametrically. The issue of selecting a smoothing parameter for orthogonal series density estimators was discussed by Hall (1987).

Early references on smoothing splines include Schoenberg (1964), Reinsch (1967) and Kimeldorf and Wahba (1970, 1971). Practical implementations of spline techniques are given in de Boor (1978). The choice of smoothing parameters was discussed by Utreras (1980) and Li (1985, 1986). Asymptotic theory for smoothing splines can also be found in Rice and Rosenblatt (1983). Confidence intervals can be constructed by using the Bayesian method described in Nychka (1988). Multivariate spline approximations were discussed by Wong (1984) and Gu (1990). Eubank (1984, 1985) studied the hat matrix for smoothing splines and diagnostics based on it. The use of splines in inverse problems was dealt with in detail by O'Sullivan (1986). Nonparametric estimation of relative risk using spline techniques was discussed by O'Sullivan (1988). Recent work on smoothing spline density estimation can be found in Gu (1993), Gu and Qiu (1993) and references therein.

Smoothing curves under shape restrictions has been studied by Grenander (1956), Wegman (1969, 1970), Groeneboom (1985), Birgé (1987a, b) and Mammen (1991a, b), among others.

CHAPTER 3

Framework for local polynomial regression

3.1 Introduction

In the previous chapter we gave a brief overview of the most important methods for estimating a regression function or a density function. We now start the exploration of the method of local polynomial fitting. We introduce the framework for this particular smoothing technique in the case of one-dimensional explanatory variables $X_1, \cdots, X_n$. This lays out the theoretical foundation and provides insights into the local polynomial approximation method. The framework established here will serve as a starting basis for the following chapters. The case of multivariate explanatory variables is discussed in Chapter 7.

Consider the bivariate data $(X_1, Y_1), \cdots, (X_n, Y_n)$, which form an independent and identically distributed sample from a population (X, Y). Of interest is to estimate the regression function $m(x_0) = E(Y|X = x_0)$ and its derivatives $m'(x_0), m''(x_0), \cdots,$ $m^{(p)}(x_0)$. To help us understand the estimation methodology, we can regard the data as being generated from the model

$$Y = m(X) + \sigma(X)\varepsilon, \qquad (3.1)$$

where $E(\varepsilon) = 0$, $\text{Var}(\varepsilon) = 1$, and X and ε are independent. However, this *location-scale model* assumption is not necessary for our development, but is adopted to provide intuitive understanding. We always denote the conditional variance of Y given $X = x_0$ by $\sigma^2(x_0)$ and the marginal density of X, i.e. the *design density*, by $f(\cdot)$.

Suppose that the $(p+1)^{th}$ derivative of $m(x)$ at the point x_0 exists. We then approximate the unknown regression function $m(x)$ locally by a polynomial of order p. A Taylor expansion gives, for x

in a neighborhood of x_0,

$$m(x) \approx m(x_0) + m'(x_0)(x - x_0) + \frac{m''(x_0)}{2!}(x - x_0)^2$$
$$+ \cdots + \frac{m^{(p)}(x_0)}{p!}(x - x_0)^p. \qquad (3.2)$$

This polynomial is fitted locally by a weighted least squares regression problem: minimize

$$\sum_{i=1}^{n} \{Y_i - \sum_{j=0}^{p} \beta_j (X_i - x_0)^j\}^2 K_h(X_i - x_0), \qquad (3.3)$$

where, as in Section 2.3, h is a bandwidth controlling the size of the local neighborhood, and $K_h(\cdot) = K(\cdot/h)/h$ with K a kernel function assigning weights to each datum point. Throughout this chapter we assume that K is a symmetric probability density function with bounded support, although this technical assumption can be relaxed significantly.

Denote by $\widehat{\beta}_j$, $j = 0, \ldots, p$, the solution to the least squares problem (3.3). It is clear from the Taylor expansion in (3.2) that $\widehat{m}_\nu(x_0) = \nu! \widehat{\beta}_\nu$ is an estimator for $m^{(\nu)}(x_0), \nu = 0, 1, \cdots, p$. To estimate the entire function $m^{(\nu)}(\cdot)$ we solve the above weighted least squares problem for all points x_0 in the domain of interest.

It is more convenient to work with matrix notation. Denote by **X** the design matrix of problem (3.3):

$$\mathbf{X} = \begin{pmatrix} 1 & (X_1 - x_0) & \cdots & (X_1 - x_0)^p \\ \vdots & \vdots & & \vdots \\ 1 & (X_n - x_0) & \cdots & (X_n - x_0)^p \end{pmatrix},$$

and put

$$\mathbf{y} = \begin{pmatrix} Y_1 \\ \vdots \\ Y_n \end{pmatrix} \quad \text{and} \quad \widehat{\beta} = \begin{pmatrix} \widehat{\beta}_0 \\ \vdots \\ \widehat{\beta}_p \end{pmatrix}.$$

Further, let **W** be the $n \times n$ diagonal matrix of weights:

$$\mathbf{W} = \text{diag}\{K_h(X_i - x_0)\}.$$

Then the weighted least squares problem (3.3) can be written as

$$\min_{\beta}(\mathbf{y} - \mathbf{X}\beta)^T \mathbf{W}(\mathbf{y} - \mathbf{X}\beta), \qquad (3.4)$$

with $\beta = (\beta_0, \cdots, \beta_p)^T$. The solution vector is provided by weighted

INTRODUCTION

least squares theory and is given by

$$\widehat{\beta} = (\mathbf{X}^T\mathbf{W}\mathbf{X})^{-1}\mathbf{X}^T\mathbf{W}\mathbf{y}. \tag{3.5}$$

There are several important issues which have to be discussed. First of all there is the choice of the bandwidth parameter h, which plays a rather crucial role. A too large bandwidth under-parametrizes the regression function, causing a large modelling bias, while a too small bandwidth over-parametrizes the unknown function and results in noisy estimates. Ideal theoretical choices of the bandwidth are easy to obtain, as will be shown in Section 3.2.3. However this theoretical choice is not directly practically usable since it depends on unknown quantities. Finding a practical procedure for selecting the bandwidth parameter is one of the most important tasks. This task is carried out in Chapter 4.

Another issue in local polynomial fitting is the choice of the order of the local polynomial. Since the modelling bias is primarily controlled by the bandwidth, this issue is less crucial however. For a given bandwidth h, a large value of p would expectedly reduce the modelling bias, but would cause a large variance and a considerable computational cost. It will be shown in Section 3.3 that there is a general pattern of increasing variability: for estimating $m^{(\nu)}(x_0)$, there is no increase in variability when passing from an even (i.e. $p-\nu$ even) $p = \nu + 2q$ order fit to an odd $p = \nu + 2q + 1$ order fit, but when passing from an odd $p = \nu + 2q + 1$ order fit to the consecutive even $p = \nu + 2q + 2$ order fit there is a price to be paid in terms of increased variability. Therefore, even order fits $p = \nu + 2q$ are not recommended, as will be explained in detail in Section 3.3.2. Since the bandwidth is used to control the modelling complexity, we recommend the use of the lowest odd order, i.e. $p = \nu + 1$, or occasionally $p = \nu + 3$.

Another question concerns the choice of the kernel function K. Since the estimate is based on the local regression (3.3) no negative weight K should be used. In fact it will be shown in Section 3.2.6 that for all choices of p and ν the optimal weight function is $K(z) = \frac{3}{4}(1-z^2)_+$, the Epanechnikov kernel, which minimizes the asymptotic MSE of the resulting local polynomial estimators. From our discussion it will also be clear how this *universal optimal weighting scheme* relates to the higher order kernels provided by Gasser, Müller and Mammitzsch (1985).

How good are the local polynomial estimators compared with other estimators? An answer to this question is provided by studying the efficiency of the local polynomial fit. The results of this study are presented in Sections 3.4 and 3.5, where it is basically shown that local polynomial fitting is nearly optimal in an asymptotic minimax sense.

From a computational point of view local polynomial estimators are attractive, due to their simplicity. It might be desirable however to speed up the computations especially when computing intensive procedures, such as for example bandwidth selection, are to be implemented. Carefully thought out algorithms make it possible to do local polynomial fitting in $O(n)$ operations, as was shown in Fan and Marron (1994). They report that with the implementation of such fast algorithms local polynomial fitting is at least as fast as other popular methods. This computational issue is discussed in Section 3.6.

3.2 Advantages of local polynomial fitting

Local polynomial fitting is an attractive method both from theoretical and practical point of view. Other commonly used kernel estimators, such as the Nadaraya-Watson estimator and the Gasser-Müller estimator suffer from some drawbacks, as explained in Chapter 2. In summary, the Nadaraya-Watson estimator leads to an undesirable form of the bias, while the Gasser-Müller estimator has to pay a price in variance when dealing with a random design model. Local polynomial fitting also has other advantages. The method adapts to various types of designs such as random and fixed designs, highly clustered and nearly uniform designs. Furthermore, there is an absence of boundary effects: the bias at the boundary stays automatically of the same order as in the interior, without use of specific boundary kernels. This is remarkably different from the other methods described in Chapter 2. With local polynomial fitting no boundary modifications are required, and this is an important merit especially when dealing with multidimensional situations, for which the boundary can be quite substantial in terms of the number of data points involved (see Silverman (1986) and Fan and Marron (1993)). Boundary modifications in higher dimensions are a very difficult task. Local polynomial estimators have nice minimax efficiency properties: the asymptotic minimax efficiency for commonly used orders is 100% among all linear estimators and only a small loss has to be tolerated beyond this class. The local polynomial approximation method is appealing on general scientific grounds: the least squares principle to be applied opens the way to a wealth of statistical knowledge and thus easy generalizations. The advantage of falling back on simple least squares regression theory becomes clear in the next section, which discusses bias and variance of local polynomial estimators. In the following sections we explain in more detail the features of the local polynomial approximation method.

3.2.1 Bias and variance

When dealing with the bandwidth selection problem, a key issue is to have a good insight into bias and variance of the estimators, since a tradeoff between these two quantities forms the core of many bandwidth selection criteria. In this section, we will give exact and asymptotic expressions for bias and variance.

The conditional bias and variance of the estimator $\widehat{\beta}$ are derived immediately from its definition in (3.5):

$$E(\widehat{\beta}|\mathbf{X}) = (\mathbf{X}^T\mathbf{W}\mathbf{X})^{-1}\mathbf{X}^T\mathbf{W}\mathbf{m}$$
$$= \beta + (\mathbf{X}^T\mathbf{W}\mathbf{X})^{-1}\mathbf{X}^T\mathbf{W}\mathbf{r}$$

$$\mathrm{Var}(\widehat{\beta}|\mathbf{X}) = (\mathbf{X}^T\mathbf{W}\mathbf{X})^{-1}(\mathbf{X}^T\Sigma\mathbf{X})(\mathbf{X}^T\mathbf{W}\mathbf{X})^{-1}, \quad (3.6)$$

where $\mathbf{m} = \{m(X_1), \cdots, m(X_n)\}^T$, $\beta = \{m(x_0), \cdots, m^{(p)}(x_0)/p!\}^T$, $\mathbf{r} = \mathbf{m} - \mathbf{X}\beta$, the vector of residuals of the local polynomial approximation, and $\Sigma = \mathrm{diag}\,\{K_h^2(X_i - x_0)\sigma^2(X_i)\}$.

These exact bias and variance expressions are not directly usable, since they depend on unknown quantities: the residual $\mathbf{r}$ and the diagonal matrix Σ. So, there is a need for approximating bias and variance. First order asymptotic expansions for the bias and the variance of the estimator $\widehat{m}_\nu(x_0) = \nu!\widehat{\beta}_\nu$ are given in the next theorem. This result was obtained by Ruppert and Wand (1994). In Section 3.7 we give a simple derivation of the necessary expansions. The following notation will be used. The moments of K and K^2 are denoted respectively by

$$\mu_j = \int u^j K(u)du \quad \text{and} \quad \nu_j = \int u^j K^2(u)du.$$

Some matrices and vectors of moments appear in the asymptotic expressions. Let

$$S = (\mu_{j+\ell})_{0\leq j,\ell\leq p} \qquad c_p = (\mu_{p+1}, \cdots, \mu_{2p+1})^T$$
$$\tilde{S} = (\mu_{j+\ell+1})_{0\leq j,\ell\leq p} \qquad \tilde{c}_p = (\mu_{p+2}, \cdots, \mu_{2p+2})^T$$
$$S^* = (\nu_{j+\ell})_{0\leq j,\ell\leq p}.$$

Further, we consider the unit vector $e_{\nu+1} = (0, \cdots, 0, 1, 0, \cdots, 0)^T$, with 1 on the $(\nu+1)^{th}$ position. Recall that $(X_1, Y_1), \cdots, (X_n, Y_n)$ is an i.i.d. sample from the population (X, Y), and that $o_P(1)$ denotes a random quantity that tends to zero in probability.

Theorem 3.1 *Assume that $f(x_0) > 0$ and that $f(\cdot)$, $m^{(p+1)}(\cdot)$ and $\sigma^2(\cdot)$ are continuous in a neighborhood of x_0. Further, assume that $h \to 0$ and $nh \to \infty$. Then the asymptotic conditional variance of $\widehat{m}_\nu(x_0)$ is given by*

$$Var\{\widehat{m}_\nu(x_0)|\mathbf{X}\} = e_{\nu+1}^T S^{-1} S^* S^{-1} e_{\nu+1} \frac{\nu!^2 \sigma^2(x_0)}{f(x_0) n h^{1+2\nu}}$$
$$+ o_P\left(\frac{1}{nh^{1+2\nu}}\right). \qquad (3.7)$$

The asymptotic conditional bias for $p - \nu$ odd is given by

$$Bias\{\widehat{m}_\nu(x_0)|\mathbf{X}\} = e_{\nu+1}^T S^{-1} c_p \frac{\nu!}{(p+1)!} m^{(p+1)}(x_0) h^{p+1-\nu}$$
$$+ o_P\left(h^{p+1-\nu}\right). \qquad (3.8)$$

Further, for $p - \nu$ even the asymptotic conditional bias is

$$Bias\{\widehat{m}_\nu(x_0)|\mathbf{X}\} = e_{\nu+1}^T S^{-1} \tilde{c}_p \frac{\nu!}{(p+2)!} \{m^{(p+2)}(x_0)$$
$$+ (p+2) m^{(p+1)}(x_0) \frac{f'(x_0)}{f(x_0)} \} h^{p+2-\nu}$$
$$+ o_P\left(h^{p+2-\nu}\right), \qquad (3.9)$$

provided that $f'(\cdot)$ and $m^{(p+2)}(\cdot)$ are continuous in a neighborhood of x_0 and $nh^3 \to \infty$.

A deeper result than Theorem 3.1, specifying higher order terms in the asymptotic bias and variance expressions, can be found in Fan, Gijbels, Hu and Huang (1996).

From the above result it is already clear that there is a theoretical difference between the cases $p - \nu$ odd and $p - \nu$ even. For $p - \nu$ even, the leading term $O(h^{p+1})$ in the bias expression is zero due to the symmetry of the kernel K and hence the second order term is presented in (3.9). For $p - \nu$ odd the asymptotic bias has a simpler structure and does not involve $f'(x_0)$, a factor appearing in the asymptotic bias when $p - \nu$ is even. This remark is in fact a generalization of what has already been observed for the special case of the local constant fit ($p = 0$) used for estimating the regression function ($\nu = 0$). Indeed, the estimator resulting from such a fit, the Nadaraya-Watson estimator, has an additional term in the asymptotic bias, as was reported in Table 2.1. So, here we have some theoretical grounds for a distinction between the cases $p - \nu$ odd and $p - \nu$ even. It will turn out clearly later on that polynomial fits with $p - \nu$ odd outperform those with $p - \nu$ even. This issue is elaborated in detail in Section 3.3. There we give an intuitive explanation for this important observation, and justify why the case $p - \nu$ odd should be referred to as the natural one. Hence, throughout the rest of this Section 3.2 we take $p - \nu$ odd.

3.2.2 Equivalent kernels

In this section we show how the local polynomial approximation method assigns weights to each datum point. This provides further insight into the method and serves as a technical tool for understanding and deriving its properties, such as asymptotic bias and variance and minimax results.

Recall the notation

$$S_{n,j} = \sum_{i=1}^{n} K_h(X_i - x_0)(X_i - x_0)^j \qquad (3.10)$$

and put $S_n \equiv \mathbf{X}^T \mathbf{W} \mathbf{X}$, the $(p+1) \times (p+1)$ matrix $(S_{n,j+\ell})_{0 \leq j, \ell \leq p}$. Note first of all that the estimator $\widehat{\beta}_\nu$ can be written as

$$\begin{aligned} \widehat{\beta}_\nu = e_{\nu+1}^T \widehat{\beta} &= e_{\nu+1}^T S_n^{-1} \mathbf{X}^T \mathbf{W} \mathbf{y} \\ &= \sum_{i=1}^{n} W_\nu^n \left(\frac{X_i - x_0}{h} \right) Y_i, \qquad (3.11) \end{aligned}$$

where $W_\nu^n(t) = e_{\nu+1}^T S_n^{-1} \{1, th, \cdots, (th)^p\}^T K(t)/h$. The above expression reveals that the estimator $\widehat{\beta}_\nu$ is very much like a conventional kernel estimator except that the 'kernel' W_ν^n depends on the design points and locations. This explains why the local polynomial fit can adapt automatically to various designs and to boundary estimation. It is easy to show (see Section 3.7) that the weights W_ν^n satisfy the following discrete moment conditions:

$$\sum_{i=1}^{n} (X_i - x_0)^q W_\nu^n \left(\frac{X_i - x_0}{h} \right) = \delta_{\nu,q} \quad 0 \leq \nu, q \leq p. \qquad (3.12)$$

A direct consequence of this relation is that the finite sample bias when estimating polynomials up to order p is zero. This point was also made clear in Ruppert and Wand (1994). This nice feature highlights a contrast with methods based on higher order kernels, such as the Nadaraya-Watson estimator and the Gasser-Müller kernel method, for which the above moment conditions and the respective zero bias only hold true asymptotically.

We continue our investigation of the weight function W_ν^n by noting that
$$S_{n,j} = nh^j f(x_0)\mu_j\{1 + o_P(1)\} \tag{3.13}$$
which is easy to show, and of which a short proof is given in Section 3.7. From this one obtains immediately that
$$S_n = nf(x_0)HSH\{1 + o_P(1)\}, \tag{3.14}$$
where $H = \operatorname{diag}(1, h, \cdots, h^p)$. Hence, substituting this into the definition of W_n^ν, we find that
$$W_\nu^n(t) = \frac{1}{nh^{\nu+1}f(x_0)}e_{\nu+1}^T S^{-1}(1, t, \cdots, t^p)^T K(t)\{1 + o_P(1)\}$$
and therefore
$$\widehat{\beta}_\nu = \frac{1}{nh^{\nu+1}f(x_0)}\sum_{i=1}^n K_\nu^*\left(\frac{X_i - x_0}{h}\right) Y_i\{1 + o_P(1)\}, \tag{3.15}$$
where
$$K_\nu^*(t) = e_{\nu+1}^T S^{-1}(1, t, \cdots, t^p)^T K(t) = (\sum_{\ell=0}^p S^{\nu\ell}t^\ell)K(t), \tag{3.16}$$
with $S^{-1} = (S^{j\ell})_{0 \le j,\ell \le p}$.

We refer to K_ν^* as the *equivalent kernel*. This kernel satisfies the following moment conditions:
$$\int u^q K_\nu^*(u)du = \delta_{\nu,q} \quad 0 \le \nu, q \le p, \tag{3.17}$$
which are an asymptotic version of the discrete moment conditions presented in (3.12). The derivations of the moment conditions in (3.12) and (3.17) are easy and details are provided in Section 3.7. The moment conditions in (3.17) reveal also that, up to normalizing constants, the equivalent kernel K_ν^* is a kernel of order $(\nu, p+1)$. See the definition of these kernels in Gasser, Müller and Mammitzsch (1985). Equivalent kernels have been previously used for analyzing polynomial fitting by Lejeune (1985) and Müller (1987) in a slightly different way. Table 3.1 gives the forms of some equivalent kernel functions. To emphasize the dependence of p, we use $K_{\nu,p}^*$ to denote the equivalent kernel function given by (3.16). As an illustration we plot in Figure 3.1 the Epanechnikov kernel $K(z) = \frac{3}{4}(1-z^2)_+$, as well as the equivalent kernel $K_{\nu,p}^*$ for some values of p and ν.

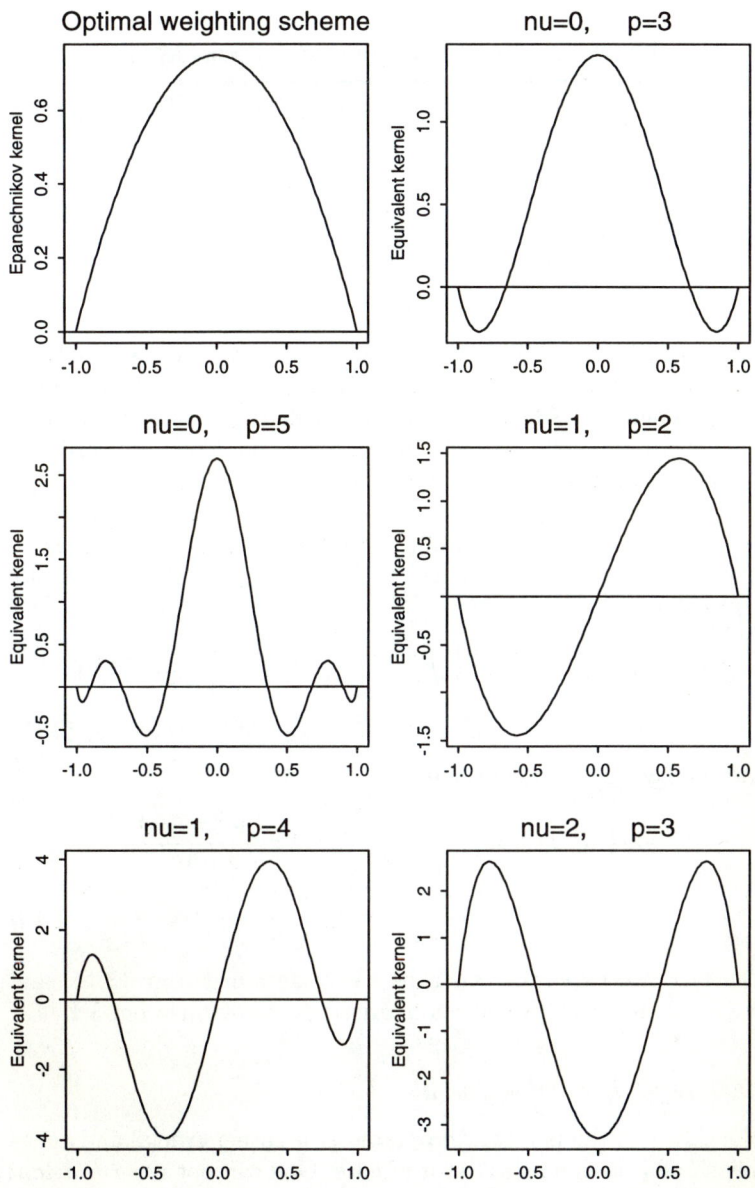

Figure 3.1. *Epanechnikov kernel and its equivalent kernel for some values of p and ν. These equivalent kernels are similar to those given by Gasser, Müller and Mammitzsch (1985).*

Table 3.1. *The equivalent kernel functions $K^*_{\nu,p}$.*

ν	p	Equivalent kernel function $K^*_{\nu,p}(t)$
0	1	$K(t)$
0	3	$(\mu_4 - \mu_2^2)^{-1}(\mu_4 - \mu_2 t^2)K(t)$
1	2	$\mu_2^{-1} t K(t)$
2	3	$(\mu_4 - \mu_2^2)^{-1}(t^2 - \mu_2)K(t)$

From (3.11) and (3.15) it now becomes clear that local polynomial fitting creates automatically a higher order weighting scheme which depends on the unknown design density f. Thus, this weighting scheme not only corrects bias up to a polynomial of order p, but also adapts automatically to all design densities.

The conditional bias and variance of the estimator $\widehat{m}_\nu(x_0)$, given in respectively (3.8) and (3.7), can equally well be re-expressed in terms of the equivalent kernel K^*_ν, leading to the asymptotic expression

$$\text{Bias}\{\widehat{m}_\nu(x_0)|\mathbb{X}\} = \left\{\int t^{p+1} K^*_\nu(t) dt\right\} \frac{\nu!}{(p+1)!} m^{(p+1)}(x_0)$$
$$\times h^{p+1-\nu} + o_P\left(h^{p+1-\nu}\right), \qquad (3.18)$$

and its asymptotic variance equals

$$\text{Var}\{\widehat{m}_\nu(x_0)|\mathbb{X}\} = \int K^{*2}_\nu(t) dt \frac{\nu!^2 \sigma^2(x_0)}{f(x_0) n h^{1+2\nu}}$$
$$+ o_P\left(\frac{1}{nh^{1+2\nu}}\right). \qquad (3.19)$$

The last two expressions can be easily obtained from (3.15) and (3.17). Hence, this provides a heuristic proof of Theorem 3.1.

3.2.3 Ideal choice of bandwidth

The choice of the bandwidth parameter is rather crucial and hence this should be done with a lot of care. One can opt for a *constant bandwidth* – also referred to as a *global bandwidth* – or a *variable bandwidth*. For a variable bandwidth the distinction should be made between a *local variable bandwidth* and a *global variable bandwidth*. A local variable bandwidth $h(x_0)$ varies with the location point x_0, whereas a global variable bandwidth changes with the data points,

ADVANTAGES OF LOCAL POLYNOMIAL FITTING

Table 3.2. The constants $C_{\nu,p}(K)$.

ν	p	Gaussian	Uniform	Epanechnikov	Biweight	Triweight
0	1	0.776	1.351	1.719	2.036	2.312
0	3	1.161	2.813	3.243	3.633	3.987
1	2	0.884	1.963	2.275	2.586	2.869
2	3	1.006	2.604	2.893	3.208	3.503

and is of the form $h(X_i)$. A variable bandwidth is introduced to allow for different degrees of smoothing, resulting in a possible reduction of the bias at peaked regions and of the variance at flat regions. This enhances the flexibility of the local polynomial fitting so that it can adapt to spatially inhomogeneous curves.

The theoretical choice of a global variable bandwidth was discussed in detail in Fan and Gijbels (1992). Here we will focus on the choice of a constant and local (variable) bandwidth. A theoretical optimal local bandwidth for estimating $m^{(\nu)}(x_0)$ is obtained by minimizing the conditional Mean Squared Error (MSE) given by

$$[\text{Bias}\{\widehat{m}_\nu(x_0)|\boldsymbol{X}\}]^2 + \text{Var}\{\widehat{m}_\nu(x_0)|\boldsymbol{X}\}.$$

This ideal choice of a local bandwidth can be approximated by the asymptotically optimal local bandwidth, i.e. the bandwidth which minimizes the asymptotic MSE. The relative error of this approximation can be found in Fan, Gijbels, Hu and Huang (1996). An expression for this asymptotic MSE is provided via (3.18) and (3.19), and minimization of it leads to

$$h_{\text{opt}}(x_0) = C_{\nu,p}(K) \left[\frac{\sigma^2(x_0)}{\{m^{(p+1)}(x_0)\}^2 f(x_0)} \right]^{1/(2p+3)} n^{-1/(2p+3)},$$

(3.20)

where

$$C_{\nu,p}(K) = \left[\frac{(p+1)!^2 (2\nu+1) \int K_\nu^{*2}(t) dt}{2(p+1-\nu)\{\int t^{p+1} K_\nu^*(t) dt\}^2} \right]^{1/(2p+3)}$$

The latter constants are easy to calculate, and Table 3.2 lists some of them. These values will be used in Section 4.2 which deals with bandwidth selection. A commonly used, simple measure of global loss is the weighted Mean Integrated Squared Error (MISE). Minimization of the conditional weighted MISE

$$\int \left([\text{Bias}\{\widehat{m}_\nu(x)|\boldsymbol{X}\}]^2 + \text{Var}\{\widehat{m}_\nu(x)|\boldsymbol{X}\} \right) w(x) dx,$$

with $w \geq 0$ some weight function, leads to a theoretical optimal constant bandwidth. Using again the asymptotic expressions in (3.18) and (3.19) we find an asymptotically optimal constant bandwidth given by

$$h_{\text{opt}} = C_{\nu,p}(K) \left[\frac{\int \sigma^2(x) w(x)/f(x) dx}{\int \{m^{(p+1)}(x)\}^2 w(x) dx} \right]^{1/(2p+3)} n^{-1/(2p+3)}.$$
(3.21)

It is understood that the integrals are finite and that the denominator does not vanish.

These asymptotically optimal bandwidths depend on unknown quantities such as the design density $f(\cdot)$, the conditional variance $\sigma^2(\cdot)$ and the derivative function $m^{(p+1)}(\cdot)$, and hence further work is needed for achieving practical bandwidth selection procedures. One possible approach is to substitute the unknown quantities by pilot estimators, leading to so-called 'plug-in' type bandwidth selectors. An alternative, appealing practical approach for selecting optimal bandwidths is introduced and discussed in Chapter 4.

3.2.4 Design adaptation property

The bias and variance expressions in (3.8) and (3.7) are obtained under the random design model, but remain valid for fixed designs. Hence, local polynomial estimators adapt to both random and fixed designs. This is in contrast with the Gasser-Müller estimator which cannot adapt to random designs: the unconditional variance is higher by a factor 1.5 for random designs (see also Table 2.1). More explanation about this statement can be found in Mack and Müller (1989) and Chu and Marron (1991).

Recall that for $p - \nu$ even, additional terms arise in the asymptotic conditional bias. For example, when estimating the regression function $m(x_0)$ ($\nu = 0$), an extra term $m'(x_0)f'(x_0)/f(x_0)$ appears in the asymptotic bias of the Nadaraya-Watson estimator. The bias of this estimator depends on the intrinsic part $m''(x_0)$ interplaying with the artifact $m'(x_0)f'(x_0)/f(x_0)$. Keeping $m''(x_0)$ fixed, we first remark that in the highly clustered (asymmetric) design where $|f'(x_0)/f(x_0)|$ is large, the bias of the Nadaraya-Watson estimator can be large. See Figure 2.2. Thus this estimator cannot adapt to highly clustered designs. Similar artifacts hold true for polynomial fits of an even order $p - \nu$. Local polynomial fitting with $p - \nu$ odd however rules out such artifacts and results in design-adaptive estimators. A neat illustration of this can be found in Fan (1992) who discusses in detail the local linear fit in comparison with the local constant fit.

3.2.5 Automatic boundary carpentry

In applications design points always have a bounded support. For estimating $m^{(\nu)}(x_0)$, with x_0 a point close to the boundary, the local neighborhood $x_0 \pm h$ can lie outside the design region. See Figure 2.2. Hence, certain symmetric moment conditions, valid for all interior points, are no longer valid for x_0 in a boundary region, causing a large boundary bias for most of the smoothing techniques. If a bandwidth is chosen to be 25% of the data range, then for about 50% of the data range the local neighborhood will lie partly outside the design region. Hence the boundary region is about 50% of the whole data range. In higher dimensions these figures are even more striking, reflecting even more severe problems. Since many of the smoothing techniques show the aforementioned bias problem at the boundary, considerable efforts have been devoted to methods for correcting this boundary bias. Two popular approaches are *boundary kernel methods* and *reflection methods*. But none of these methods are as simple and as efficient as the automatic boundary correction when using local polynomial fitting. In the following we will explain this in detail. To investigate theoretically what is happening at the boundary, we adapt the formulation of Gasser and Müller (1979). Without loss of generality we assume that the design density has a bounded support $[0, 1]$. A *left boundary point* is thought of as being of the form $x = ch$, with $c \geq 0$, whereas a *right boundary point* is of the form $x = 1 - ch$.

The behavior of the estimator $\widehat{m}_\nu(x_0)$ for points x_0 at the interior of the support has been studied in the previous sections. In this section we address the question of how local polynomial estimators behave at boundary points. For most regression smoothers the rate of convergence at boundary points is slower than that at points in the interior. In the literature, one refers to this problem as *boundary effects* or *edge effects*. These effects are visually very disturbing in practice, and in addition they can play a dominant role in theoretical analysis. Hence, in the case of boundary effects there is a strong request for boundary modifications, in order to overcome the problem.

The main purpose of this section is to demonstrate that local polynomial estimators do not suffer from boundary effects: they adapt automatically to estimation at the boundaries, and hence no extra efforts in terms of boundary modifications are needed. Moreover, unlike most other methods the local polynomial approximation method does not require knowledge of the location of the endpoints of the support. Fan and Gijbels (1992) proved that theoretically the local linear regression estimator adapts automatically to the boundary. This was also empirically observed by Tibshirani and Hastie (1987). Ruppert and Wand (1994) extended the results of Fan and Gijbels (1992) to the case of local polynomial estimators.

The aforementioned *automatic boundary carpentry* can be easily seen from the representation of the local polynomial estimator in terms of an equivalent kernel, presented in (3.16). Consider a point at the left boundary $x = ch$. A derivation similar to (3.13)–(3.16) can be made for these points. The only difference is that now the finite sample moments behave as

$$S_{n,j} = nh^j f(0+)\mu_{j,c}\{1 + o_P(1)\}$$

where $\mu_{j,c} = \int_{-c}^{\infty} u^j K(u) du$ (note the difference with (3.13)). This leads to the following equivalent kernel at the boundary

$$K_{\nu,c}^*(t) = e_{\nu+1}^T S_c^{-1}(1, t, \ldots, t^p)^T K(t) \quad \text{with} \quad S_c = (\mu_{j+\ell,c})_{0 \leq j, \ell \leq p}, \tag{3.22}$$

which should be compared with (3.16). This equivalent kernel differs from K_ν^* only in the matrix S, and satisfies the boundary moment conditions of Gasser, Müller and Mammitzsch (1985). This reflects the automatic adaptation to the boundary. This point was already illustrated in Figure 2.2. In Figure 3.2. we show the Epanechnikov kernel and some of its equivalent kernels for some boundary points and for various values of p and ν.

Theorem 3.2 *Assume that $f(0+) > 0$ and that $f(\cdot)$, $m^{(p+1)}(\cdot)$ and $\sigma^2(\cdot)$ are right continuous at the point 0. Then, the conditional MSE of the estimator $\widehat{m}_\nu(x)$ at the left boundary point $x = ch$ is given by*

$$\left[\left\{\int_{-c}^{\infty} t^{p+1} K_{\nu,c}^*(t) dt\right\}^2 \left\{\nu! \frac{m^{(p+1)}(0+)}{(p+1)!}\right\}^2 h^{2(p+1-\nu)} \right.$$
$$\left. + \int_{-c}^{\infty} K_{\nu,c}^{*2}(t) dt \frac{\nu!^2 \sigma^2(0+)}{f(0+)nh^{1+2\nu}}\right]\{1 + o_P(1)\}. \tag{3.23}$$

Assume momentarily that the kernel function K has support $[-1, 1]$. Then, a point $x = ch$ is a real boundary point if $c <$

ADVANTAGES OF LOCAL POLYNOMIAL FITTING

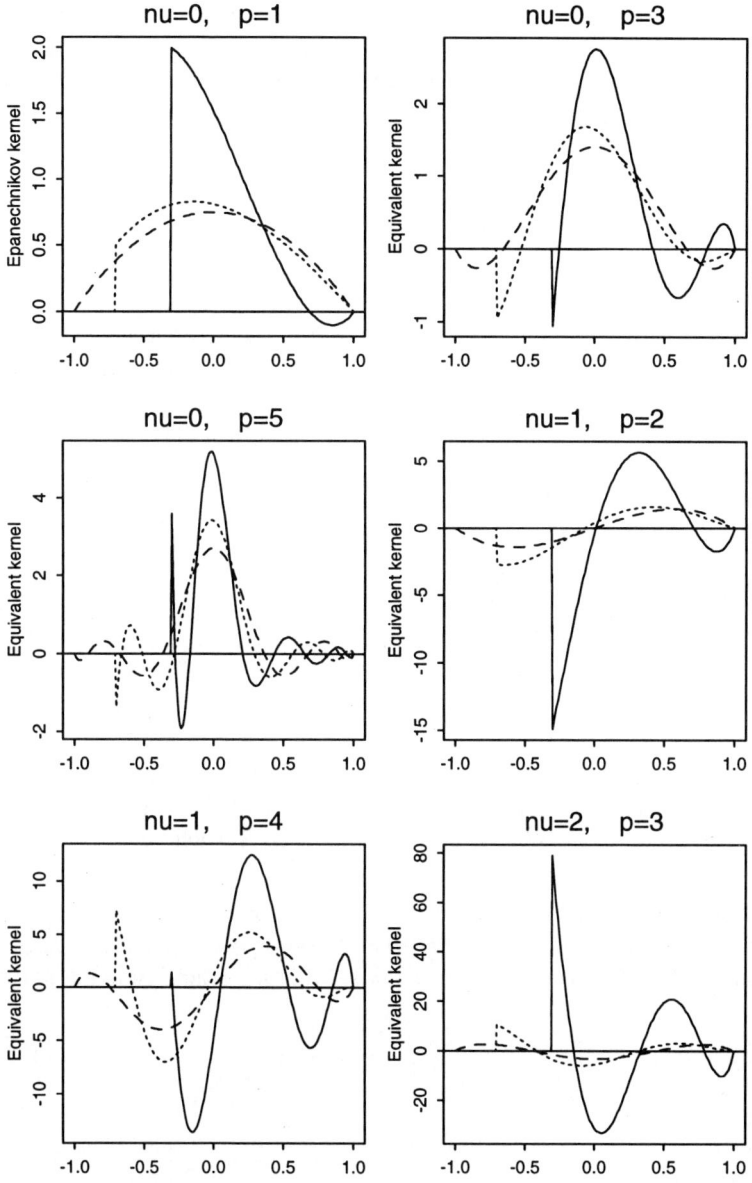

Figure 3.2. *The Epanechnikov kernel and its equivalent kernels at the boundary points $c = 0.3$ (solid line) and $c = 0.7$ (dotted line) and interior points $c \geq 1$ (dashed line) for various values of p and ν.*

1, and corresponds to an interior point if $c \geq 1$. It is easy to see that, when $p - \nu$ is odd, the conditional MSE is a continuous function in c, which implies that the risk changes continuously from a boundary point to an interior point. This is a theoretical justification of why the local polynomial fit with $p - \nu$ odd does not require boundary modifications. However, when $p - \nu$ is even, the bias at the boundary is of a larger order. Compare Theorem 3.1 with Theorem 3.2. Thus, a boundary modification is needed.

Analogously, one can obtain expressions for the conditional MSE of the estimator at right boundary points. The integrations in (3.23) are now from $-\infty$ to c (including those in the definition of $K_{\nu,c}^*$) and the functions $m^{(p+1)}$, σ^2 and f are now evaluated at 1.

To get a deeper insight into the boundary property of the local polynomial estimator, we focus now on the special case of the local linear regression estimator ($p = 1$) used for estimating the regression function ($\nu = 0$). In this particular case it is easy to see from (3.22) that the equivalent kernel $K_{0,c}^*(t)$ at a left boundary point $x = ch$ equals

$$K_{0,c}^*(t) = \frac{\mu_{2,c}}{\mu_{0,c}\mu_{2,c} - \mu_{1,c}^2} K(t) - \frac{\mu_{1,c}}{\mu_{0,c}\mu_{2,c} - \mu_{1,c}^2} tK(t).$$

This immediately leads to an expression for the conditional MSE of the local linear regression estimator at left boundary points. This result was established by Fan and Gijbels (1992) in the more general setting of a global variable bandwidth, and under general conditions on K (not necessarily requiring kernels with bounded support).

Theorem 3.3 *Assume that $f(\cdot)$, $m''(\cdot)$ and $\sigma(\cdot)$ are right continuous at the point 0. Suppose that $\limsup_{u \to -\infty} |K(u)u^5| < \infty$. Then, the conditional MSE of the estimator at the boundary point $x = ch$ is given by*

$$\frac{1}{4}\left\{m''(0+)\frac{\mu_{2,c}^2 - \mu_{1,c}\mu_{3,c}}{\mu_{2,c}\mu_{0,c} - \mu_{1,c}^2}\right\}^2 h^4$$

$$+ \frac{\int_{-c}^{\infty}(\mu_{2,c} - u\mu_{1,c})^2 K^2(u)du}{(\mu_{2,c}\mu_{0,c} - \mu_{1,c}^2)^2} \frac{\sigma^2(0+)}{f(0+)nh}$$

$$+ o_P(h^4 + \frac{1}{nh}). \tag{3.24}$$

It is demonstrated clearly now that the rate of the estimator is not influenced by the position of the point under considera-

ADVANTAGES OF LOCAL POLYNOMIAL FITTING

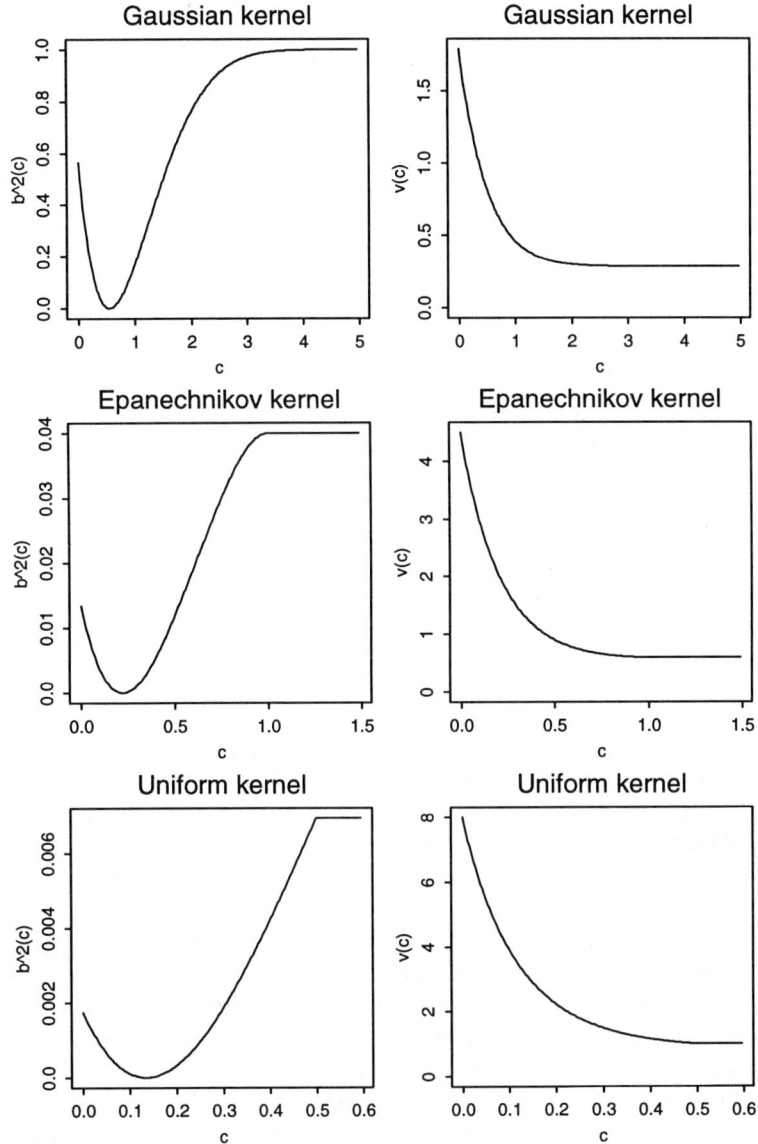

Figure 3.3. *Behavior of the constant factors in the squared bias and the variance for a Gaussian kernel (top panel); Epanechnikov kernel (middle panel); and uniform kernel (bottom panel). Adapted from Fan and Gijbels (1992).*

tion. What can we say about the constant factors appearing in the asymptotic expression for the conditional MSE? There is the constant factor

$$b^2(c) = \left(\frac{\mu_{2,c}^2 - \mu_{1,c}\mu_{3,c}}{\mu_{2,c}\mu_{0,c} - \mu_{1,c}^2}\right)^2$$

in the squared bias, and the constant factor in the variance

$$v(c) = \frac{\int_{-c}^{\infty}(\mu_{2,c} - u\mu_{1,c})^2 K^2(u)du}{(\mu_{2,c}\mu_{0,c} - \mu_{1,c}^2)^2}.$$

How do these factors change with c, where c measures how many effective bandwidths h the point $x = ch$ is away from the left boundary? To gain an insight into this, we plot both functions $b^2(\cdot)$ and $v(\cdot)$ for three commonly used kernels:

the Gaussian kernel : $K(z) = \frac{1}{\sqrt{2\pi}} \exp\left(-z^2/2\right)$,
the Epanechnikov kernel : $K(z) = \frac{3}{4}\left(1 - z^2\right)_+$,
the uniform kernel : $K(z) = 1_{[-0.5,+0.5]}(z)$.

All three pictures, presented in Figure 3.3, show the same behavior. A first feature is that with $[-s_0, s_0]$ being the support of K,

$$\lim_{c \to s_0} b^2(c) = \mu_2^2 \quad \text{and} \quad \lim_{c \to s_0} v(c) = \nu_0,$$

and these limits are exactly the constant factors appearing in respectively the squared bias and variance for an interior point. Furthermore, it is clear from the pictures that $b^2(c)$ is smaller than μ_2^2 and that $v(c)$ is larger than ν_0, for all values of c. This implies that the squared bias of the local linear regression estimator is smaller at a boundary point than at an interior point, at least if the value of m'' is the same and the same amount of smoothing is used. This is due to the fact that one uses a linear approximation on a smaller interval around the boundary point. The variance however tends to be larger, because on a smaller interval fewer observations contribute in computing the estimator.

From the above discussion we conclude that the local polynomial estimators do not suffer from boundary effects and this is an important gain in comparison with many other methods.

3.2.6 Universal optimal weighting scheme

In Section 3.2.3 we dealt with the question of how to choose an optimal bandwidth. Next the question arises of which weight function to use for different choices of ν and p. When using the optimal bandwidths (3.20) and (3.21) both the asymptotically MSE and MISE depend on the kernel function K through

$$T_{\nu,p}(K) \equiv \left|\int t^{p+1} K_\nu^*(t) dt\right|^{2\nu+1} \left\{\int K_\nu^{*2}(t) dt\right\}^{p+1-\nu}. \quad (3.25)$$

Note that when $p - \nu$ is even, $T_{\nu,p}(K) = 0$ for any symmetric kernel K. We assume throughout this section that $p - \nu$ is odd. *Optimal kernels* K_ν^*, minimizing the right hand side of (3.25), have been derived by Gasser, Müller and Mammitzsch (1985) and Granovsky and Müller (1991), assuming a minimal number of sign changes to avoid degeneracy. The object function $T_{\nu,p}(K)$ depends on K in a complicated manner. Nevertheless, in the context of local polynomial fitting the solution to the above minimization problem is surprisingly simple as is shown in the next theorem.

Occasionally one would use an estimator in an undersmoothed form, so that the bias is negligible and one would see a finer structure. If this smoothing scheme is to be performed, which kernel function should one choose? In this case only the asymptotic variance contributes to the MSE or the MISE, and a possibility is to select the kernel which minimizes this variance. This approach leads to the *minimum variance kernel*, i.e. the kernel minimizing $\int K_\nu^{*2}(t) dt$. Finding this kernel is also an easy task in the framework of local polynomial fitting, as is evidenced by the next result, which is obtained by Fan, Gasser, Gijbels, Brockmann and Engel (1995).

Recall that $p - \nu$ is assumed to be odd, which is what we recommend to use in practice. The reason behind this recommendation will be further elaborated in the next section.

Theorem 3.4 *The Epanechnikov weight function $K(z) = 3/4(1-z^2)_+$ is the optimal kernel in the sense that it minimizes $T_{\nu,p}(K)$ over all nonnegative, symmetric, and Lipschitz continuous functions. The minimum variance kernel minimizing $\int K_\nu^{*2}(t) dt$ is given by the uniform density $1/2 I_{\{|z| \leq 1\}}$.*

Denote by $K_{\nu,p}^{\text{opt}}(z)$ the equivalent kernel induced by the optimal Epanechnikov kernel. It has been shown that this kernel equals

$$(K^*)^{opt} = K_{\nu,p}^{\text{opt}}(z) = \sum_{j=0}^{p+1} \lambda_j z^j, \quad (3.26)$$

where

$$\lambda_j = \begin{cases} 0 & \text{if } j+p+1 \text{ odd} \\ \frac{(-1)^{(j+\nu)/2}C_p(p+1-\nu)(p+1+j)!}{j!(j+\nu+1)2^{2p+3}(\frac{p+1-j}{2})!(\frac{p+1+j}{2})!} & \text{if } j+p+1 \text{ even,} \end{cases}$$
(3.27)

with $C_p = (p+\nu+2)!/\{(\frac{p+1+\nu}{2})!(\frac{p+1-\nu}{2})!\}$. Moreover, we have the following explicit formulae for its $(p+1)^{th}$ moment and L_2-norm:

$$|\int t^{p+1}K_{\nu,p}^{\text{opt}}(t)dt| = \frac{C_p\{(p+1)!\}^2}{(2p+3)!},$$

$$\int \{K_{\nu,p}^{\text{opt}}(t)\}^2 dt = \frac{C_p^2(p+1-\nu)^2}{(2\nu+1)(2p+3)(p+\nu+2)2^{2p+2}}. \qquad (3.28)$$

The question then arises of whether a similar kind of *universal optimal weighting scheme* can be given at the boundary? For example, if one wants to estimate $m(x_0)$, with $x_0 = ch$ which kernel function should be used? In this case the optimality criterion is as in (3.25) but with K_ν^* replaced by $K_{\nu,c}^*$. When using a local polynomial fit with the uniform kernel, the weighting scheme still possesses the minimum variance property. The behavior is thus optimal in this sense for the interior as well as for the boundary. This follows immediately from a characterization of minimum variance kernels given by Müller (1991). The situation is not as simple when taking MSE instead of variance as a criterion for optimality. See Fan, Gasser, Gijbels, Brockmann and Engel (1995) for a preliminary discussion on this. Cheng, Fan and Marron (1993) study this problem in detail and show that the kernel function $K(z) = (1-z)I_{[0,1]}(z)$ induces optimal kernels at the left boundary point $x_0 = 0$, whereas $K(z) = (1+z)I_{[-1,0]}(z)$ induces optimal kernels at the right boundary point $x_0 = 1$. For other boundary points the solution has not yet been obtained.

It is known that the choice of the kernel function K is not very important for the performance of the resulting estimators, both theoretically and empirically. However, since the Epanechnikov kernel is optimal in minimizing MSE and MISE at an interior point, we recommend using this kernel function. The structure of this kernel also enables us to implement fast computing algorithms, which is another reason behind our recommendation.

3.3 Which order of polynomial fit to use?

We now address the important question of which order of polynomial fit that should be used. Fitting polynomials of higher order

WHICH ORDER OF POLYNOMIAL FIT TO USE? 77

leads to a possible reduction of the bias, but on the other hand also to an increase of the variability, caused by introducing more local parameters. These points will be illustrated in Section 3.3.1. In Section 3.3.2 we will also argue that odd order polynomial fits outperform even order ones at the interior of design points.

Intuitively it is clear that in a flat non-sloped region a local constant or linear fit is recommendable, whereas at peaks and valleys local quadratic and cubic fits are preferable. Thus, for a very spatially inhomogeneous curve, the order of the polynomial approximation should be adjusted to the curvature of the unknown regression function. How do we develop a procedure which accomplishes this job of spatial adaptation? In Section 3.3.3 we propose such an order selection procedure.

We would like to mention, however, that for many applications the choice $p = \nu + 1$ suffices. A variable order selection carries a possible price including the stochastic element introduced by the selection procedure and computational costs. Such an order selection procedure is mainly proposed for recovering spatially inhomogeneous curves.

3.3.1 Increases of variability

We will first of all make a comparison of gains and losses of different order approximations. Suppose we fit a local polynomial of order p in order to estimate the derivative $m^{(\nu)}(x_0)$. The bias of such a fit will be of order $h^{p+1-\nu}$ (for $p - \nu$ odd) or of order $h^{p+2-\nu}$ (for $p - \nu$ even) as can be seen from Theorem 3.1. So, higher order polynomial approximations result in a smaller order of the bias. But let's see what happens to the variance if we increase the order of the approximation. The asymptotic variance of the estimator $\widehat{m}_\nu(x_0)$ is given by

$$\begin{aligned}\text{Var}\{\widehat{m}_\nu(x_0)|\mathbb{X}\} &= e_{\nu+1}^T S^{-1} S^* S^{-1} e_{\nu+1} \frac{\nu!^2 \sigma^2(x_0)}{f(x_0)nh^{1+2\nu}} \\ &\quad \times \{1 + o_P(1)\} \\ &= \int K_\nu^{*2}(t)dt \frac{\nu!^2 \sigma^2(x_0)}{f(x_0)nh^{1+2\nu}}\{1 + o_P(1)\}\end{aligned}$$

as established in Theorem 3.1 and (3.19). The order of this asymptotic variance is $n^{-1}h^{-(1+2\nu)}$, which is not affected by the order of the polynomial fit. But what about the constant terms? To investigate this further we concentrate for simplicity on the problem of estimating the regression function ($\nu = 0$). This is by no means a restriction; conclusions drawn for the case of estimating the regression function carry over to the estimation of its derivative functions. The asymptotic variance of the estimator for the regression function is of the form

$$V_p \frac{\sigma^2(x_0)}{f(x_0)nh}\{1 + o_P(1)\}, \qquad (3.29)$$

where V_p is the $(1,1)^{th}$ element of the matrix $S^{-1}S^*S^{-1}$. For $p = 0, 1, 2, 3$, the constant term V_p can be explicitly written as

$$V_0 = V_1 = \nu_0, \qquad V_2 = V_3 = \frac{\mu_4^2 \nu_0 - 2\mu_2\mu_4\nu_2 + \mu_2^2\nu_4}{(\mu_4 - \mu_2^2)^2}.$$

For a Gaussian kernel, we have that

$$\mu_{2j} = (2j-1)(2j-3)\cdots 3 \cdot 1 \quad \text{and} \quad \nu_{2j} = 2^{-j-1}\mu_{2j}/\sqrt{\pi},$$

and for the symmetric beta kernel defined in (2.5) we find

$$\mu_{2j} = \frac{\text{Beta}(j+1/2, \gamma+1)}{\text{Beta}(1/2, \gamma+1)} \quad \text{and} \quad \nu_{2j} = \frac{\text{Beta}(j+1/2, 2\gamma+1)}{\{\text{Beta}(1/2, \gamma+1)\}^2}.$$

Recall that the choices $\gamma = 0, 1, 2$ and 3 lead respectively to the uniform, the Epanechnikov, the biweight and the triweight kernel function. Table 3.3 shows how much the variance increases with the order of the approximation, relative to the variance of the Nadaraya-Watson estimator (local constant fit, $p = 0$). We summarize in Table 3.3 the values for V_p/V_0 for various commonly used kernel functions. The calculations can be carried through easily by using the definition of V_p in (3.29) and the above moment expressions.

Note that there is no loss in terms of asymptotic variance by doing a local linear instead of a local constant fit. This remark applies to the comparison of any even order approximation with its consecutive odd order approximation. This was also discussed in Ruppert and Wand (1994). However, the asymptotic variance increases when moving from an odd order approximation to its consecutive even order approximation. For example in the case of the Epanechnikov kernel, the variance increases by a factor of 2.0833 when a local quadratic instead of a local linear fit is used. Regarding the Gaussian kernel function as the limit of the symmetric beta kernel functions (see Section 2.2.1), the longer the effective support of the kernel function is, the more the increase in variability. The uniform kernel shows the highest increase. We also provide

WHICH ORDER OF POLYNOMIAL FIT TO USE?

Table 3.3. *Increase of the variability with the order of the polynomial approximation p.*

p	Gaussian	Uniform	Epanechnikov	Biweight	Triweight
1	1	1	1	1	1
2	1.6876	2.2500	2.0833	1.9703	1.9059
3	1.6876	2.2500	2.0833	1.9703	1.9059
4	2.2152	3.5156	3.1550	2.8997	2.7499
5	2.2152	3.5156	3.1550	2.8997	2.7499
6	2.6762	4.7852	4.2222	3.8133	3.5689
7	2.6762	4.7852	4.2222	3.8133	3.5689
8	3.1224	6.0562	5.2872	4.7193	4.3753
9	3.1224	6.0562	5.2872	4.7193	4.3753
10	3.5704	7.3281	6.3509	5.6210	5.1744

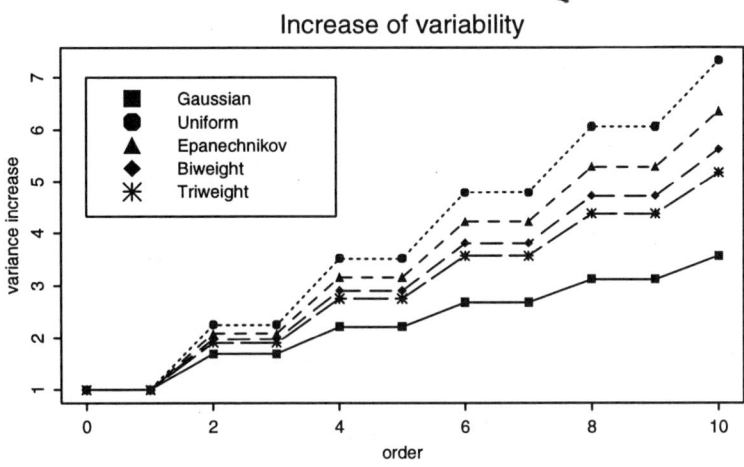

Figure 3.4. *Increase of the variability with the order of the polynomial approximation p. Results taken from Fan and Gijbels (1995b).*

a graphical presentation of these results in Figure 3.4.

3.3.2 It is an odd world

It now becomes clear that odd order fits are preferable. A fit of odd order $2q + 1$ introduces an extra parameter in comparison with a fit of even order $2q$, but there is no increase of variability caused by this. With this extra parameter an opportunity is created for a significant bias reduction especially in the boundary regions and in highly clustered design regions. These conclusions are in fact along the same lines as those evidenced by Fan (1992) who argues strongly in favor of local linear fits above local constant fits. Moreover, even order fits suffer from low efficiency, as was established by Fan (1993a) for the local constant fit. In addition serious boundary effects appear when using even order fits; this contrasts with odd order fits which have the nice boundary adaptive property described in Section 3.2.5.

The above asymptotic considerations demonstrate that it is 'an odd world': odd order polynomial fits are preferable to even order polynomial fits. An odd world indeed, and that is why we say that $p - \nu$ odd is 'natural'.

In local parametric fitting in nonparametric density estimation Hjort and Jones (1996a) find that an even number of parameters is preferable to an odd number of parameters. This coincides with the preference for odd order polynomials (involving an even number of parameters) when estimating a regression function.

3.3.3 Variable order approximation

From the previous two sections we know that preference should be given to odd order polynomial approximations. So, when focusing on estimation of the regression function, one can essentially restrict attention to local linear fits, local cubic fits or any odd order fit. Local constant fits and local quadratic fits for example do not have to be considered since they are asymptotically outperformed by respectively local linear and local cubic fits. In general, there is no direct comparison between two odd order fits: a lower order fit results in a larger bias, but has a smaller variance. Therefore, it is desirable to choose the order of the approximation adaptively: let the local curvature of the regression curve decide which order of fit should be used. In the following we explain how such a data-driven choice of the order of the polynomial approximation can be established.

Suppose that a bandwidth h is given, and hence that the local neighborhood of the point x_0 is fixed. A purely theoretical procedure to choose the order p of the local polynomial would be: choose the order p for which the MSE of the estimator is minimized. Since the MSE is unknown we replace it by an appropriate estimator $\widehat{\text{MSE}}(x_0; h)$. Then the basic idea for selecting an order of polynomial approximation in that fixed neighborhood of x_0, is to choose that order for which the estimated mean squared error is the smallest. A particular estimator $\widehat{\text{MSE}}(x_0; h)$ with good finite sample performance is proposed in Section 4.3. We omit the details here.

We now describe the order selection procedure for estimating $m^{(\nu)}(x_0)$. Denote by $\widehat{\text{MSE}}_{\nu,p}(x_0; h)$ any estimator of the MSE when fitting a polynomial of degree p. The estimated curve is usually evaluated at grid points of the form

$$x_j = x_L + j\Delta; \qquad j = 0, \cdots, n_{\text{grid}},$$

with x_L the first grid point. Suppose that we are interested only in polynomial approximations up to order R. For a given bandwidth h the algorithm of adaptive order approximation consists of the following steps:

Step 1. For each order p ($\nu < p \leq R$) and for each grid point obtain $\widehat{\text{MSE}}_{\nu,p}(x_j; h)$;

Step 2. For each order p, and for each grid point calculate the smoothed estimated MSE by taking the weighted local average of the estimated MSE in the neighboring $2[h/\Delta]+1$ grid points;

Step 3. For each grid point x_j choose the order p_j which has the smallest smoothed estimated MSE, and use a p_j order polynomial approximation to estimate $m^{(\nu)}(x_j)$.

We first of all mention that smoothing the estimated MSE in Step 2 is obtained by averaging the estimated MSE in a local neighborhood of width h. This step is not crucial, and can simply be replaced by taking the estimated MSE itself. In principle even order fits do not have to be included, but we did not exclude them in the above algorithm since they can possibly perform well at finite sample sizes. A detailed description of the variable order selection procedure based on the particular estimator for the MSE elaborated in Chapter 4 can be found in Fan and Gijbels (1995b).

We here illustrate briefly the benefits of the proposed variable order selection rule. The estimator for the MSE is taken to be the one proposed in Chapter 4. Consider the model

$$Y = X + 2\exp(-16X^2) + N(0, 0.5^2),$$

where $X \sim \text{Uniform}(-2, 2)$. Of interest is to estimate the regression curve $m(x)$. We simulated 1000 samples of size 200 from this model. For the estimation we use the Epanechnikov kernel. Figure 3.5 (a) shows the true regression curve and a typical simulated

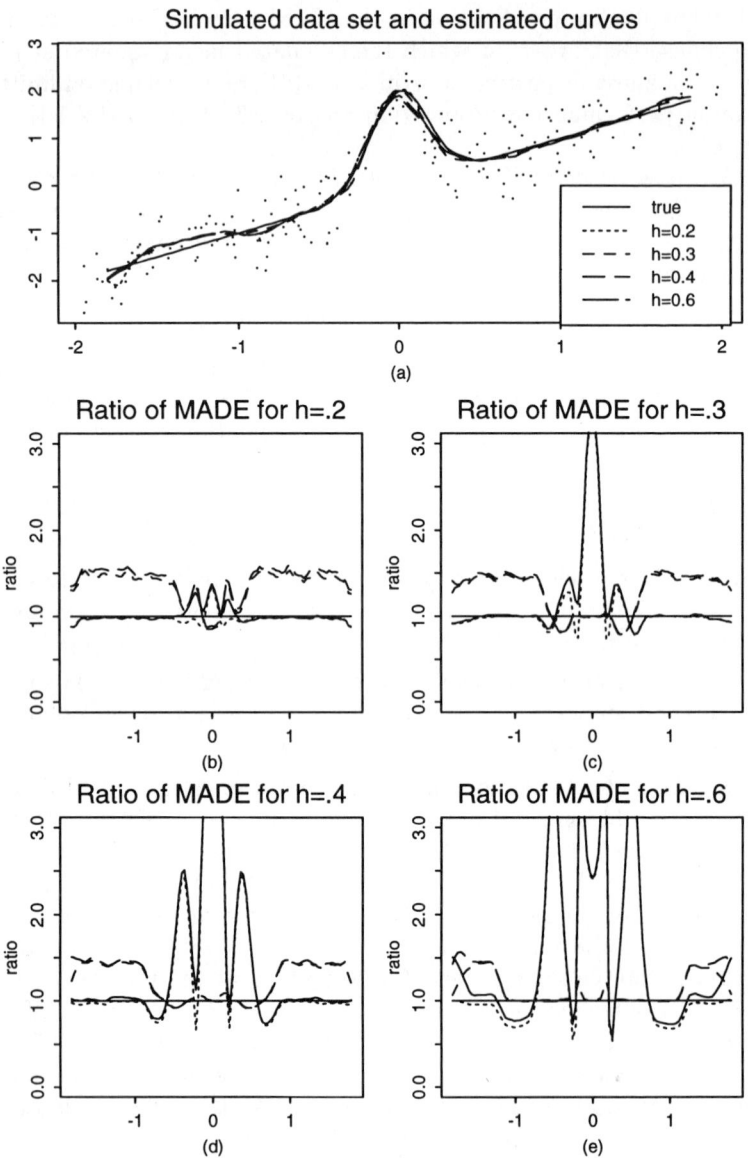

Figure 3.5. *Variable order selection. (a) The true regression curve, a typical simulated data set, and its estimated curves using the adaptive order procedure for a wide range of bandwidths. (b)–(e) Ratios of Mean Absolute Deviation Errors (MADE) for fixed orders to that for the adaptive order at grid points; solid line – local constant; short-dashed line – local linear; dashed line – local quadratic; long-dashed line – local cubic.*

dataset. Here the estimation task consists of recovering the linear structure and the bump. Clearly, it is desirable to use a local linear fit in both ends and a local quadratic or cubic fit for the middle part of the curve. Interesting questions are: 'How does the variable order approximation perform in comparison with fixed order approximations?', 'Does it select the right order of fits?', and 'How does the variable order approximation perform for different values of the fixed bandwidth h?' To answer these questions we implemented fixed order approximations together with the variable order approximation, obtained via the above algorithm, for different values of h, namely $h = 0.2, 0.3, 0.4$ and 0.6. The estimated regression curve was calculated in the grid points $x_j = -1.8 + 0.036j$, $j = 0, \cdots, 100$.

In Figure 3.5 (a) we also present typical estimated curves, using the adaptive order approximation, for the four different choices of h. All estimates seem to do a good job in tracking back the structure of the unknown curve. Although the bandwidth has been increased by a factor of 3 (from 0.2 to 0.6), the quality of the estimate does not change. So, the estimate based on the adaptive order selection procedure is *not very sensitive* to the choice of bandwidth. This can be explained as follows. When the bandwidth is too large for a local linear fit, the adaptive method tends to choose a higher order approximation. On the other hand, when the bandwidth is too small for a local cubic fit, the procedure opts for a lower order approximation. This highlights a nice feature of the order selection procedure: it is 'robust' for the choice of the bandwidth.

In order to compare the various fixed order and the variable order approximations, we calculate for each fitted order the *Mean Absolute Deviation Error* (MADE) in each grid point based on 400 simulations. For each fixed order, we plot the ratio of the MADE for that fixed order to the MADE for the adaptive order. These plots are depicted in Figures 3.5 (b)–(e), for respectively $h = 0.2, 0.3, 0.4$ and $h = 0.6$. All pictures reveal that the adaptive order has small risk at all locations. This means that the variable order selection procedure correctly selected order 1 at both ends and order 2 or 3 in the middle, which is what we intuitively expected. Moreover, the procedure selects the right order, even at locations such as $-0.7 \leq x \leq -0.3$ where even humans are not sure which order to use.

3.4 Best linear smoothers

Most regression estimators studied in the literature are weighted averages of the responses and hence of the form

$$\widehat{m}(x) = \sum_{j=1}^{n} W_j(x, X_1, \cdots, X_n) Y_j.$$

Such an estimator is called a *linear smoother*. The Nadaraya-Watson estimator and the Gasser-Müller estimator are both linear estimators, as can be seen easily from (2.4) and (2.6) respectively. Also local polynomial estimators belong to the class of linear smoothers as is obvious from expression (3.11). The class of linear smoothers includes also orthogonal series estimators and spline smoothers. How do local polynomial estimators perform in comparison with other linear smoothers? A way to answer this question is to study the linear minimax risk. This study is the main objective of the next section. It turns out that local polynomial estimators have high linear minimax efficiency and some of them can achieve 100% efficiency.

3.4.1 Best linear smoother at interior: optimal rates and constants

When investigating which linear smoother is the best one, we should specify the class of unknown regression functions which is under consideration. We will concentrate here on the estimation of the functional $S_\nu(m) = m^{(\nu)}(x_0)$, where x_0 represents an arbitrary interior point. Let

$$C_{p+1} = \{m : |m(z) - \sum_{j=0}^{p} \frac{m^{(j)}(x_0)}{j!}(z - x_0)^j | \leq C \frac{|z - x_0|^{p+1}}{(p+1)!} \}.$$
(3.30)

This class basically consists of all regression functions whose $(p+1)^{th}$ derivative is bounded by C in a neighborhood of x_0. Some basic conditions are needed in order to obtain the minimax results. These conditions read as follows:

(i) $f(\cdot)$ is continuous at the point x_0 with $f(x_0) > 0$.

(ii) $\sigma(\cdot)$ is continuous at the point x_0.

BEST LINEAR SMOOTHERS

The *linear minimax risk* is defined by

$$R_{\nu,L}(n, \mathcal{C}_{p+1}) = \inf_{\widehat{S}_\nu \text{ linear}} \sup_{m \in \mathcal{C}_{p+1}} E\left[\{\widehat{S}_\nu - m^{(\nu)}(x_0)\}^2 | \mathbf{X}\right], \quad (3.31)$$

and is a benchmark to efficiency of linear estimation. The best linear smoother is the one that achieves this linear minimax risk. The linear minimax efficiency of local linear regression estimators was studied by Fan (1992). For getting insight into linear minimax efficiency properties we concentrate first of all on this special case ($p = 1$ and $\nu = 0$). Here we consider the class $\mathcal{C}_2$, which consists basically of regression functions having a bounded second derivative. We first compute the linear minimax risk and point out its implications later on.

Theorem 3.5 *The linear minimax risk is given by*

$$R_{0,L}(n, \mathcal{C}_2) = \frac{3}{4} 15^{-1/5} \left\{ \frac{\sqrt{C}\sigma^2(x_0)}{nf(x_0)} \right\}^{4/5} \{1 + o_P(1)\}. \quad (3.32)$$

This linear minimax risk is achieved by the local linear regression estimator $\widehat{m}^*(x_0)$ with Epanechnikov kernel and bandwidth:

$$h = \left\{ \frac{15\sigma^2(x_0)}{f(x_0)C^2 n} \right\}^{1/5}. \quad (3.33)$$

The proof of this result, which is simple and elementary, is sketched in Section 3.7. So, the linear smoother $\widehat{m}^*(x_0)$ is the best linear smoother, having the optimal rate as well as the optimal constant; or equivalently,

$$\frac{R_{0,L}(n, \mathcal{C}_2)}{\sup_{m \in \mathcal{C}_2} E\left[\{\widehat{m}^*(x_0) - m(x_0)\}^2 | \mathbf{X}\right]} \xrightarrow{P} 1.$$

The linear minimax risk plays an important role in evaluating the minimax efficiency of linear smoothers among the class of all linear smoothers. For a linear smoother $\widehat{m}_L$, this *minimax efficiency* can be defined as

$$\text{efficiency of } \widehat{m}_L = \left(\frac{R_{0,L}(n, \mathcal{C}_2)}{\sup_{m \in \mathcal{C}_2} E\left[\{\widehat{m}_L(x_0) - m(x_0)\}^2 | \mathbf{X}\right]} \right)^{5/4}. \quad (3.34)$$

Hence, an 80% efficient estimator uses only about 80% of the available data. In other words, such an estimator based on a sample of size 100 works as well as the best linear smoother with sample size

Table 3.4. *Minimax efficiency for kernel regression smoothers (in percentage).*

Kernel	Local linear fit	Gasser-Müller	Nadaraya-Watson
Epanechnikov	100	66.67	0
Gaussian	95.12	63.41	0
Uniform	92.95	61.97	0

Adapted from Fan (1992).

only 80. Consider the local linear regression estimator, and recall that its conditional MSE is given by

$$\left[\frac{1}{4}\mu_2^2\{m''(x_0)\}^2 h^4 + \frac{1}{nh}\frac{\nu_0 \sigma^2(x_0)}{f(x_0)}\right]\{1 + o_P(1)\}, \quad (3.35)$$

which over the class $\mathcal{C}_2$ is bounded by

$$\left[\frac{1}{4}\mu_2^2 C^2 h^4 + \frac{1}{nh}\frac{\nu_0 \sigma^2(x_0)}{f(x_0)}\right]\{1 + o_P(1)\}. \quad (3.36)$$

The rigorous proof of this step is given in Fan (1993a). Now choose the bandwidth such that this risk is minimized. Then, it is easy to see that its asymptotic minimax efficiency is determined by

$$0.268(\mu_2 \nu_0^2)^{-1/2},$$

where $0.268 = (3/5 \times 15^{1/5})^{5/4}$. The Gasser-Müller estimator has, for random designs, the efficiency

$$\frac{2}{3}0.268(\mu_2 \nu_0^2)^{-1/2}.$$

The Nadaraya-Watson estimator depends on the derivative $m'(x_0)$, and hence its maximum risk over $\mathcal{C}_2$ is infinite, unless $f'(x_0) = 0$, and hence its asymptotic minimax efficiency is 0. Table 3.4 gives an overview of the minimax efficiencies of those three estimators for some commonly used kernels.

The above minimax properties can be extended to the general case of local polynomial estimators and derivative estimation. This was done by Fan, Gasser, Gijbels, Brockmann and Engel (1995). They established the following general theorems (recall definitions (3.30) and (3.31)).

Let $K_{\nu,p}^{\text{opt}}$ be the equivalent kernel induced by the Epanechnikov kernel, defined as in (3.26). Define $r = 2(p + 1 - \nu)/(2p + 3)$,

$s = (2\nu + 1)/(2p + 3)$ and

$$B_{\nu,p} = r^{-r}s^{-s}\left(\frac{C}{(p+1)!}\right)^{2s}\left(\frac{\sigma^2(x_0)}{nf(x_0)}\right)^r.$$

Theorem 3.6 *The linear minimax risk is bounded from above and below by*

$$R_{\nu,p}\{1 + o_P(1)\} \geq R_{\nu,L}(n, C_{p+1}) \geq r_{\nu,p}\{1 + o_P(1)\}, \quad (3.37)$$

where

$$R_{\nu,p} = B_{\nu,p}\left(\int |t^{p+1} K_{\nu,p}^{\text{opt}}(t)| dt\right)^{2s} \left(\int \{K_{\nu,p}^{\text{opt}}(t)\}^2 dt\right)^r$$

and

$$r_{\nu,p} = B_{\nu,p}\left(\left|\int t^{p+1} K_{\nu,p}^{\text{opt}}(t) dt\right|\right)^{2s} \left(\int \{K_{\nu,p}^{\text{opt}}(t)\}^2 dt\right)^r.$$

Using expressions (3.26) and (3.28), we have the following explicit formula:

$$r_{\nu,p} = \left(\frac{2p+3}{2\nu+1}\right)\left\{\frac{(p+\nu+2)!}{\left(\frac{p+1+\nu}{2}\right)!\left(\frac{p+1-\nu}{2}\right)!}\right\}^2 \left\{\frac{r}{(p+\nu+2)4^{p+2}}\right\}^r$$

$$\times \left\{\frac{(p+1)!C}{(2p+3)!}\right\}^{2s} \left\{\frac{\sigma^2(x_0)}{nf(x_0)}\right\}^r. \quad (3.38)$$

The lower and upper bounds are not sharp, except for some commonly used cases: $p = 1, \nu = 0$ and $p = 2, \nu = 1$ (see Table 3.5). We conjecture that the sharp risk is obtained by replacing $|K_{\nu,p}^{\text{opt}}|$ in the definition of $R_{\nu,p}$ by $|K_{\nu,p}^{\text{opt}*}|$ which minimizes

$$T_{\nu,p}^*(K) = \left\{\int |t^{p+1} K_\nu^*(t)| dt\right\}^{2\nu+1} \left\{\int K_\nu^{*2}(t) dt\right\}^{p+1-\nu}$$

(compare with (3.25)). How far are the lower bound and upper bound apart? Table 3.5 gives the ratio between $r_{\nu,p}$ and $R_{\nu,p}$, which is

$$E_{\nu,p} = \left(\frac{|\int t^{p+1} K_{\nu,p}^{\text{opt}}(t) dt|}{\int |t^{p+1} K_{\nu,p}^{\text{opt}}(t)| dt}\right)^{2s}.$$

Let $\widehat{m}_\nu^*(x_0) = \nu! \widehat{\beta}_\nu$ be the estimator resulting from a local polynomial fit of order p with the Epanechnikov kernel K_0 (and its corresponding equivalent kernel $K_{0\nu}^* = K_{\nu,p}^{\text{opt}}$ defined as in (3.16)

Table 3.5. *Ratio of the minimax lower and upper bounds.*

$p-\nu$ \ p	1	2	3	4	5
1	1	1	.9805	.8325	.5807
3			.8984	.6604	.4371
5					.8070

and given by (3.26)) and with the bandwidth

$$h = \left[\frac{(2\nu+1)(p+1)!^2 \int \{K_{\nu,p}^{opt}(t)\}^2 \, dt \, \sigma^2(x_0)}{2(p+1-\nu)\{\int |t^{p+1} K_{\nu,p}^{opt}(t)| dt\}^2 f(x_0) C^2 n} \right]^{1/(2p+3)}. \tag{3.39}$$

Note that these choices of the kernel function and the bandwidth reduce to the choices mentioned in Theorem 3.5 for the special case that $p = 1$ and $\nu = 0$. This is easily seen from (3.28). With the above choices of the kernel and the bandwidth the local polynomial estimator has a high linear minimax efficiency, as established in the next theorem.

Theorem 3.7 *The local polynomial estimator $\widehat{m}_\nu^*(x_0)$ has high linear minimax efficiency for estimating $m^{(\nu)}(x_0)$ in the sense that*

$$\frac{R_{\nu,L}(n, \mathcal{C}_{p+1})}{\sup_{m \in \mathcal{C}_{p+1}} E\left[\{\widehat{m}_\nu^*(x_0) - m^{(\nu)}(x_0)\}^2 | \mathbf{X}\right]} \geq E_{\nu,p}\{1 + o_P(1)\}. \tag{3.40}$$

Thus, for commonly used local polynomial fitting with $p = 1, \nu = 0$ or $p = 2, \nu = 1$, the estimator $\widehat{m}_\nu^*(x_0)$ is the best linear estimator. For other values, the local polynomial fit is nearly the best, as evidenced by Table 3.5. In fact, if our conjecture after Theorem 3.6 is true, the efficiency of $\widehat{m}_\nu^*(x_0)$ is far higher than $E_{\nu,p}$. Further, the local polynomial regression estimator is the best linear smoother in a weak minimax sense, namely it is a linear minimax estimator when the 'parameter space' $\mathcal{C}$ consists only of any finite number of regression functions whose $(p+1)^{th}$ derivative is bounded by C:

$$\mathcal{C} = \{\frac{Cx^{p+1}}{(p+1)!}, m^*, -m^*, m_1, m_2, \cdots, m_L : |m_j^{(p+1)}(x_0)| \leq C\},$$

where $m^*(\cdot)$ and $-m^*(\cdot)$ are the least favorable pair of the regression functions that Fan, Gasser, Gijbels, Brockmann and Engel (1995) used to construct the linear minimax lower bound (3.37).

3.4.2 Best linear smoother at boundary

In the previous section we showed that when dealing with estimation at the interior local polynomial estimators are nearly best linear estimators and are the best for some commonly used values of p and ν. Do these minimax properties also hold when one considers estimation at the boundary? The answer to this question is affirmative, as was pointed out by Cheng, Fan and Marron (1993). Assume that the design density has a support $[0,1]$. We only discuss here the case of the left boundary point $x_0 = 0$, and remark that similar discussions hold true for the right boundary point $x_0 = 1$. The major difference between the results for the point $x_0 = 0$ and an interior point is that the optimal Epanechnikov kernel for an interior point should be replaced by the optimal kernel for the left boundary point $x_0 = 0$, namely $K(z) = (1-z)I_{[0,1]}(z)$.

Suppose that $f(\cdot)$ and $\sigma(\cdot)$ are right continuous at the point 0 with $f(0+) > 0$ and $\sigma(0+) < \infty$. Here we consider the class of regression functions

$$\mathcal{C}_{p+1}^0 = \{m : |m(z) - \sum_{j=0}^{p} \frac{m^{(j)}(0+)}{j!} z^j| \leq C \frac{|z|^{p+1}}{(p+1)!} \text{ for } z > 0\}.$$

The linear minimax risk is defined by

$$R_{\nu,L}^0(n, \mathcal{C}_{p+1}^0) = \inf_{\widehat{S}_\nu \text{ linear}} \sup_{m \in \mathcal{C}_{p+1}^0} E\left[\{\widehat{S}_\nu - m^{(\nu)}(0+)\}^2 | \mathbf{X}\right].$$

We discuss first the special case of a local linear regression estimator for the regression function ($p=1$, $\nu=0$). It turns out that the local linear regression estimator, with an appropriate choice of the kernel and the bandwidth, is nearly the best linear smoother. The next theorem formulates the result as proven by Cheng, Fan and Marron (1993).

Theorem 3.8 *The linear minimax risk is bounded from above and below by*

$$\frac{3}{2}\left(\frac{15}{2}\right)^{1/5} \left\{\frac{\sqrt{C}\sigma^2(0+)}{nf(0+)}\right\}^{4/5} \{1 + o_P(1)\}$$

$$\geq R_{0,L}^0(n, \mathcal{C}_2^0) \geq 3 \cdot 15^{-1/5} \left\{\frac{\sqrt{C}\sigma^2(0+)}{nf(0+)}\right\}^{4/5} \{1 + o_P(1)\}, \quad (3.41)$$

and the upper bounded is achieved by the local linear regression estimator with kernel $(1-z)I_{[0,1]}(z)$ *and bandwidth* $h = \left\{\frac{2048\sigma^2(0+)}{15f(0+)C^2 n}\right\}^{\frac{1}{5}}$.

The proof of this result is simple, and follows exactly the same lines as that of Theorem 3.5. Note that the constant factors in the lower and upper bounds are respectively 1.7454 and 2.2444. The lower bound is not sharp. The sharp lower bound has a constant factor 2.1242 and hence the local linear regression estimator given by Theorem 3.8 has linear minimax efficiency 94.64%. More details can be found at the end of this section.

Using similar arguments it was established that the local linear regression estimator with kernel $(1+z)I_{[-1,0]}(z)$ and bandwidth $h = \left\{ \frac{2048\sigma^2(1-)}{15f(1-)C^2 n} \right\}^{1/5}$ is nearly the best linear smoother for estimating $m(1-)$.

The above minimax theory can be developed also for the general case of local polynomial fitting and derivative estimation. We only give the main result. All details of this study can be found in Cheng, Fan and Marron (1993).

Consider again the estimation at the left boundary point 0. Then, by Theorem 3.2, the MSE of the local polynomial estimator depends on the kernel function through $T^0_{\nu,p}(K)$ which is defined analogously to (3.25) except now the integration takes place from 0 to ∞. The optimal kernel that minimizes $T^0_{\nu,p}(K)$ among $K \geq 0$ is the triangular kernel $K_0(z) = (1-z)I_{[0,1]}(z)$. Let $K^{\text{opt},0}_{\nu,p}$ denote the corresponding equivalent kernel at the point 0. Denote by $\hat{m}^*_\nu(0)$ the estimator resulting from a local polynomial fit of order p with kernel $K_0(z) = (1-z)I_{[0,1]}(z)$ and bandwidth

$$h = \left[\frac{(2\nu+1)(p+1)!^2 \int \{K^{\text{opt},0}_{\nu,p}(t)\}^2 dt \, \sigma^2(0+)}{2(p+1-\nu) \left\{ \int |t^{p+1} K^{\text{opt},0}_{\nu,p}(t)| dt \right\}^2 f(0+) C^2 n} \right]^{1/(2p+3)}.$$

Then, we have the following results.

Theorem 3.9 *The linear minimax risk for estimating the νth derivative of the regression function at its left endpoint is*

$$R^0_{\nu,p}\{1+o_P(1)\} \geq R^0_{\nu,L}(n, C^0_{p+1}) \geq r^0_{\nu,p}\{1+o_P(1)\}, \quad (3.42)$$

where with r and s as in Theorem 3.6,

$$R^0_{\nu,p} = B^0_{\nu,p} \left(\int |t^{p+1} K^{\text{opt},0}_{\nu,p}(t)| dt \right)^{2s} \left(\int \{K^{\text{opt},0}_{\nu,p}(t)\}^2 dt \right)^r,$$

$$r^0_{\nu,p} = B^0_{\nu,p} \left(\left| \int t^{p+1} K^{\text{opt},0}_{\nu,p}(t) dt \right| \right)^{2s} \left(\int \{K^{\text{opt},0}_{\nu,p}(t)\}^2 dt \right)^r$$

and
$$B^0_{\nu,p} = r^{-r}s^{-s}\left(\frac{C}{(p+1)!}\right)^{2s}\left(\frac{\sigma^2(0+)}{nf(0+)}\right)^r.$$

Moreover, the upper bound is attained by the local polynomial estimator $\widehat{m}^*_\nu(0)$.

It can be shown that

$$\begin{aligned}r^0_{\nu,p} &= \left(\frac{2p+3}{2\nu+1}\right)\left\{\frac{(p+\nu+2)!}{(p-\nu+1)!\nu!}\right\}^2\left\{\frac{r}{2(p+\nu+2)}\right\}^r \\ &\quad \times \left\{\frac{(p+1)!C}{(2p+3)!}\right\}^{2s}\left\{\frac{\sigma^2(0+)}{nf(0+)}\right\}^r.\end{aligned} \qquad (3.43)$$

However, this lower bound is not sharp. We conjecture that the linear minimax risk is $R^0_{\nu,p}$ with the kernel function $K^{\text{opt},0}_{\nu,p}(t)$ replaced by $K^{\text{opt},0,*}_{\nu,p}(t)$, which minimizes

$$T^{0*}_{\nu,p}(K) = \left\{\int_0^\infty |t^{p+1}K^*_\nu(t)|dt\right\}^{2\nu+1}\left\{\int_0^\infty K^{*2}_\nu(t)dt\right\}^{p+1-\nu}.$$

Sacks and Ylvisaker (1981a) characterize the solution. The minimum value of $T^{0*}_{1,0}(K)$ is 3.764336 and this implies that

$$R^0_{0,L}(n, \mathcal{C}^0_{p+1}) = 2.1242\left\{\frac{\sqrt{C}\sigma^2(0+)}{nf(0+)}\right\}^{4/5}\{1+o_P(1)\}.$$

Thus, the local linear regression estimator has 94.64% linear minimax efficiency in terms of the risk. In other words, it is nearly the best boundary correction method. Further, it is the best boundary correction method in a weak minimax sense. See Cheng, Fan and Marron (1993).

3.5 Minimax efficiency of local polynomial fitting

In the previous section it was shown that local polynomial estimators are nearly 100% asymptotic minimax efficient among the class of linear smoothers, both in the interior and at the boundary. The practical performance of kernel regression smoothers depends very much on the choice of the bandwidth. In practical implementation a data-driven choice of the bandwidth is recommended. With such a data-driven choice of the bandwidth, the resulting estimator is of a more complex structure, and can no longer be considered as a linear smoother. Therefore, it is of interest to investigate the overall minimax risk, i.e. when looking at the class of all estimators, linear and nonlinear ones. The overall minimax risk has been studied for estimation at the interior. In this section we give an overview of the results which were obtained. The investigation of the minimax efficiency at the boundary has not been accomplished yet, but can be obtained using arguments similar to those in Fan (1993a).

Denote by $\widehat{T}_\nu(x_0)$ any estimator of $m^{(\nu)}(x_0)$, where x_0 is a point at the interior. Then, the *minimax risk* is defined as

$$R_\nu(n, \mathcal{C}_{p+1}) = \inf_{\widehat{T}_\nu} \sup_{m \in \mathcal{C}_{p+1}} E\left[\{\widehat{T}_\nu - m^{(\nu)}(x_0)\}^2 | \boldsymbol{X}\right]. \quad (3.44)$$

An upper bound of the minimax risk is easily obtained from the asymptotic expression for the MSE. For example, the linear minimax risk (3.37) provides an upper bound.

A key tool for deriving lower bounds for minimax risks is a geometric quantity, called the *modulus of continuity* (see Donoho and Liu (1991a, b) and Fan (1993a)). In the next section we give a brief introduction to this quantity and its usefulness in deriving minimax risks.

3.5.1 Modulus of continuity

Consider the problem of estimating a functional $T(m)$ of the regression function. For example $T(m) = m^{(\nu)}(x_0)$. The modulus of continuity of $T(m)$ over a class $\mathcal{C}_m$ is defined as

$$\omega_{\mathcal{C}_m}(\varepsilon) = \sup\{|T(m_1) - T(m_0)| : m_0, m_1 \in \mathcal{C}_m, \|m_1 - m_0\| \leq \varepsilon\}, \quad (3.45)$$

where $\|\cdot\|$ denotes the usual L_2-norm on $L_2(-\infty, +\infty)$.

What is the relationship between this modulus of continuity and a lower bound for the minimax risk? This relationship was described thoroughly by Donoho and Liu (1991a, b) for the Gaussian white noise model and by Fan (1993a) for the current nonparametric regression model. Here, we indicate only the main ideas. In nonparametric applications, the modulus of continuity typically behaves as follows

$$\omega_{\mathcal{C}_m}(\varepsilon) = A\varepsilon^\alpha\{1 + o(1)\} \qquad \text{as } \varepsilon \to 0, \alpha \in (0,1),$$

and the extremal pair is attained at $m_0(\cdot)$ and $m_1(\cdot)$ satisfying

$$m_1(x) - m_0(x) = \varepsilon^\alpha H\left(\frac{x_0 - x}{\varepsilon^{2(1-\alpha)}}\right)\{1 + o(1)\}$$

uniformly in x as $\varepsilon \to 0$,

where $H(\cdot)$ is a bounded continuous function. A functional $T(m)$ is said to be *regular* on $\mathcal{C}_m$ with exponent α if the extremal pair of the modulus of continuity exists and has the above form.

Consider the following class of joint density functions $f_{X,Y}$, imposing constraints on the marginal density of X and the conditional variance:

$$\mathcal{C}_{b,B} = \{f_{X,Y}(\cdot,\cdot) : f(x_0) \geq b, \sigma^2(x_0) \leq B, |f(x)-f(y)| \leq c|x-y|^\eta\},$$

with $0 < \eta < 1$. The Lipschitz condition on the marginal density is used only for technical arguments. In fact, the constants c and η are not related to the upper and lower bound of the minimax risk. We would also like to mention that the constraint $\mathcal{C}_{b,B}$ is not directly related to the functional $T(m)$. It is allowed that $\mathcal{C}_{b,B}$ has only one member of f and σ, as was the case in the previous section.

The next result, established by Fan (1993a), provides the key tool for obtaining bounds for the minimax risks and the linear minimax risks, using knowledge of the modulus of continuity. Indeed, the derivations of Theorems 3.5–3.9 are all based on the following theorem. We say that the class $\mathcal{C} = \mathcal{C}_m \cap \mathcal{C}_{b,B}$ of joint densities is *rich*, if there exists a bounded density g with $g(x_0) = b$ such that the normal submodel

$$f_\theta(x,y) = \frac{1}{\sqrt{2\pi B}} \exp(-(y - m_\theta(x))^2/B) g(x)$$

is in the class.

Theorem 3.10 *Let $\mathcal{C}_m$ be convex and $\mathcal{C} = \mathcal{C}_m \cap \mathcal{C}_{b,B}$ be rich. If $T(m)$ is regular on $\mathcal{C}_m$ with exponent α, then the minimax risk $R(n,\mathcal{C})$ for estimation of $T(m)$ is bounded by*

$$R(n,\mathcal{C}) \geq \varepsilon_\alpha \frac{\alpha^\alpha (1-\alpha)^{1-\alpha}}{4} \omega_{\mathcal{C}_m}^2 \left(2\sqrt{\frac{B}{nb}}\right)\{1 + o(1)\},$$

with $0.8 \leq \varepsilon_\alpha \leq 1$. Moreover, a lower bound for the linear minimax risk $R_L(n,\mathcal{C})$ is provided by

$$R_L(n,\mathcal{C}) \geq \frac{\alpha^\alpha (1-\alpha)^{1-\alpha}}{4} \omega_{\mathcal{C}_m}^2 \left(2\sqrt{\frac{B}{nb}}\right)\{1 + o(1)\}.$$

Hence, the main task in finding a lower bound for the minimax risk in (3.44) is to investigate the modulus of continuity. Once this is accomplished, one then simply applies the above theorem.

3.5.2 Best rates and nearly best constant

The minimax risk over the class of regression functions $\mathcal{C}_{p+1}$ behaves as follows.

Theorem 3.11 *The minimax risk (3.44) is asymptotically bounded by*
$$R_{\nu,p} \geq R_\nu(n, \mathcal{C}_{p+1}) \geq (0.894)^2 r_{\nu,p}, \qquad (3.46)$$
with $r_{\nu,p}$ as in (3.38).

We now show that a local polynomial fit with appropriate optimal choices of the bandwidth and the kernel comes asymptotically fairly close to the minimax risk. Recall that $\widehat{m}_\nu^*(x_0)$ denotes the estimator resulting from a local polynomial fit of order p with the Epanechnikov kernel K_0, and with the bandwidth as in (3.39). This local polynomial estimator is efficient in the rate, and nearly efficient in the constant factor, as is established in the following theorem.

Theorem 3.12 *The local polynomial regression estimator $\widehat{m}_\nu^*(x_0)$ has an asymptotic efficiency of at least 89.4 $E_{\nu,p}$ % among all estimators:*
$$\frac{R_\nu(n, \mathcal{C}_{p+1})}{\sup_{m \in \mathcal{C}_{p+1}} E\left[\{\widehat{m}_\nu^*(x_0) - m^{(\nu)}(x_0)\}^2 | \boldsymbol{X}\right]}$$
$$\geq (0.894)^2 E_{\nu,p} + o_P(1), \qquad (3.47)$$
with $E_{\nu,p}$ as in Theorem 3.7.

The minimax theory provided in this and the preceding section is an additional justification of the intuitively appealing local polynomial approximation method.

3.6 Fast computing algorithms

In the recent literature some fast implementations of kernel-based nonparametric curve estimators have been proposed. In comparison with naive direct implementations, those fast implementations can increase the speed by a factor up into the hundreds. We discuss in this section how to implement these fast computing algorithms for local polynomial estimators. After implementing such fast algorithms, a local polynomial fit can be carried out in a few seconds. Consequently, local polynomial fitting with the use of fast computing algorithms competes clearly with other fast methods.

Recent proposals for fast implementations of nonparametric curve estimators include the *binning methods* and the *updating meth-*

ods. See the bibliographic notes for references. Fan and Marron (1994) give careful speed comparisons of these two fast implementations and naive direct implementations, under a variety of settings and using various machines and software. Both fast methods turn out to be much faster than the direct implementations, but neither of the two dominates the other. In terms of accuracy the difference among the two fast methods and direct implementation is very small, and is mostly negligible from a practical point of view. The updating method occasionally seems to have some problems with numerical instability. In the following we describe briefly how the binning and updating algorithms are used to speed up the computation of local polynomial estimators.

Note that the local polynomial estimator $\hat{\beta} = (\mathbf{X}^T \mathbf{W} \mathbf{X})^{-1} \mathbf{X}^T \mathbf{W} \mathbf{y}$ can be written as

$$\begin{pmatrix} S_{n,0}(x_0) & S_{n,1}(x_0) & \cdots & S_{n,p}(x_0) \\ S_{n,1}(x_0) & S_{n,2}(x_0) & \cdots & S_{n,p+1}(x_0) \\ \vdots & \vdots & \ddots & \vdots \\ S_{n,p}(x_0) & S_{n,p+1}(x_0) & \cdots & S_{n,2p}(x_0) \end{pmatrix}^{-1} \begin{pmatrix} T_{n,0}(x_0) \\ T_{n,1}(x_0) \\ \vdots \\ T_{n,p}(x_0) \end{pmatrix} \tag{3.48}$$

where $S_{n,\ell}(x_0) = S_{n,\ell}$ is as defined in (3.10), and

$$T_{n,\ell}(x_0) = \sum_{i=1}^{n} K_h(X_i - x_0)(X_i - x_0)^\ell Y_i, \quad \ell = 0, 1, \ldots, p.$$

For example, it is easily seen that the local linear regression estimator for $m(x_0)$ can be expressed as

$$\widehat{m}(x_0) = \frac{S_{n,2}(x_0) T_{n,0}(x_0) - S_{n,1}(x_0) T_{n,1}(x_0)}{S_{n,2}(x_0) S_{n,0}(x_0) - S_{n,1}^2(x_0)},$$

requiring five different quantities to be calculated. Hence, calculation of local polynomial estimators reduces to that of the quantities $S_{n,\ell}(x_0)$ and $T_{n,\ell}(x_0)$, where the number of these quantities involved is $3p + 2$, as is easily seen from (3.48). Computation of kernel density estimators and the Nadaraya-Watson estimator also results in calculating these kinds of quantities. So, the major task now is how to calculate such quantities quickly.

Estimated curves are commonly presented in graphical form, and therefore the estimators are evaluated in grid points $x_1, \cdots, x_{n_{\text{grid}}}$, with n_{grid} the number of grid points. Note that direct calculation of the quantities $S_{n,\ell}(\cdot)$ and $T_{n,\ell}(\cdot)$ in all grid points, consists of carrying through $O(n \cdot n_{\text{grid}})$ kernel evaluations. These kernel evaluations are actually the most time-consuming part in direct implementations, and it is exactly this part which needs speeding up.

3.6.1 Binning implementation

This approach results basically in an approximation, obtained after binning the data. The key idea is to reduce the number of kernel evaluations considerably, exploiting the fact that many of these evaluations are nearly the same. This kind of approach requires that the grid $x_1, \cdots, x_{n_\text{grid}}$ is *equally spaced*. The data are also approximated by 'equally spaced data'. This is achieved by *binning* the data, i.e. by moving each datum point (X_i, Y_i) to its nearest grid point $(x_{j(i)}, Y_i)$. Let I_j be the index set, recording observations X_i moved to x_j, namely $I_j = \{i : X_i \mapsto x_j\}$. Then the modified data set of length n can clearly be summarized by the *binned data*

$$\{(x_j, \overline{Y}_j, c_j) : j = 1, \cdots, n_\text{grid}\},$$

where

$$\overline{Y}_j = \text{average}\{Y_i : i \in I_j\}, \qquad c_j = \#\{X_i : i \in I_j\} = \#\{I_j\}$$

are respectively the *bin averages* and the *bin counts*. Note that the only approximation in the binning process is that X_i is replaced by its nearest grid point $x_{j(i)}$.

The estimation is then carried out based on the modified data set. Hence the original quantity $T_{n,\ell}(x_{j'})$, $x_{j'}$ a grid point, is now approximated (via approximating X_i by $x_{j(i)}$) by

$$\begin{aligned}
\overline{T}_{n,\ell}(x_{j'}) &\equiv \sum_{i=1}^{n} K_h(x_{j(i)} - x_{j'})(x_{j(i)} - x_{j'})^\ell Y_i \\
&= \sum_{j=1}^{n_\text{grid}} \sum_{i \in I_j} K_h(x_j - x_{j'})(x_j - x_{j'})^\ell Y_i \\
&= \sum_{j=1}^{n_\text{grid}} K_h(x_j - x_{j'})(x_j - x_{j'})^\ell Y_j^\Sigma, \qquad (3.49)
\end{aligned}$$

where $Y_j^\Sigma = \sum_{i \in I_j} Y_i = c_j \overline{Y}_j$.

Similarly the quantity $S_{n,\ell}(x_{j'})$ is approximated by

$$\overline{S}_{n,\ell}(x_{j'}) \equiv \sum_{j=1}^{n_\text{grid}} K_h(x_j - x_{j'})(x_j - x_{j'})^\ell c_j. \qquad (3.50)$$

FAST COMPUTING ALGORITHMS

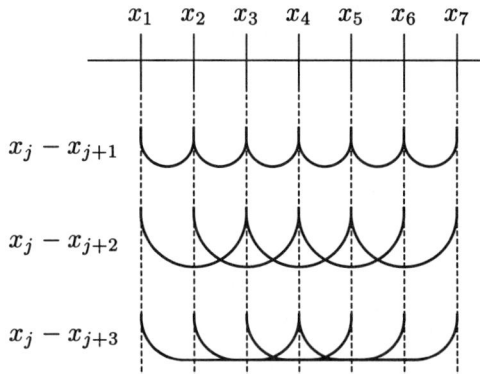

Figure 3.6. *Pairwise differences in the case of equally spaced grid points. The idea is from Fan and Marron (1994).*

It is clear that implementation of (3.49) and (3.50) would reduce the number of kernel evaluations from $O(n \cdot n_{\text{grid}})$ to $O(n_{\text{grid}}^2)$. But, even a further reduction can be achieved by studying closely (3.49) and (3.50). A careful look at these quantities reveals that for equally spaced grid points many of the pairwise differences $x_j - x_{j'}$ appearing in the calculations are the same, as is illustrated clearly in Figure 3.6. Consequently many kernel evaluations are the same, and an important gain in terms of computation results from this.

More precisely, define $\Delta = x_j - x_{j-1}$ as the *binwidth*. Then $x_j - x_{j-k} = k\Delta$, for each j. Hence, $\overline{T}_{n,\ell}(x_{j'})$ can be simply calculated via

$$\overline{T}_{n,\ell}(x_{j'}) = \sum_{j=1}^{n_{\text{grid}}} K_h(\Delta(j-j'))(\Delta(j-j'))^\ell Y_j^\Sigma$$

$$= \sum_{j=1}^{n_{\text{grid}}} \kappa_{\ell, j-j'} Y_j^\Sigma, \qquad (3.51)$$

where $\kappa_{\ell,j} = K_h(\Delta j)(\Delta j)^\ell$. A similar simplification holds for the calculation of $\overline{S}_{n,\ell}(x_{j'})$. In this way, the number of kernel evaluations is further reduced to $O(n_{\text{grid}})$.

Linear binning is a refinement of the *simple binning* procedure described above. Here each datum point (X_i, Y_i) is split into two 'fractional points', assigned to the two nearest grid points (bin centers). The assigned fraction is proportional to the distance of X_i

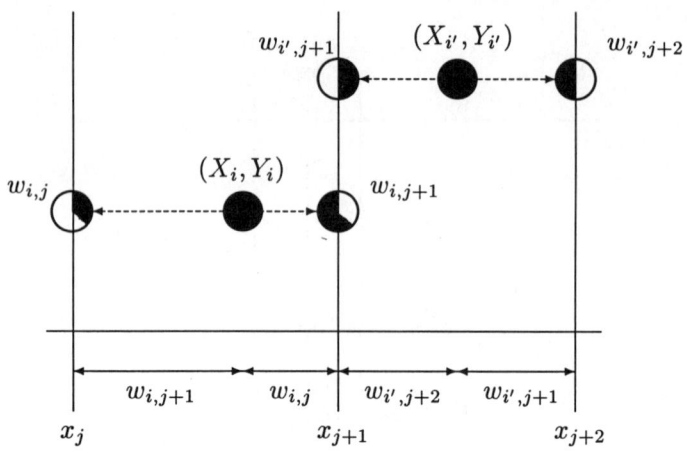

Figure 3.7. *The principle of linear binning. The idea is adapted from Fan and Marron (1994).*

to the nearest grid point on the opposite side. Figure 3.7 illustrates this linear binning procedure.

More precisely, with

$$w_{i,j} = \left(1 - \frac{|X_i - x_j|}{\Delta}\right)_+ \qquad i = 1, \cdots, n; \ j = 1, \cdots, n_{\text{grid}},$$

the fraction $w_{i,j}$ is put into the bin centered at x_j, whereas the fraction $w_{i,j+1}$ is put into the bin centered at x_{j+1}. Hence, evaluation of $\overline{T}_{n,\ell}$ and $\overline{S}_{n,\ell}$ now requires storing

$$c_j = \sum_{i=1}^n w_{i,j} \quad \text{and} \quad Y_j^\Sigma = \sum_{i=1}^n w_{i,j} Y_i \quad j = 1, \cdots, n_{\text{grid}}.$$

Both simple and linear binning of the data require only $O(n)$ operations. To appreciate how, consider the transformation $L(x) = (1/\Delta)(x - x_1) + 1$, which maps the set of grid points $\{x_1, \cdots, x_{n_{\text{grid}}}\}$ into the index set $\{1, \cdots, n_{\text{grid}}\}$. For each observation $X_i, i = 1, \cdots, n$, we calculate $L(X_i)$. Its integer part, denoted by $\ell(X_i)$, indicates the two nearest bin centers to X_i. This calculation already suffices for simple binning. For linear binning the fractional part $L(X_i) - \ell(X_i)$ should be computed, leading to the weights assigned to each of the two nearest bin centers, namely $w_{i,\ell(X_i)} = 1 - \{L(X_i) - \ell(X_i)\}$ and $w_{i,\ell(X_i)+1} = L(X_i) - \ell(X_i)$. Once the data have been binned, we use direct implementation of $\overline{T}_{n,\ell}$ and $\overline{S}_{n,\ell}$ which can be calculated now with only $O(n_{\text{grid}})$ kernel evaluations.

Note that the quantities $\overline{T}_{n,\ell}(x_{j'})$ and $\overline{S}_{n,\ell}(x_{j'})$ admit a *convolution form*. For example, $\overline{T}_{n,\ell}$ is the convolution of $\{\kappa_{\ell,j}\}$ and $\{Y_j^\Sigma\}$. When n_{grid} is large *fast Fourier transform methods* can be used to further speed up the computations. Fan and Marron (1994) report that for $n_{\text{grid}} \leq 400$, however, no significant gain can be made by using fast Fourier transform methods.

3.6.2 Updating algorithm

This fast method relies on computing averages recursively, by updating averages previously computed. The supersmoother of Friedman (1984) is also based on this kind of updating idea. The fast smoother LOWESS introduced by Cleveland (1979) uses a different kind of updating idea to determine its local neighborhoods. When using smooth, nonuniform, kernels one has to deal with weighted local averages, where the weights change with the location, and hence updating is not so straightforward. Gasser and Kneip (1989) have adapted updating ideas for polynomial kernels. These ideas were implemented for local polynomial fitting by Fan and Marron (1994) and Seifert, Brockmann, Engel and Gasser (1994). The main principle is to expand the polynomials into terms which can be computed quickly using updating ideas. We explain this briefly.

Suppose that the kernel function is supported on $[-1, 1]$. For a location point x, data points contributing to the quantities $S_{n,\ell}(x)$ and $T_{n,\ell}(x)$ are those belonging to the set

$$I_x = \left\{ i : \left| \frac{X_i - x}{h} \right| \leq 1 \right\}.$$

Now consider a kernel of the symmetric beta family (see (2.5)), i.e. $K(t) = C_\gamma (1 - t^2)_+^\gamma$, where C_γ denotes the normalizing factor in (2.5), and γ is a nonnegative integer. Then

$$K(t) = C_\gamma \sum_{k=0}^{\gamma} \binom{\gamma}{k} (-t^2)^k$$

and therefore

$$\begin{aligned} T_{n,\ell}(x) &= \sum_{i=1}^{n} K_h(X_i - x)(X_i - x)^\ell Y_i \\ &= C_\gamma \sum_{i \in I_x} \sum_{k=0}^{\gamma} \binom{\gamma}{k} h^{-2k} (-1)^k (X_i - x)^{2k+\ell} Y_i \end{aligned}$$

$$
\begin{aligned}
&= C_\gamma \sum_{i \in I_x} \sum_{k=0}^{\gamma} \binom{\gamma}{k} h^{-2k}(-1)^k \\
&\quad \times \sum_{q=0}^{2k+\ell} \binom{2k+\ell}{q} x^q (-1)^q X_i^{2k+\ell-q} Y_i \\
&= C_\gamma \sum_{k=0}^{\gamma} \binom{\gamma}{k} (-1)^k \sum_{q=0}^{2k+\ell} \binom{2k+\ell}{q} x^q \\
&\quad \times \sum_{i \in I_x} h^{-2k}(-1)^q X_i^{2k+\ell-q} Y_i.
\end{aligned}
$$
(3.52)

The summations of the form

$$\sum_{i \in I_x} h^{-2k}(-1)^q X_i^{2k+\ell-q} Y_i, \qquad \text{for } x = x_1, \cdots, x_{n_{\text{grid}}}$$

can be calculated using updating ideas. Assume that the data are sorted according to $X_1, \cdots, X_n$. Then, by differencing the cumulative sums $U_m \equiv \sum_{i=1}^{m} h^{-2k}(-1)^q X_i^{2k+\ell-q} Y_i$ we obtain

$$\sum_{i \in I_{x_j}} h^{-2k}(-1)^q X_i^{2k+\ell-q} Y_i = U_{\overline{I}(x_j)} - U_{\underline{I}(x_j)},$$

for appropriate choices of indices $\overline{I}(x_j)$ and $\underline{I}(x_j)$.

In a similar way, $S_{n,\ell}(x)$ can be computed using updating ideas. Calculating the cumulative sums requires $O(n)$ operations, and the remaining calculations are independent of n. Hence, the updating procedure results in a $O(n)$ algorithm (ignoring the time needed for presorting the data). An advantage of this updating method is that it allows for bandwidth variation, as explained by Fan and Marron (1994). The disadvantage is that updating has the potential for numerical instability due to rounding errors. See Seifert, Brockmann, Engel and Gasser (1994) for a possible solution.

3.7 Complements

In this section we give the derivations of some of the results presented in the previous sections of this chapter. These derivations might give the reader an idea of the technical work under the framework of local polynomial fitting. We keep the derivations on a simple level. Details about all presented results can be found in the literature (see references mentioned in the text).

Derivation of asymptotic bias and variance given in Theorem 3.1

We first of all show how to derive the asymptotic expression for the conditional variance given in (3.7). Recall the notation $S_n = \mathbf{X}^T \mathbf{W} \mathbf{X}$ and denote by $S_n^* \equiv \mathbf{X}^T \Sigma \mathbf{X}$ the $(p+1) \times (p+1)$ matrix $(S_{n,j+\ell}^*)_{0 \le j, \ell \le p}$, with $S_{n,j}^* = \sum_{i=1}^n (X_i - x_0)^j K_h^2(X_i - x_0) \sigma^2(X_i)$. Then, the conditional variance in (3.6) can be re-expressed as

$$S_n^{-1} S_n^* S_n^{-1}, \qquad (3.53)$$

and the task is now to find approximations for the two matrices S_n and S_n^*. Recall definition (3.10) and note that

$$\begin{aligned}
S_{n,j} &= E S_{n,j} + O_P\{\sqrt{\operatorname{Var}(S_{n,j})}\} \\
&= nh^j \int u^j K(u) f(x_0 + hu) du \\
&\quad + O_P\left(\sqrt{n E\{(X_1 - x_0)^{2j} K_h^2(X_1 - x_0)\}}\right) \\
&= nh^j \{f(x_0) \mu_j + o(1) + O_P(1/\sqrt{nh})\} \\
&= nh^j f(x_0) \mu_j \{1 + o_P(1)\}, \qquad (3.54)
\end{aligned}$$

provided that $h \to 0$ and $nh \to \infty$. From this we obtain immediately that

$$S_n = n f(x_0) H S H \{1 + o_P(1)\}, \qquad (3.55)$$

as was already noted in (3.14). Using similar arguments, we find that

$$S_{n,j}^* = nh^{j-1} f(x_0) \sigma^2(x_0) \nu_j \{1 + o_P(1)\}, \qquad (3.56)$$

and therefore,

$$S_n^* = nh^{-1} f(x_0) \sigma^2(x_0) H S^* H \{1 + o_P(1)\}, \qquad (3.57)$$

with S^* as in Theorem 3.1. Now, starting from (3.53) and using (3.55) and (3.57) we find that

$$\operatorname{Var}(\widehat{\beta} | \mathbf{X}) = \frac{\sigma^2(x_0)}{f(x_0) nh} H^{-1} S^{-1} S^* S^{-1} H^{-1} \{1 + o_P(1)\}, \qquad (3.58)$$

and since $\widehat{m}_\nu(x_0) = \nu! e_{\nu+1}^T \widehat{\beta}$ this leads directly to the asymptotic expression for the conditional variance.

For the bias we have to distinguish between the case that $p - \nu$ is odd and $p - \nu$ is even. We first deal with the simplest situation occurring when $p - \nu$ is odd, and indicate afterwards how to obtain the asymptotic expression for the conditional bias when

$H = \text{diag}(1, h, \cdots, h^p)$

$p - \nu$ is even. By using the Taylor expansion the conditional bias $S_n^{-1} \mathbf{X}^T \mathbf{W} \mathbf{r}$ of $\widehat{\beta}$ can be written as

$$S_n^{-1} \mathbf{X}^T \mathbf{W} \left[\beta_{p+1}(X_i - x_0)^{p+1} + o_P\{(X_i - x_0)^{p+1}\} \right]_{1 \le i \le n}$$
$$= S_n^{-1} \{\beta_{p+1} c_n + o_P(n h^{p+1})\}, \tag{3.59}$$

where $c_n = (S_{n,p+1}, \cdots, S_{n,2p+1})^T$ and $\beta_{p+1} = m^{(p+1)}(x_0)/(p+1)!$. Applying (3.54) and (3.55) we obtain from (3.59) that

$$\text{Bias}(\widehat{\beta}|\mathbb{X}) = H^{-1} S^{-1} c_p \beta_{p+1} h^{p+1} \{1 + o_P(1)\}, \tag{3.60}$$

where we recall that $c_p = (\mu_{p+1}, \cdots, \mu_{2p+1})^T$. This immediately leads to the asymptotic expression for the conditional bias as given in (3.8).

The above derivation of course holds for any value of $p - \nu$, but the problem is that for $p - \nu$ even the $(\nu + 1)^{th}$ element of the vector $S^{-1} c_p$ is zero. This can be easily seen by writing out the structure of the matrix S and the vector c_p, and recalling that the odd order moments of a symmetric kernel are zero. Hence, the $(\nu + 1)^{th}$ element of the main term in expansion (3.60) is zero, and one clearly has to proceed to higher order expansions. This essentially means that in all the expansions we derived to obtain (3.60) some extra terms have to be taken along. We now indicate these detailed expansions. The justifications of these expansions are similar to those given above, but now relying on the more stringent conditions on $f(\cdot)$ and $m^{(p+2)}(\cdot)$. It is easily seen that the expansion in (3.54) can be extended to

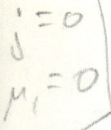

$$S_{n,j} = n h^j \{ f(x_0) \mu_j + h f'(x_0) \mu_{j+1} + O_P(a_n) \}, \tag{3.61}$$

where $a_n = h^2 + 1/\sqrt{nh}$. Hence, we get

$$S_n = nH \{ f(x_0) S + h f'(x_0) \widetilde{S} + O_P(a_n) \} H, \tag{3.62}$$

with $\widetilde{S} = (\mu_{j+\ell+1})_{0 \le j, \ell \le p}$. Using a higher order Taylor expansion we can write the conditional bias as

$$S_n^{-1} \mathbf{X}^T \mathbf{W} \left[\beta_{p+1}(X_i - x_0)^{p+1} + \beta_{p+2}(X_i - x_0)^{p+2} \right.$$
$$\left. + o_P\{(X_i - x_0)^{p+2}\} \right]_{1 \le i \le n}$$
$$= S_n^{-1} \{\beta_{p+1} c_n + \beta_{p+2} \widetilde{c}_n + o_P(n h^{p+2})\}, \tag{3.63}$$

where $\widetilde{c}_n = (S_{n,p+2}, \cdots, S_{n,2p+2})^T$. Then, substituting (3.61) and (3.62) into (3.63), we obtain

COMPLEMENTS

$$\text{Bias}(\widehat{\beta}|\mathbf{X}) = H^{-1}\{f(x_0)S + hf'(x_0)\tilde{S} + O_P(a_n)\}^{-1}$$
$$\times h^{p+1}[\beta_{p+1}f(x_0)c_p + h\{f'(x_0)\beta_{p+1}$$
$$+ \beta_{p+2}f(x_0)\}\tilde{c}_p + O_P(a_n)],$$

(3.64)

where $\tilde{c}_p = (\mu_{p+2}, \cdots, \mu_{2p+2})^T$. Finally, by noting that

$$(A + hB)^{-1} = A^{-1} - hA^{-1}BA^{-1} + O(h^2), \quad (3.65)$$

and using some simple algebra, we find the following asymptotic expansion for the bias term

$$\text{Bias}(\widehat{\beta}|\mathbf{X}) = h^{p+1}H^{-1}\{\beta_{p+1}S^{-1}c_p + hb^*(x_0) + O_P(a_n)\} \quad (3.66)$$

where

$$b^*(x_0) = \frac{f'(x_0)\beta_{p+1} + \beta_{p+2}f(x_0)}{f(x_0)}S^{-1}\tilde{c}_p - \frac{f'(x_0)}{f(x_0)}\beta_{p+1}S^{-1}\tilde{S}S^{-1}c_p.$$

Taking the $(\nu + 1)^{th}$ element of this bias vector, and keeping in mind that the $(\nu + 1)^{th}$ element of $S^{-1}c_p$ and of $S^{-1}\tilde{S}S^{-1}c_p$ is zero, we obtain the result in (3.9).

Moment conditions for the equivalent kernel

In Section 3.2.2 we established an equivalent kernel representation for local polynomial estimators. Here we derive the moment conditions for this kernel.

We first show that the weights W_ν^n defined in (3.11) satisfy the discrete moment conditions (3.12). Note that, using the definition of those weights,

$$\sum_{i=1}^n (X_i - x_0)^q W_\nu^n \left(\frac{X_i - x_0}{h}\right)$$

$$= e_{\nu+1}^T S_n^{-1} \sum_{i=1}^n (X_i - x_0)^q \begin{pmatrix} 1 \\ X_i - x_0 \\ \vdots \\ (X_i - x_0)^p \end{pmatrix} K_h(X_i - x_0)$$

$$= e_{\nu+1}^T S_n^{-1} S_n e_{q+1} = \delta_{\nu,q},$$

where the last two equalities follow from the definition of S_n. This proves the statement.

The moment conditions for the equivalent kernel stated in (3.17) are obtained via similar arguments. Indeed, using definition (3.16)

we find
$$\int u^q K_\nu^*(u)du = e_{\nu+1}^T S^{-1} S e_{q+1} = \delta_{\nu,q},$$
using the definition of the matrix S.

Linear minimax efficiency at interior points

In the following we prove the linear minimax result stated in Theorem 3.5. Note first of all that for any linear smoother
$$\widehat{m}_L(x_0) = \sum_{j=1}^n W_j(x_0, X_1, \cdots, X_n) Y_j$$
the conditional risk is given by
$$E\left[\{\widehat{m}_L(x_0) - m(x_0)\}^2 | \mathbf{X}\right]$$
$$= \{\sum_{j=1}^n W_j m(X_j) - m(x_0)\}^2 + \sum_{j=1}^n W_j^2 \sigma^2(X_j)$$
$$\geq \frac{m^2(x_0)}{1 + \sum_{j=1}^n m^2(X_j)/\sigma^2(X_j)}, \qquad (3.67)$$

where the latter inequality follows from applying the next lemma, established by Fan (1992).

Lemma 3.1 *Let* $\mathbf{a} = (a_1, \cdots, a_n)^T$ *and* $\mathbf{w} = (w_1, \cdots, w_n)^T$ *be n-dimensional real vectors. Then,*
$$\min_{\mathbf{w}}\{(\mathbf{w}^T \mathbf{a} - b)^2 + \sum_{i=1}^n c_i w_i^2\} = \frac{b^2}{1 + \sum_{i=1}^n a_i^2/c_i},$$
where the minimum is attained at
$$w_j = \frac{b}{1 + \sum_{i=1}^n a_i^2/c_i} a_j/c_j.$$

Since the random variables $X_1, \cdots, X_n$ are i.i.d., we find that
$$\sum_{j=1}^n m^2(X_j)/\sigma^2(X_j)$$
$$= nE\{m^2(X_1)/\sigma^2(X_1)\} + O_P\left(\sqrt{nE\{m^4(X_1)/\sigma^4(X_1)\}}\right). \qquad (3.68)$$

Now take $m_0(y) = (b_n^2/2) \cdot \{1 - C(y - x_0)^2/b_n^2\}_+$. Clearly $m_0 \in C_2$. Here b_n is taken to be $\{15\sqrt{C}\sigma^2(x_0)/nf(x_0)\}^{1/5}$; the reason for this choice will become clear later on. Then, straightforward calculations show that

$$E\{m_0^2(X_1)/\sigma^2(X_1)\} = \frac{4f(x_0)b_n^5}{15\sqrt{C}\sigma^2(x_0)}\{1+o(1)\}, \qquad (3.69)$$

and $E\{m_0^4(X_1)/\sigma^4(X_1)\} = O(b_n^8)$. Combining (3.67), (3.68) and (3.69), we get

$$E\left[\{\widehat{m}_L(x_0) - m_0(x_0)\}^2 | \mathbf{X}\right]$$

$$\geq \frac{b_n^4/4}{1 + \frac{4f(x_0)nb_n^5}{15\sqrt{C}\sigma^2(x_0)}\{1+o_P(1)\}}$$

$$= \frac{3}{4}15^{-1/5}\left\{\frac{\sqrt{C}\sigma^2(x_0)}{nf(x_0)}\right\}^{4/5}\{1+o_P(1)\},$$

by the choice of b_n, which maximizes the expression on the right-hand side of the inequality. Hence, from the definition of the linear minimax risk,

$$R_{0,L}(n,\mathcal{C}_2) \geq \frac{3}{4}15^{-1/5}\left\{\frac{\sqrt{C}\sigma^2(x_0)}{nf(x_0)}\right\}^{4/5}\{1+o_P(1)\}. \qquad (3.70)$$

Now take $\widehat{m}(x_0)$ to be the local linear regression estimator with the Epanechnikov kernel $K(z) = \frac{3}{4}(1-z^2)_+$ and the bandwidth as indicated in (3.33). Then, it is easy to see from the asymptotic expression of the MSE in (3.35) and in particular its bound in (3.36) that

$$R_{0,L}(n,\mathcal{C}_2) \leq \frac{3}{4}15^{-1/5}\left\{\frac{\sqrt{C}\sigma^2(x_0)}{nf(x_0)}\right\}^{4/5}\{1+o_P(1)\}, \qquad (3.71)$$

which together with (3.70) implies that the linear minimax risk is given by (3.32).

From the above derivation it is also clear that this linear minimax risk is achieved by the local linear regression estimator $\widehat{m}(x_0)$ with Epanechnikov kernel and bandwidth as in (3.33).

3.8 Bibliographic notes

Locally weighted least squares regression has been around for many years. In early years it was used in time series analysis by Macauley (1931). In the context of nonparametric regression the idea was first systematically studied by Stone (1977, 1980), Cleveland (1979) and

Katkovnik (1979). Lejeune (1985) and Müller (1987) investigated the issue of effective weights in local polynomial fitting. Tsybakov (1986) studied the minimax rate of the local polynomial approximation method in a robust estimation setting. The principle of local polynomial approximation was also exploited by Tibshirani and Hastie (1987) in their study of local likelihood estimation. A variety of examples illustrating the use of local polynomial fitting, in particular local quadratic fitting, in various domains of applications, can be found in Cleveland and Devlin (1988). Almost sure convergence of local polynomial estimators has been studied by Ioffe and Katkovnik (1989, 1990). Another fresh look at local polynomial fitting was provided by Fan (1992, 1993a). In the last few years the 'old' idea of local least squares regression has received a lot of attention, resulting in deep insights into the method, its simplicity and wide applicability. Recent illuminating works on this smoothing method include Fan and Gijbels (1992, 1995a, b), Hastie and Loader (1993), Ruppert and Wand (1994), Cheng, Fan and Marron (1993) and Fan, Gasser, Gijbels, Brockmann and Engel (1995). See also Cheng and Chu (1993), Wei and Chu (1994), Cheng (1995), Cheng and Bai (1995) and Fan, Gijbels, Hu and Huang (1996). References discussing application of local polynomial fitting in various fields will be given in later chapters.

Variations on local polynomial fitting have recently been proposed and studied by Seifert and Gasser (1995) and Hall and Marron (1995), among others.

Many papers in the literature deal with possible approaches to boundary modifications, in order to reduce boundary effects. Gasser and Müller (1979), Gasser, Müller and Mammitzsch (1985), Granovsky and Müller (1991) and Müller (1991, 1993) discuss using so-called boundary kernels to correct for boundary effects, in the case of conventional kernel estimators. Schucany and Sommers (1977) and Rice (1984) suggest a linear combination of two kernel estimators with different bandwidths to reduce the bias. Schuster (1985) discusses the problem in the density estimation setup, and 'reflects' the data points beyond the support to end up with an augmented data set, on which the estimator is based. Hall and Wehrly (1991) essentially introduce a more sophisticated regression version of Schuster's approach. Djojosugito and Speckman (1992) discuss a method for boundary bias reduction based on a finite-dimensional projection in a Hilbert space. Boundary correction methods for smoothing splines have been studied by Rice and Rosenblatt (1981) and Eubank and Speckman (1991), among others.

BIBLIOGRAPHIC NOTES   107

The concept of 'global' variable bandwidth was introduced by Breiman, Meisel and Purcell (1977) in the density estimation context. Further studies on the use of global variable bandwidth in density estimation can be found in Abramson (1982), Hall and Marron (1988) and Hall (1990). Müller and Stadtmüller (1987) investigated Gasser-Müller type estimation of regression curves with a local variable bandwidth. A nice discussion of the distinction between 'global' variable bandwidth and local variable bandwidth is given by Jones (1990). More references on the issue of bandwidth selection are provided in Chapter 4.

There is a vast literature on minimax estimation in various settings. The following list of references is by no means complete, but only reflects some important historical developments. Minimax estimation rates were studied by Farrell (1972) in the setting of nonparametric density estimation. They were further investigated by Hasminskii (1979), Stone (1980, 1982) and Donoho and Johnstone (1996). Many interesting ideas on minimax estimation can be found in Le Cam (1985). Deep results on asymptotic minimax risks over Sobolov spaces were established by Pinsker (1980), Efromovich and Pinsker (1982) and Nussbaum (1985). Minimax risks over more general classes were studied by Ibragimov and Hasminskii (1984), Donoho and Liu (1991a, b), Donoho and Low (1992) and Donoho (1994) for estimating a functional, and by Donoho, Liu and MacGibbon (1990), Fan (1993b) and Low (1993), among others, for estimating an entire function. See also Sacks and Ylvisaker (1978, 1981b) and Li (1982) for early results on linear minimax risks. Most of the above studies focused on estimating a linear functional. For estimation of a quadratic functional, the minimax risks were studied by Bickel and Ritov (1988), Donoho and Nussbaum (1990), Fan (1991) and Hall and Marron (1991). Theory on adaptive minimax estimation was developed by Efromovich (1985, 1996), Lepskii (1992) and Brown and Low (1995), among others. Korostelev and Tsybakov (1993) discuss minimax theory in image reconstruction.

Binning methods for nonparametric curve estimators have been discussed by Silverman (1982), Jones and Lotwick (1984) and Härdle and Scott (1992). Silverman (1986) also discusses the implementation of fast Fourier transform methods. Jones (1989) and Hall and Wand (1996) investigated the accuracy of binned density estimators. For a variation on the local linear regression estimator, using binning ideas, see Jones (1994). Multivariate extensions of fast computation methods are in Scott (1992) and Wand (1994).

CHAPTER 4

Automatic determination of model complexity

4.1 Introduction

In Chapter 3 we outlined the framework of local polynomial fitting. This method consists of fitting a polynomial of degree p locally, where the local neighborhood is determined by a bandwidth h and a kernel function K. A bandwidth $h = 0$ basically results in interpolating the data, and hence leads to the most complex model. A bandwidth $+\infty$ corresponds to fitting *globally* a polynomial of degree p, the simplest model. So the bandwidth governs the complexity of the model. In Section 3.2.3 of Chapter 3, we obtained expressions for an asymptotically optimal constant and variable bandwidth. See respectively (3.21) and (3.20). However, those asymptotically optimal bandwidths depend on unknown quantities, and hence are not directly applicable in practice. Here we discuss some practical choices of bandwidths.

The main contribution of this chapter is to propose data-driven procedures for selecting a constant or variable bandwidth in local polynomial fitting. The basic ingredients for the procedures are simple. First we derive, in Section 4.3, good approximations of the bias and variance, not relying completely on asymptotic expressions. As a by-product of the estimated bias and variance, we discuss the construction of pointwise confidence intervals in Section 4.4. A second basic ingredient is a residual squares criterion, which is introduced in Section 4.5. The procedures developed can be applied for selecting bandwidths for estimating the regression function as well as its derivatives. The performance of the procedures is illustrated largely via simulated examples. They also illustrate the fact that local polynomial fitting with a data-driven variable bandwidth possesses the necessary flexibility for capturing complicated shapes of curves. In other words, the resulting estimators have the nice feature that they adapt readily to spatially inhomogeneous curves. This spatial adaptation property is shared by the wavelet methods. In Section 4.7 we discuss briefly the property of spatial adaptation and its importance in nonparametric regression. Local polynomial fitting with the proposed data-driven bandwidth selection procedures turns out to be an appealing smoothing technique in practice. This is demonstrated via a variety of real data examples in Section 4.8. They give readers an impression of how this particular smoothing technique works in practice.

The scenario of the local modelling idea is not restricted to the least squares problem. Indeed, it is applicable to all likelihood-based models. Section 4.9 outlines the key idea of the local modelling scheme in likelihood-based models, and shows how the bandwidth can be chosen automatically by using the data. The approach is applicable to a wide array of statistical problems and forms a *nearly universal principle* for automatic determination of the model complexity. This nearly universal principle appears to be new in the literature and opens gateways for future research. Our experiences in the least squares problem suggest that it should be very useful for a large array of statistical models.

There are many other proposals for selecting smoothing parameters in the literature. Among those, we briefly discuss in Section 4.10 the cross-validation technique, the normal-reference method, the plug-in approach, and the bandwidth selection rule proposed by Sheather and Jones (1991). Some of those were proposed only for the density estimation setting.

4.2 Rule of thumb for bandwidth selection

In many data analyses, one would like to get a quick idea about how large the amount of smoothing should be. A rule of thumb is very suitable in such a case. Such a rule is meant to be somewhat crude, but possesses simplicity and requires little programming effort that other methods are hard to compete with. In certain situations the choice of the bandwidth might be less important and a crude bandwidth selector might suffice. An example of such a situation is the adaptive order procedure in Section 3.3.3, which is less sensitive to the choice of bandwidth. Another example is to select a *pilot bandwidth* for certain bandwidth selection procedures.

With the local polynomial regression method such a crude bandwidth selector can easily be obtained as follows. We start from the asymptotically optimal constant bandwidth derived in (3.21). It contains the unknown quantities $\sigma^2(\cdot)$, $m^{(p+1)}(\cdot)$ and $f(\cdot)$, which need to be estimated. A simple way to do so is by fitting a polynomial of order $p+3$ *globally* to $m(x)$, leading to the parametric fit

$$\tilde{m}(x) = \check{\alpha}_0 + \cdots + \check{\alpha}_{p+3} x^{p+3}. \tag{4.1}$$

The standardized residual sum of squares from this parametric fit is denoted by $\check{\sigma}^2$. This choice of a global fit results in a derivative function $\tilde{m}^{(p+1)}(x)$ which is of a quadratic form, allowing for a certain flexibility in estimating the curvature. Suppose that we are interested in estimating $m^{(\nu)}(\cdot)$ and we take $w(x) = f(x)w_0(x)$ for some specific function w_0. Regarding the conditional variance $\sigma^2(x)$ as a constant, and substituting the *pilot estimates* $\tilde{m}(\cdot)$ and $\check{\sigma}^2$ into the expression for the asymptotically optimal constant bandwidth (3.21) we obtain the following expression

$$C_{\nu,p}(K) \left[\frac{\check{\sigma}^2 \int w_0(x) dx}{n \int \{\tilde{m}^{(p+1)}(x)\}^2 w_0(x) f(x) dx} \right]^{1/(2p+3)}. \tag{4.2}$$

The denominator in the above expression can be estimated by

$$\sum_{i=1}^{n} \{\tilde{m}^{(p+1)}(X_i)\}^2 w_0(X_i),$$

which leads to the *rule of thumb bandwidth selector*

$$\check{h}_{\text{ROT}} = C_{\nu,p}(K) \left[\frac{\check{\sigma}^2 \int w_0(x) dx}{\sum_{i=1}^{n} \{\tilde{m}^{(p+1)}(X_i)\}^2 w_0(X_i)} \right]^{1/(2p+3)}. \tag{4.3}$$

This simple bandwidth selector $\check{h}_{\text{ROT}}$ is derived under certain conditions. However, even in situations where these conditions are not strictly fulfilled this bandwidth selector can be applied to get an initial idea of the amount of smoothing to be used.

Table 4.1. *Simulation models.*

Example	$m(x)$	σ	$\approx$ noise / signal
1	$x + 2e^{-16x^2}$	0.4	1/2
2	$\sin(2x) + 2e^{-16x^2}$	0.3	1/3
3	$m_0(x)$ [1]	0.1	1
4	$0.4x + 1$	0.15	1/2

[1] $m_0(x) = 0.3 \exp\{-4(x+1)^2\} + 0.7 \exp\{-16(x-1)^2\}$

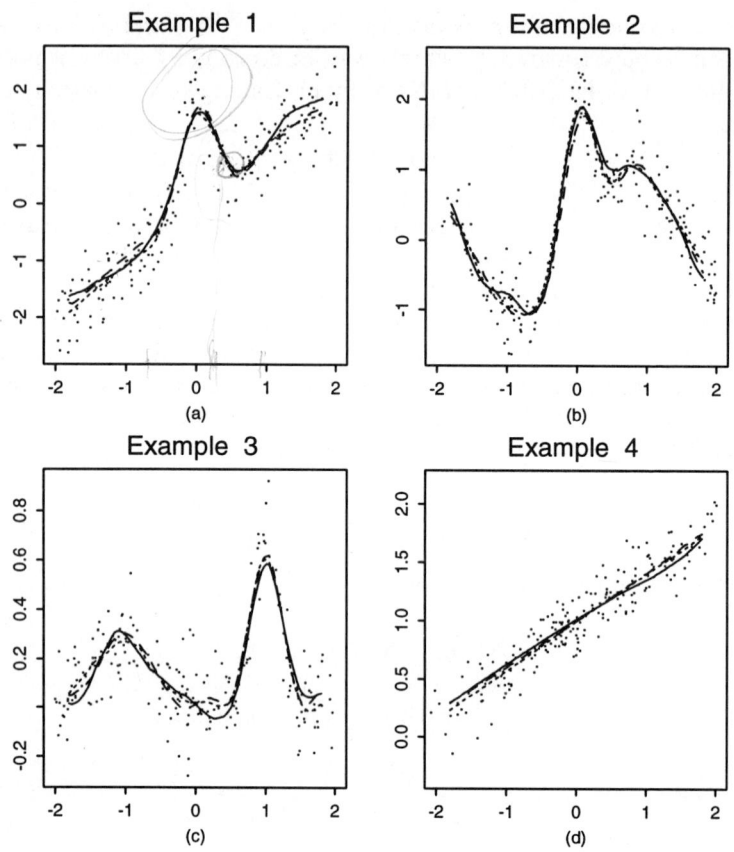

Figure 4.1. *Use of the rule of thumb bandwidth selector in compliance with the variable order selection procedure. A typical simulated data set and three typical estimated curves are presented; these correspond to the 10th (the solid line), the 50th (the dotted line) and the 90th percentile (the dashed line) among the MISE-ranked curves. Adapted from Fan and Gijbels (1995b).*

To illustrate the use of $\check{h}_{\text{ROT}}$ we now apply this bandwidth selector in conjunction with the variable order selection procedure established in Section 3.3.3. Since the procedure uses adaptively $p = 1$ and $p = 3$ when estimating m, we take the bandwidth $1.5\check{h}_{\text{ROT}}$ for $p = 1$ and $\nu = 0$ with $w_0(x)$ being the indicator function of the interval $[-1.8, 1.8]$, resulting in a fully adaptive variable order regression smoother. Details of this procedure can be found in Fan and Gijbels (1995b). To get an idea of the performance, we consider four simulated models summarized in Table 4.1. In each of these models $Y = m(X) + \sigma\varepsilon$ with $\varepsilon \sim N(0,1)$. In Examples 1–3 the covariate X has a Uniform$(-2, 2)$ distribution, and in Example 4 we took $X \sim N(0,1)$. An indicator for the difficulty of the estimation problem is the *noise to signal ratio* $\sigma^2/\text{Var}\{m(X)\}$. This ratio is very high for these examples. We draw 1000 random samples of size 200 from the above models. Throughout this chapter, we use the Epanechnikov kernel. The resulting estimated curves are ranked according to their MISE values and are summarized in Figure 4.1.

Another rule of thumb bandwidth selector is proposed in Härdle and Marron (1995), where a piecewise polynomial fitting is used to recover the curvature. Another useful reference model method, using a Box-Cox power transformation, is proposed by Doksum, Blyth, Bradlow, Meng and Zhao (1994).

4.3 Estimated bias and variance

The constant bandwidth selector introduced in the previous section provides an initial guess for the amount of smoothing, and is satisfactory in many applications. In other situations a more carefully designed selection procedure is needed. This section and Section 4.5 provide the building blocks for such a selection procedure.

Exact expressions of the conditional bias and variance of local polynomial regression estimators can be obtained immediately from their definition. However, these expressions are not directly accessible since they involve unknown quantities. A possible way out is to rely on asymptotic expressions for the conditional bias and variance, such as those established in Theorem 3.1. This leads to asymptotically optimal bandwidths, namely bandwidths that minimize the asymptotic mean (integrated) squared error. See (3.20) and (3.21). They depend on unknown functions which could be replaced by pilot estimates. This direct 'plug-in' approach will be discussed briefly in Section 4.10.

We opt for a selection procedure which does not rely completely on the asymptotic expressions, but stays 'closer' to the exact expressions for the conditional bias and variance and hence carries more information about the finite sample bias and variance.

The conditional bias $(\mathbf{X}^T \mathbf{W} \mathbf{X})^{-1} \mathbf{X}^T \mathbf{W} \mathbf{r}$, given in (3.6), contains the unknown quantity $\mathbf{r} = \mathbf{m} - \mathbf{X}\beta$. Using Taylor's expansion of order a, this conditional bias can be approximated by

$$S_{m,j} = \sum_{i=1}^{m} K_h(X_i - x_o)(X_i - x_o)^j$$

$(\mathbf{X}^T\mathbf{W}\mathbf{X})^{-1}\mathbf{X}^T\mathbf{W}\tau$, where τ is a $n \times 1$ vector with as i^{th} element

$$\beta_{p+1}(X_i - x_0)^{p+1} + \cdots + \beta_{p+a}(X_i - x_0)^{p+a}.$$

Which order of approximation a should one take? The choice of a has an influence on the performance of the resulting bandwidth selector. Suppose that a local linear fit ($p = 1$) is carried through. Then, the choice $a = 3$ entails that the resulting bandwidth selector is $\sqrt{n}$-consistent in relative rate. See Hall, Sheather, Jones and Marron (1991) for the density estimation setting, and see Huang (1995) for a proof of the above consistency statement in the local linear regression setup. The choice $a = 2$ would reduce the computational costs, and results in a bandwidth selector which is not far from being $\sqrt{n}$-consistent – the rate of convergence for the local linear fit is expected to be $n^{2/5}$ following the argument in Huang (1995). We took $a = 2$ in all of the simulations and experienced a good performance of the proposed data-driven bandwidth selection procedures in a variety of examples we considered.

Recall the definition of $S_{n,j}$ given in (3.10), and recall the notation $S_n = \mathbf{X}^T\mathbf{W}\mathbf{X}$. The approximated bias can be written as

$$S_n^{-1} \begin{pmatrix} \beta_{p+1}S_{n,p+1} + \cdots + \beta_{p+a}S_{n,p+a} \\ \vdots \\ \beta_{p+1}S_{n,2p+1} + \cdots + \beta_{p+a}S_{n,2p+a} \end{pmatrix}, \quad (4.4)$$

and this can be estimated by

$$S_n^{-1} \begin{pmatrix} \widehat{\beta}_{p+1}S_{n,p+1} + \cdots + \widehat{\beta}_{p+a}S_{n,p+a} \\ \vdots \\ \widehat{\beta}_{p+1}S_{n,2p+1} + \cdots + \widehat{\beta}_{p+a}S_{n,2p+a} \end{pmatrix}, \quad (4.5)$$

where $\widehat{\beta}_{p+1}, \cdots, \widehat{\beta}_{p+a}$ are the estimated regression coefficients obtained by fitting a polynomial of degree $p + a$ locally. For this $(p + a)^{th}$ order fit, one needs a pilot bandwidth h^*, and the question is then how to choose this pilot bandwidth. We postpone this discussion until Section 4.5.

The difference between our approach and conventional asymptotic approaches (such as direct 'plug-in' techniques) appears clearly in the above derivation. When conventional asymptotic approaches are employed, one would approximate the quantities $S_{n,j}$ by their asymptotic counterparts provided in (3.54) and (3.61). Note that these asymptotic counterparts depend on for example the design density and its derivatives, and introduce more unknown quantities. However, there is no real need for carrying through such an approximation since $S_{n,j}$ is fully known from the data. The estimated bias (4.5) makes no more use of asymptotics than needed, and is expected to be close to the finite sample bias.

A slight modification is needed in order to improve upon the finite sample behavior, especially in the case of a higher order fit ($p \geq 2$). A satisfactory modification is to put

$$S_{n,p+a+1} = 0, \ S_{n,p+a+2} = 0, \ \cdots \ S_{n,2p+a} = 0 \qquad (4.6)$$

in the vector in (4.5). The effect of this operation is to reduce the collinearity among the terms $\{(X_i - x_0)^j\}$, such as $\{(X_i - x_0)^2\}$ and $\{(X_i - x_0)^4\}$. The operation has however no effect on the asymptotic properties, since it concerns only the higher order terms in (4.5).

We next approximate the conditional variance in (3.6). Using local homoscedasticity, this conditional variance can be approximated by

$$(\mathbf{X}^T\mathbf{W}\mathbf{X})^{-1}(\mathbf{X}^T\mathbf{W}^2\mathbf{X})(\mathbf{X}^T\mathbf{W}\mathbf{X})^{-1}\sigma^2(x_0). \qquad (4.7)$$

The unknown quantity $\sigma^2(x_0)$ can be estimated by the normalized weighted residual sum of squares

$$\widehat{\sigma}^2(x_0) = \frac{\sum_{i=1}^n (Y_i - \widehat{Y}_i)^2 K_{h^*}(X_i - x_0)}{\text{tr}\{\mathbf{W}^* - \mathbf{W}^*\mathbf{X}^*(\mathbf{X}^{*T}\mathbf{W}^*\mathbf{X}^*)^{-1}\mathbf{X}^{*T}\mathbf{W}^*\}}. \qquad (4.8)$$

from the $(p+a)^{th}$ order polynomial fit using the pilot bandwidth h^*. Here $\mathbf{X}^*$ and $\mathbf{W}^*$, similar to $\mathbf{X}$ and $\mathbf{W}$, denote respectively the design matrix and the weight matrix for this local $(p+a)^{th}$ order polynomial fit. The bias of this estimator is of order h^{p+a+1}, and is of order $h^{2(p+a+1)}$ when $\sigma^2(\cdot)$ is constant (see (4.52)). This estimator has smaller bias than the corresponding estimator resulting from the local polynomial fit (3.3) of order p, and its bias can be negligible when $a = 2$ and the pilot bandwidth is not too large. An estimator for the conditional variance is provided by

$$(\mathbf{X}^T\mathbf{W}\mathbf{X})^{-1}(\mathbf{X}^T\mathbf{W}^2\mathbf{X})(\mathbf{X}^T\mathbf{W}\mathbf{X})^{-1}\widehat{\sigma}^2(x_0). \qquad (4.9)$$

Note that all the estimators $\widehat{\beta}_{p+1}, \cdots, \widehat{\beta}_{p+a}$ and $\widehat{\sigma}^2(x_0)$ are obtained from the $(p+a)^{th}$ order polynomial fit using the initial bandwidth h^*. Section 4.5 discusses the choice of a pilot bandwidth.

With the estimated conditional bias and variance, given in (4.5) and (4.9), we obtain the following estimator for the mean squared error of $\widehat{\beta}_\nu = \widehat{m}_\nu(x_0)/\nu!$:

$$\widehat{\text{MSE}}_{\nu,p}(x_0; h) = \widehat{b}_{\nu,p}^2(x_0) + \widehat{V}_{\nu,p}(x_0), \qquad (4.10)$$

where $\widehat{b}_{\nu,p}(x_0)$ denotes the $(\nu+1)^{th}$ element of the estimated bias vector in (4.5) with the slight modification mentioned in (4.6). Further, the $(\nu+1)^{th}$ diagonal element of the matrix in (4.9) is denoted by $\widehat{V}_{\nu,p}(x_0)$. The above estimator for the mean squared error was used in Section 3.3.3 in the variable order estimation procedure.

4.4 Confidence intervals

The estimated bias vector (4.5) and variance matrix (4.9) can be used not only to select bandwidths but also to construct pointwise confidence intervals. Under certain regularity conditions, it can be shown that asymptotically

$$\{V_\nu(x_0)\}^{-1/2}\left\{\widehat{m}_\nu(x_0) - m^{(\nu)}(x_0) - b_\nu(x_0)\right\} \xrightarrow{D} N(0,1), \quad (4.11)$$

where $b_\nu(x_0)$ and $V_\nu(x_0)$ are respectively the asymptotic bias and variance of $\widehat{m}_\nu(x_0)$, given in for example (3.18) and (3.19). The asymptotic normality in (4.11) is a special case of the more general result given in Theorem 5.2 of Chapter 5. With the estimated bias and variance defined in (4.10), an approximate $(1-\alpha)100\%$ confidence interval for $m^{(\nu)}(x_0)$ is

$$\widehat{m}_\nu(x_0) - \nu!\widehat{b}_{\nu,p}(x_0) \pm z_{1-\alpha/2}\nu!\{\widehat{V}_{\nu,p}(x_0)\}^{1/2}, \quad (4.12)$$

where $z_{1-\alpha/2}$ denotes the $(1-\alpha/2)^{th}$ quantile of the standard Gaussian distribution. This approximate confidence interval is valid if $\widehat{V}_{\nu,p}(x_0)/V_\nu(x_0) \xrightarrow{P} 1$ and $\widehat{b}_{\nu,p}(x_0)/b_\nu(x_0) \xrightarrow{P} 1$. This in turn requires a different bandwidth h^* used in assessing the bias and variance, as we have done in Section 4.3.

The bias estimator in (4.12) involves indirectly estimating some higher order derivatives of the regression function. Hence, the estimator $\widehat{b}_{\nu,p}(x_0)$ itself can be unreliable. To stabilize the estimation, one can use a weighted average of $\widehat{b}_{\nu,p}(x)$ for x in a neighborhood of x_0, namely, use

$$\widehat{B}_{\nu,p}(x_0) = \int \widehat{b}_{\nu,p}(x) K_h(x - x_0) dx.$$

The integration is practically implemented as the weighted average at grid points.

The confidence interval (4.12) relies on the asymptotic normality result stated in (4.11). However, in many nonparametric smoothing situations, there may be only a few local data points in the neighborhood of x_0 and the asymptotic normality is probably not close enough. An obvious modification of the confidence interval is to replace the normal table value $z_{1-\alpha/2}$ by a Student's t-table value with the degrees of freedom to be computed as follows.

Assume that the bias in (4.8) is negligible and that the error distribution is Gaussian. Then, it can easily be shown that

$$\text{Var}\{\hat{\sigma}^2(x_0)|\mathbb{X}\} \approx 2\sigma^4(x_0)/d_n(x_0)$$

where

$$d_n(x_0) = \{\sum_{i=1}^{n} K_{h^*}(X_i - x_0)\}^2 / \sum_{i=1}^{n} K_{h^*}^2(X_i - x_0). \qquad (4.13)$$

A correction of the confidence interval is now obtained by replacing the normal table value $z_{1-\alpha/2}$ by the t-table value with degrees of freedom $[d_n(x_0)]$, the integer part of $d_n(x_0)$. This kind of idea for computing the degrees of freedom can be found in Cleveland (1979).

The pilot bandwidth h^* can be determined by, for example, the *residual squares criterion* $\hat{h}_{p+1,p+2}^{\text{RSC}}$ given in (4.19).

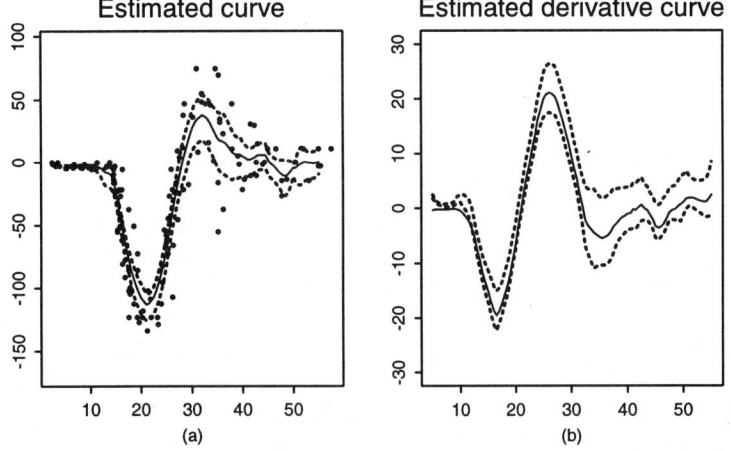

Figure 4.2. *Approximate 95% confidence intervals for (a) curve estimation, (b) derivative curve estimation for the motorcycle data. The solid curves are the estimated ones and the dashed curves are the confidence limits.*

We illustrate the proposed confidence limits (4.12) using the motorcycle data set described in Section 1.1 of Chapter 1. The curve in Figure 4.2 (a) is fitted using a local linear fit with bandwidth $h = 3.3$ and the derivative curve in Figure 4.2 (b) is fitted using a local quadratic fit with bandwidth $h = 5.3$. These two bandwidths were selected by data according to (4.21). Their associated 95% confidence limits are also plotted in Figure 4.2.

4.5 Residual squares criterion

In Section 4.3 we derived an estimator for the mean squared error of the local polynomial regression estimator. This estimated MSE is based on an initial fitting of a higher order polynomial, using a pilot bandwidth. We now propose a bandwidth selection rule which will be used in this pilot estimation stage.

4.5.1 Residual squares criterion

Recall the asymptotically optimal variable bandwidth given in (3.20), depending on unknown quantities and hence not directly accessible. Our aim is now to come up with a statistic for which the minimizer leads to an estimator for this theoretically optimal bandwidth. The *Residual Squares Criterion* (RSC) is defined as

$$\text{RSC}(x_0; h) = \hat{\sigma}^2(x_0)\{1 + (p+1)V\}, \qquad (4.14)$$

where $\hat{\sigma}^2(\cdot)$ is the normalized weighted residual sum of squares after fitting locally a pth order polynomial (see (4.8) for an analogous expression), and V is the first diagonal element of the matrix $(\mathbf{X}^T\mathbf{W}\mathbf{X})^{-1}(\mathbf{X}^T\mathbf{W}^2\mathbf{X})(\mathbf{X}^T\mathbf{W}\mathbf{X})^{-1}$. From (4.9) it is easily seen that V in fact reflects the effective number of local data points. Intuitively one can understand the statistic (4.14) as follows. When the bandwidth h is too large, the polynomial does not fit well, the bias is large and hence also the residual sum of squares $\hat{\sigma}^2(x_0)$. When the bandwidth h is too small, the variance of the fit will be large and hence also V. Since both factors $\hat{\sigma}^2(x_0)$ and V are incorporated in the RSC this quantity will prevent both extreme choices of the bandwidth. The following theorem, established by Fan and Gijbels (1995a), provides a theoretical justification for the RSC quantity. An outline of the proof is given in Section 4.11.

Theorem 4.1 *Suppose that* $\sigma^2(x) = \sigma^2(x_0)$ *in a neighborhood of* x_0. *If* $h_n \to 0$ *and* $nh_n \to \infty$, *then*

$$E\{RSC(x_0; h_n)|\mathbb{X}\}$$

$$= \sigma^2(x_0) + C_p \beta_{p+1}^2 h_n^{2p+2} + (p+1)a_0 \frac{\sigma^2(x_0)}{nh_n f(x_0)}$$
$$+ o_P\{h_n^{2p+2} + (nh_n)^{-1}\},$$

where a_0 denotes the first diagonal element of the matrix $S^{-1}S^*S^{-1}$, i.e. $\int K_0^{*2}(t)dt$, and $C_p = \mu_{2p+2} - c_p^T S^{-1} c_p$ with $c_p = (\mu_{p+1}, \cdots, \mu_{2p+1})^T$.

The above result reveals that the minimizer of $E\{RSC(x_0; h_n)|\boldsymbol{X}\}$ is approximately equal to

$$h_o(x_0) = \left\{ \frac{a_0 \sigma^2(x_0)}{2 C_p \beta_{p+1}^2 n f(x_0)} \right\}^{1/(2p+3)}. \quad (4.15)$$

Comparing this minimizer with the asymptotically optimal bandwidth in (3.20) we find the simple relationship

$$h_{\text{opt}}(x_0) = \text{adj}_{\nu,p} h_o(x_0), \quad (4.16)$$

where

$$\text{adj}_{\nu,p} = \left[\frac{(2\nu+1) C_p \int K_\nu^{*2}(t) dt}{(p+1-\nu)\{\int t^{p+1} K_\nu^*(t) dt\}^2 \int K_0^{*2}(t) dt} \right]^{1/(2p+3)}. \quad (4.17)$$

Hence the RSC quantity estimates a quantity of which the minimizer leads to estimating the target optimal bandwidth $h_{\text{opt}}(x_0)$ within a constant factor $\text{adj}_{\nu,p}$. Note that the adjusting constants $\text{adj}_{\nu,p}$ depend only on the kernel function K. Table 4.2 presents some of these adjusting constants for the Epanechnikov and the Gaussian kernel. Note that all the entities are slightly smaller than 1. This is a property which an adjusting constant should possess: in order to choose a smoothing parameter, the learning of the local curvature and variance is done from a somewhat larger interval. A similar idea of using adjusting constants was given in Müller, Stadtmüller and Schmitt (1987).

In the next two sections, we explain how to apply the RSC to select a constant or variable bandwidth. The proposed bandwidth selection rules will be refined later on in Section 4.6 using the material presented in Section 4.3.

4.5.2 Constant bandwidth selection

Suppose our interest is to estimate $m^{(\nu)}(\cdot)$ on the interval $[a, b]$ using a pth order polynomial fit. Appropriate choices would be

Table 4.2. *Adjusting constants for the Epanechnikov and Gaussian kernel. From Fan and Gijbels (1995a).*

	Epanechnikov kernel						
p	1	2	3	4	5	6	7
$p-\nu$							
1	.8941	.7643	.7776	.7639	.7827	.7835	.7989
3			.8718	.8324	.8384	.8297	.8392
5					.8819	.8639	.8679
7							.8932
	Gaussian kernel						
1	1.000	.8403	.8285	.8085	.8146	.8098	.8159
3			.9554	.8975	.8846	.8671	.8652
5					.9495	.9165	.9055
7							.9470

$p = \nu + 1$ or $p = \nu + 3$, and for simplicity we take $p = \nu + 1$. Then we find $\widehat{h}$, the minimizer of the integrated version of the residual squares criterion

$$\text{IRSC}(h) = \int_{[a,b]} \text{RSC}(y; h) dy, \qquad (4.18)$$

and obtain the *RSC constant bandwidth selector*

$$\widehat{h}_{\nu,p}^{\text{RSC}} = \text{adj}_{\nu,p} \widehat{h}. \qquad (4.19)$$

This RSC constant bandwidth selector is then used to fit a polynomial of order p, leading to the estimator $\widehat{m}_\nu(x_0)$.

The performance of the RSC constant bandwidth selector is investigated via Examples 2 and 4 in Table 4.1. Samples of size 50, 200 and 800 were drawn from these models and the number of simulations was 400. For each of the examples, we obtain $\widehat{h}_{\nu,p}^{\text{RSC}}$ and compare this with the theoretical optimal constant bandwidth $h_{\nu,p}^{\text{opt}}$, computed via the exact expression for the conditional integrated mean squared error (3.6). We report on the distribution of the 400 relative errors

$$\frac{\widehat{h}_{\nu,p}^{\text{RSC}} - h_{\nu,p}^{\text{opt}}}{h_{\nu,p}^{\text{opt}}}, \qquad (4.20)$$

by means of a kernel density estimate. For this kernel density esti-

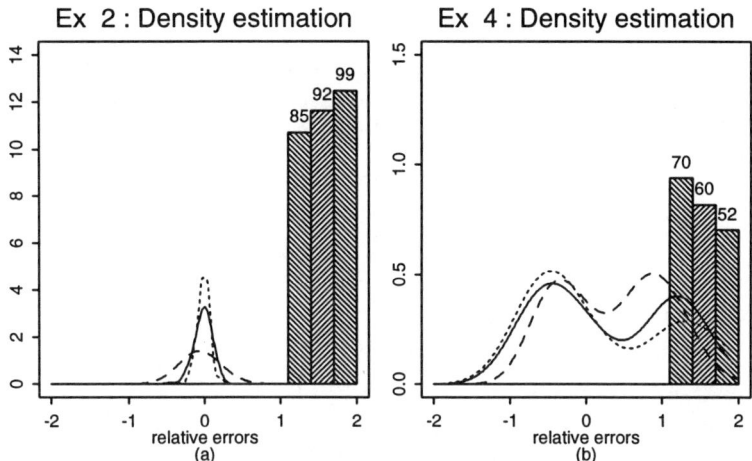

Figure 4.3. *Performance of the RSC constant bandwidth selector for Examples 2 and 4. The three curves present a density estimate for the relative errors in (4.20) for sample sizes 50, 200 and 800 ($n = 50$, dashed curve; $n = 200$, solid curve; $n = 800$, dotted curve). The bars indicate the percentage of relative errors less than 20%. Adapted from Fan and Gijbels (1995a).*

mate we use a Gaussian kernel with bandwidth $h = 1.06 s_n n^{-0.2}$, where s_n is the standard deviation of the data (see Section 2.7.1 and Silverman (1986)). The bars in Figure 4.3 (a) represent the percentage of relative errors (4.20) less than 20% for samples of sizes 50, 200 and 800. In Example 4 we took $h_{\nu,p}^{\text{opt}} = 1$, and report the percentage of selected bandwidths using more than 40% of the data, i.e. $\hat{h}_{\nu,p}^{\text{RSC}} > 0.8$. The results for Examples 1 and 3 are not included since they contain a message similar to this from Example 2. Figure 4.3 reveals that the RSC constant bandwidth selector performs well, but that the rate of convergence is rather slow. This can be seen from the estimated densities, which show a concentration peak around the theoretical optimal bandwidth. A fast rate of convergence would be manifested by peaks which get 'spiky' fast. Because of this slow rate of convergence we propose in Section 4.6 a refined bandwidth selection rule which uses RSC as a preliminary bandwidth selector in the pilot estimation stage.

4.5.3 Variable bandwidth selection

We now discuss briefly how to apply RSC to develop a variable bandwidth selector. A possible approach is to split up the interval $[a,b]$ into subintervals, say I_k, and proceed as follows:

For each interval I_k, find a bandwidth $\widehat{h}_k$ that minimizes $\mathrm{IRSC}(h) = \int_{I_k} \mathrm{RSC}(y;h)dy$. This results in a bandwidth step function $\widehat{h}_0(x) = \mathrm{adj}_{\nu,p}\widehat{h}_k$, when $x \in I_k$. Smooth this bandwidth step function by averaging locally, using the same smoothing parameter as for the initial partition (i.e. the length of I_k). Denote this smoothed bandwidth function by $\widehat{h}_1(x)$. Finally, for each x, fit a polynomial of order p using bandwidth $\widehat{h}_1(x)$ to obtain $\widehat{m}_\nu(x)$.

The smoothing step in the above procedure leads to smoother estimated curves. In our simulations we split up the interval $[a,b]$ into $[n/(10\log n)]$ pieces, which gave satisfactory results.

4.5.4 Computation and related issues

The practical implementation of the estimation procedure needs careful thought too. As already mentioned on previous occasions, the estimated curves are in practice only evaluated at grid points x_j, $j = 1, \cdots, n_{\mathrm{grid}}$. The integrals appearing in an estimation procedure are implemented as averages over appropriate grid points.

The bandwidth selection procedures developed in Sections 4.5 and 4.6 involve minimization problems. The functions of h which have to be minimized are of a very complicated form, and hence it is almost impossible to find a minimum by using the Newton-Raphson method. A feasible approach is to compare function values at grid points, which are typically of geometric type, i.e. $h_j = C^j h_{\min}$ with $h_{\min}$ the first grid point and C the grid span. Suppose we want to minimize a certain function $M(h)$ over an interval $[h_{\min}, h_{\max}]$. For example $M(h)$ could be $\mathrm{IRSC}(h)$ or any of the other functions appearing in the selection procedures of the next section. The minimum of a function $M(h)$ can be found as follows. Start from $h = h_{\min}$, keep inflating h by a factor C and compute $M(h)$ at the geometric grid points. (We compute $M(h)$ only when h is large enough so that each local neighborhood contains at least $p+2$ data points.) When the function values increase consecutively a certain number of times, denoted by IUP, or when $h > h_{\max}$ we stop, and the minimizer of $M(h)$ is chosen as the geometric grid point having the smallest computed $M(h)$ value. Note that with this minimization procedure one avoids computing $M(h)$ at a large bandwidth, unless absolutely necessary. A local polynomial fit with a large bandwidth is computationally costly, and hence the above algorithm is designed to save in this respect. In our simulated examples we took $h_{\min} = (X_{(n)} - X_{(1)})/n$, and $h_{\max} = (X_{(n)} - X_{(1)})/2$, with $X_{(1)}$ and $X_{(n)}$ respectively the first and the last order statistic. Further we took IUP = 3 and $C = 1.1$. With this implementation the computation of the estimated curve with a data-driven bandwidth is reasonably fast. It should be mentioned that fast implementation methods, such as binning and updating, can be used to speed up the computation. For a brief discussion of these methods, see Section 3.6.

4.6 Refined bandwidth selection

4.6.1 Improving rates of convergence

The RSC bandwidth selection rules introduced in Section 4.5 perform reasonably well, but the visual impression from Figure 4.3 suggests that the rate of convergence is not great, and there is evidently room for improvement. Here we discuss how to refine these bandwidth selectors, using both the estimated MSE introduced in Section 4.3 and the RSC developed in Section 4.5 as building blocks. The key ideas of the refined procedures are as follows. We select the bandwidth which minimizes the estimated MSE (or its integrated version). In order to calculate this MSE we need however a pilot bandwidth and this will be obtained via the RSC. The resulting more sophisticated estimation procedures clearly outperform the ones relying only on RSC.

4.6.2 Constant bandwidth selection

In Section 4.3 we developed an estimator for the conditional MSE of a local polynomial regression estimator. This development forms the core of the following refinement of the RSC constant bandwidth selector, which reads as follows:

Pilot estimation. Fit a polynomial of order $p+2$, use $\widehat{h}_{p+1,p+2}^{\text{RSC}}$ in (4.19) as a pilot bandwidth h^* and obtain the estimates $\widehat{\beta}_{p+1}$, $\widehat{\beta}_{p+2}$ and $\widehat{\sigma}^2(x_0)$.

Bandwidth selection. Find the bandwidth which minimizes the estimated integrated mean squared error:

$$\widehat{h}^{R}_{\nu,p} = \arg\min_{h} \int_{[a,b]} \widehat{\mathrm{MSE}}_{\nu,p}(y;h)dy, \qquad (4.21)$$

and use this *refined bandwidth selector* to fit a polynomial of order p, where $\widehat{\mathrm{MSE}}_{\nu,p}(y;h)$ is given by (4.10).

This refinement indeed leads to an important improvement upon the RSC constant bandwidth selector. For comparison purposes, we reconsider Examples 2 and 4. Examples 1 and 3 carry messages similar to this from Example 2. For each of the two examples we present two pictures in Figure 4.4. The first one shows a typical simulated data set along with some representatives of the 400 estimated curves (resulting from the 400 simulations) chosen as in Figure 4.1. The three representative curves already give a good visual impression of the performance of the estimation procedure. In the second picture we report on the kernel density estimate of the relative errors

$$\frac{\widehat{h}^{R}_{\nu,p} - h^{\mathrm{opt}}_{\nu,p}}{h^{\mathrm{opt}}_{\nu,p}}. \qquad (4.22)$$

This picture is similar to Figure 4.3. Note the improvement in the relative rate of convergence.

4.6.3 Variable bandwidth selection

The refined ideas established for the constant bandwidth selection rule are now adapted for variable bandwidth selection. As in Section 4.5.3 the main difference is that the interval of estimation $[a,b]$ is split up into subintervals I_k, and in each of these subintervals the refined ideas are used. The detailed estimation procedure reads as follows:

Pilot estimation. For each interval I_k, fit a polynomial of order $p+2$ and select the optimal bandwidth for estimating β_{p+1} by minimizing $\mathrm{IRSC}(h) = \int_{I_k} \mathrm{RSC}(y;h)dy$ and multiplying the resulting minimizer with $\mathrm{adj}_{p+1,p+2}$. Smooth the resulting bandwidth step function by averaging locally, using the same smoothing parameter as for the initial partition. Use this smoothed bandwidth function to fit a polynomial of order $p+2$, and obtain the pilot estimates $\widehat{\beta}_{p+1}, \widehat{\beta}_{p+2}$, and $\widehat{\sigma}^2(x_0)$.

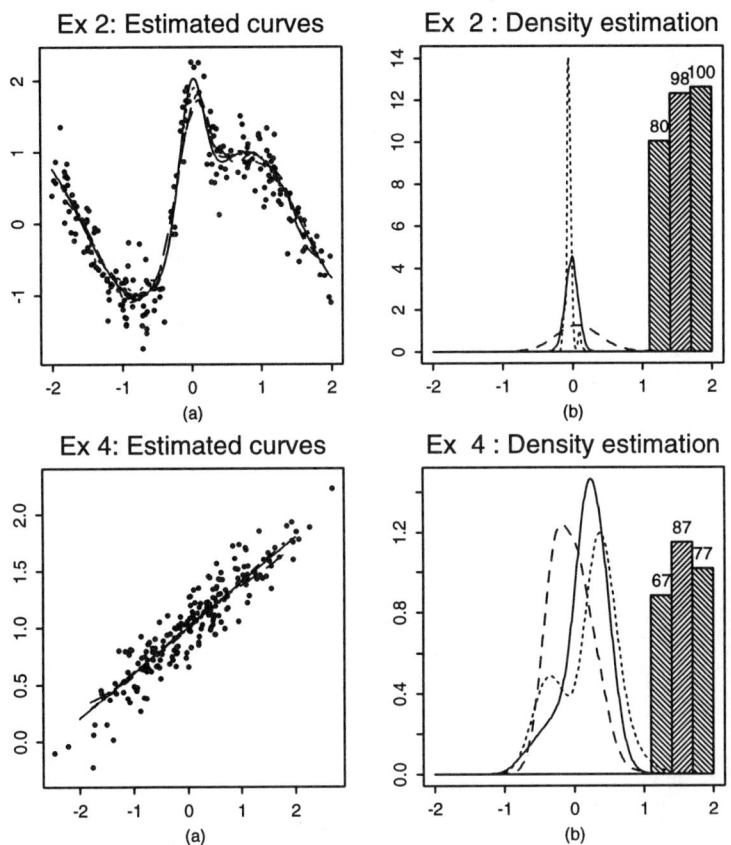

Figure 4.4. *Performance of the refined constant bandwidth selector. (a): Typical simulated data and three representatives of the estimated curves. (b): Density estimate of the relative errors in (4.22) for sample sizes 50, 200 and 800 (respectively dashed, solid and dotted curve). The bars indicate the percentage of relative errors less than 20%. From Fan and Gijbels (1995a).*

Curve estimation. For each interval I_k, minimize the estimated integrated mean squared error $\int_{I_k} \widehat{\mathrm{MSE}}_{\nu,p}(y;h) dy$. Smooth the resulting bandwidth step function, again using the length of I_k, fit a polynomial of order p and obtain the estimated curve.

We now illustrate the performance of this data-driven variable bandwidth selection procedure for Examples 1–4. Figure 4.5 shows

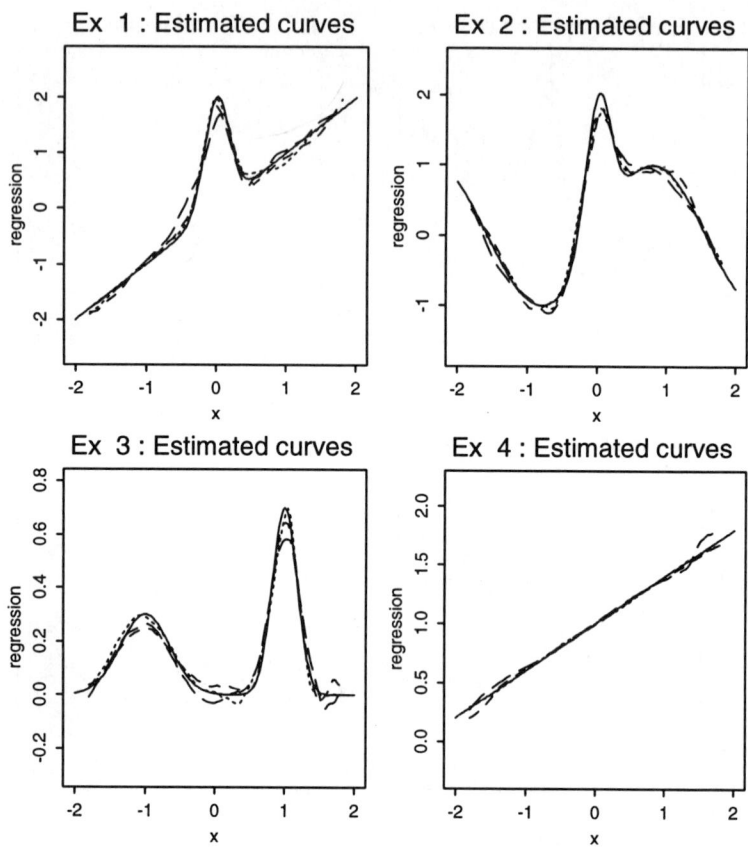

Figure 4.5. *Performance of the refined variable bandwidth selector for Examples 1–4. Solid curve – true curve. Dashed curves – three representatives of the estimated curves. Adapted from Fan and Gijbels (1995a).*

three representative estimated regression curves, chosen as before via MISE percentile ranks. Note in particular the performance of the estimation procedure in Example 4. There we have a simple linear regression model, where a constant bandwidth is needed. But even here the data-driven bandwidth selector does a good job.

How does the proposed methodology perform for examples where the curves show many alterations and are very irregular? In Section 4.7 we give some additional examples addressing this question. It turns out that the estimation procedure developed adapts neatly to the complexity of such curves. This is not surprising, since the carefully developed data-driven choice of a variable bandwidth deals with these alterations and irregularities.

So far we have only investigated the performance of the data-driven bandwidth selection procedures for estimation of the regression function. But the procedures can also be applied to estimation of derivatives of the regression function. We illustrate this by estimating the derivative curve $m'(\cdot)$ of Example 2 in Table 4.1 using the refined variable bandwidth selector. For this derivative estimation problem we carry through local quadratic fits ($p = 2$). We took samples of sizes 400 and 800. Figure 4.6 reports the results.

The above simulation study reveals that the refined bandwidth selector has good practical performance. Some theoretical results on this fully automatic bandwidth selection procedure can be found in Fan, Gijbels, Hu and Huang (1996). They established rates of convergence of the variable bandwidth selector, and obtained the asymptotic distribution of its error relative to the theoretical optimal variable bandwidth given in (3.20).

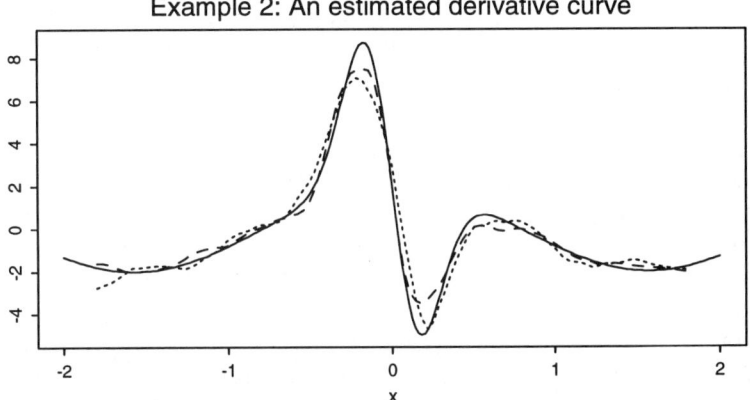

Figure 4.6. *Performance of the refined variable bandwidth selector for the derivative curve of Example 2 in Table 4.1. The true derivative curve (solid line) and typical estimated curves (50th percentile) are presented based on samples of size 400 (dotted line) and 800 (dashed line). Adapted from Fan and Gijbels (1995a).*

Due to the intrinsic connection between density estimation and nonparametric regression (see Section 2.7.2), the bandwidth selection procedures are also applicable to the density estimation setup.

128   AUTOMATIC DETERMINATION OF MODEL COMPLEXITY

4.7 Variable bandwidth and spatial adaptation

In Section 2.5 we explained that wavelet-based methods can adapt readily to spatially inhomogeneous curves, i.e. they possess what is called in the literature the spatial adaptation property. We will offer evidence in this section that local polynomial regression estimators with the proposed data-driven choice of the variable bandwidth possess the necessary flexibility for capturing complicated shapes of curves. They also perform well in estimating curves which are quite unsmooth or show many alterations, as will be seen from the simulated examples with a large spatial variability.

4.7.1 Qualification of spatial adaptation

Suppose that one wants to estimate a function g, which possesses different degrees of smoothness at different locations. See for example Figure 4.8. A way to describe this is to consider the following form of the target function

$$g(x) = g_0(x) + \sum_{j=1}^{N} \gamma_j \{\omega_j (x - x_j)\}, \qquad (4.23)$$

where γ_j are fixed functions with compact supports, $x_1, \cdots, x_N$ are fixed distinct points, and where $\omega_j \geq 1$ represents the degree of irregularity at x_j. These ω_j's may even depend on the sample size n and diverge to infinity as n increases. This allows the irregularity of the episode in g at x_j to depend on the sample size, and the smoothness of the function g at locations $x_1, \cdots, x_N$ is governed by the sizes of $\omega_1, \cdots, \omega_N$. So, the smoothness of the function is allowed to vary with the sample size, reflecting the fact that with large samples one might hope to be able to capture more erratic episodes in a curve. Assume that the functions g_0 and γ_j have r bounded derivatives. Then, the function g admits different degrees of smoothness: at a point x_0 between $\{x_1, \cdots, x_N\}$, the curve g is smooth and at a vicinity of x_j, the curve is not so smooth, depending on the value of ω_j. A spatial adaptation procedure is one that recovers the short and sharp aberrations at locations x_j and returns a smooth curve at smooth regions. Mathematically, it should obtain the optimal rate $(\omega_j n^{-1})^{2r/(2r+1)}$ of convergence for estimating $g(x_j)$, uniformly in $x_0, \cdots, x_N$, where $\omega_0 = 1$.

Fan, Hall, Martin and Patil (1993) showed in the density estimation setting that locally adaptive kernel methods possess the spatial adaptation property. This gives a theoretical endorsement of the refined variable bandwidth estimation procedure introduced in Section 4.6.3.

4.7.2 Comparison with wavelets

As already mentioned in our introduction of wavelet thresholding in Chapter 2, wavelet transforms are a device for representing functions $g(x)$ in a way that is local in both time and frequency domains (namely, in both the argument x and the roughness of g). As a consequence, wavelet-based smoothing methods adapt automatically to local variations in the roughness of g as x changes. This *local adaptability* has been expressed in terms of the *global performance* of the wavelet estimates over a large class of functions. This was the case in the papers by Donoho, Johnstone, Kerkyacharian and Picard (1993), Donoho and Johnstone (1994, 1995a) and Donoho (1995). A nice overview of these results is given by Donoho, Johnstone, Kerkyacharian and Picard (1995). But what about the performance of wavelet methods in a local setting such as mathematically modelled in (4.23)? It was shown in Fan, Hall, Martin and Patil (1993) that wavelet methods, up to a logarithmic factor, adjust optimally to varying smoothness. More precisely, if the primary resolution (see the end of Section 2.5.2) is taken to be of order $\{n(\log n)^{2r}\}^{1/(2r+1)}$ or smaller, which includes the case of a fixed primary resolution not depending on n, then, the wavelet estimators (2.27) and (2.28) achieve the mean squared convergence rate $(\omega_j n^{-1} \log n)^{2r/(2r+1)}$ simultaneously at each respective point x_j, and at the points between the x_j's the mean squared convergence rate is $(n^{-1} \log n)^{2r/(2r+1)}$. Hence, this behavior characterizes the spatial adaptability of wavelet estimators. The logarithmic factor can be removed from the convergence rate if the threshold parameter δ (see Section 2.5 for its definition) is chosen adaptively at each point x_j, such that it can alter with the frequency at the point. A level-dependent thresholding parameter selection, using Stein's unbiased risk estimation method, is proposed by Donoho and Johnstone (1995b).

From the above discussions we conclude that variable bandwidth methods, such as local polynomial regression estimators with a data-driven variable bandwidth as proposed in Section 4.6.3, perform comparably with wavelet methods. Both possess the capability of spatial adaptation.

We now illustrate the spatial adaptation property of the wavelet-based methods and the local polynomial regression estimates with

Table 4.3. *Simulation models.*

Example	$m(x)$	σ	$\approx$ noise/ signal
5	$m_0(x)$ [1]	1.0	1/6
6	blocks	1.0	1/6
7	bumps	1.0	1/6
8	heavisine	1.0	1/6

[1] $m_0(x) = 24\sqrt{x(1-x)}\sin\{2\pi 1.05/(x+0.05)\}$

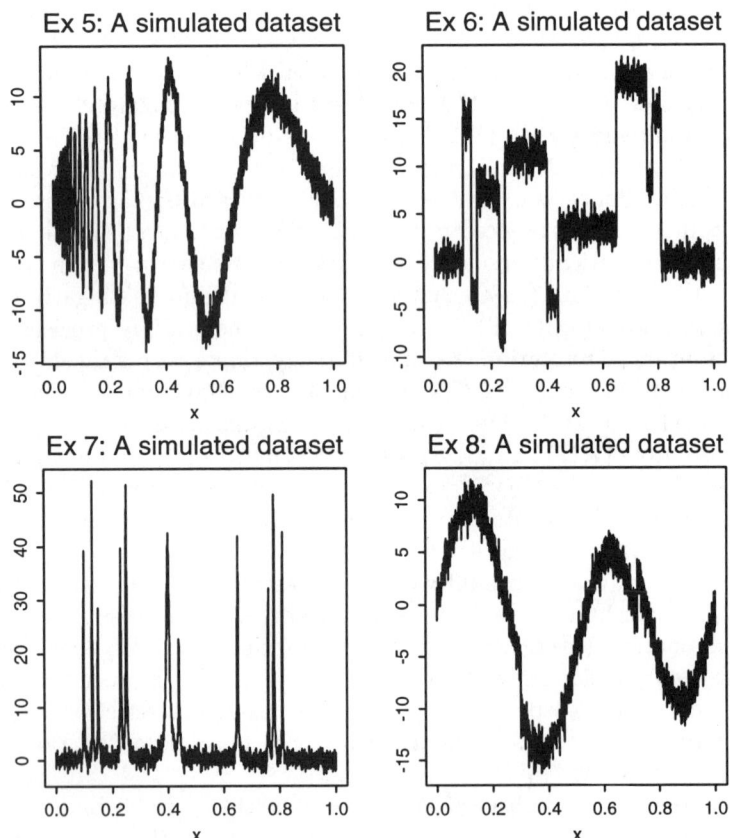

Figure 4.7. *Typical simulated data sets for the examples in Table 4.3.*

a data-driven variable bandwidth, via the set of examples listed in Table 4.3. Explicit formulae for Examples 6–8 are similar to those in Donoho and Johnstone (1994) and are given here:

Blocks : $m(x) = \sum h_j K(x - x_j), K(x) = \{1 + \text{sgn}(x)\}/2$
$(x_j) = (10, 13, 15, 23, 25, 40, 44, 65, 76, 78, 81)/100$
$(h_j) = (40, -50, 30, -40, 50, -42, 21, 43, -31, 21, -42)$
$\times 0.37$

Bumps : $m(x) = \sum h_j K\{(x - x_j)/w_j\}, K(x) = (1 + 25|x|^2)^{-1}$
$(t_j) = t_{\text{Blocks}}$
$(h_j) = (40, 50, 30, 40, 50, 42, 21, 43, 31, 51, 42)/10$
$(w_j) = (5, 5, 6, 10, 10, 30, 10, 10, 5, 8, 5)/1000$

Heavisine : $m(x) = 4\sin(4\pi x) - \text{sgn}(x - 0.3) - \text{sgn}(0.72 - x)$.

The target regression curves of the models in Table 4.3 are quite unsmooth or show many alterations. Donoho and Johnstone (1994) used these examples to illustrate the performance of their wavelet thresholding methods. They form a challenging set of examples to test the performance of local polynomial regression estimators with a data-driven variable bandwidth. Note that the noise to signal ratio for the examples in Table 4.3 is lower than for the examples in Table 4.1.

The design is a uniform fixed design $x_i = i/n$. The sample size is 2048 for each of the examples, and we do 31 simulations. Figure 4.7 indicates typical simulated data sets. We depict typical estimated regression curves (the curves with median MISE value among the 31 MISE values) using respectively the local linear fit with the proposed data-driven variable bandwidth, and the wavelet hard-thresholding estimator. To reproduce the wavelet thresholding estimates we used the S-software developed by Nason and Silverman (1994), which implements the recipe of Donoho and Johnstone (1994). Figure 4.8 demonstrates clearly the spatial adaptability of both estimation methods.

What about estimation of derivative curves for these examples? As we have already emphasized our data-driven estimation procedure presented in this chapter applies to derivative estimation as well. As an illustration we estimated the derivative $m'(\cdot)$ of Example 5 via a local quadratic fit ($p = 2$). Figure 4.9 depicts the true derivative curve and a typical estimated curve. At the very high

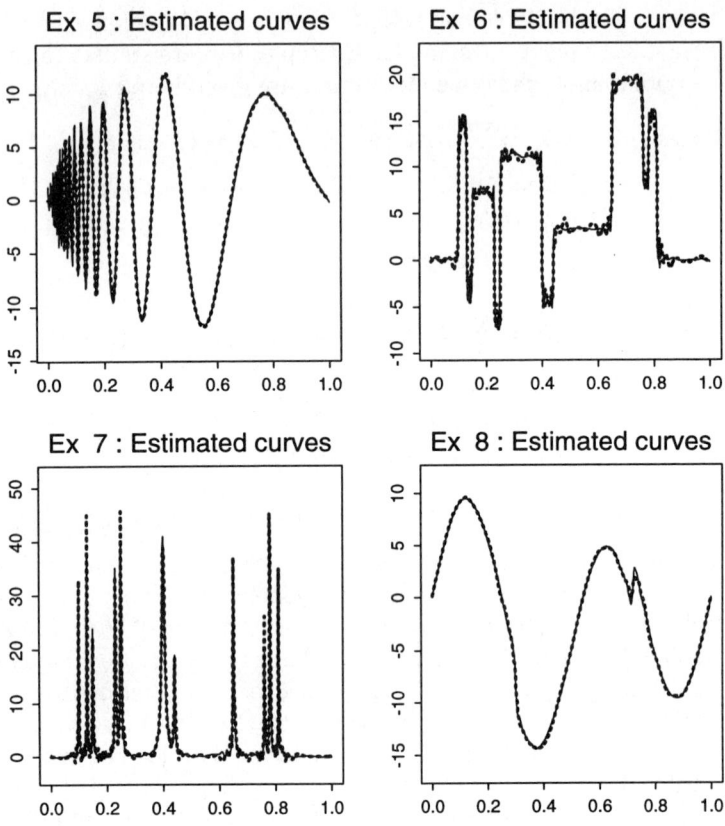

Figure 4.8. *Typical estimated curves for Examples 5–8. Solid curve – local linear fit with refined variable bandwidth; dashed curve – wavelet hard-thresholding estimator. Results from Fan and Gijbels (1995a).*

frequency region the estimated curve is not that good, but from $x = 0.1$ on it performs very neatly.

4.8 Smoothing techniques in use

Experiences with simulation studies and the theoretical foundations seem to indicate that the fully automatic estimation procedures developed in the previous sections work reasonably well. However, the reason for developing smoothing techniques is their use in real life examples. In Section 4.4 we already showed one such

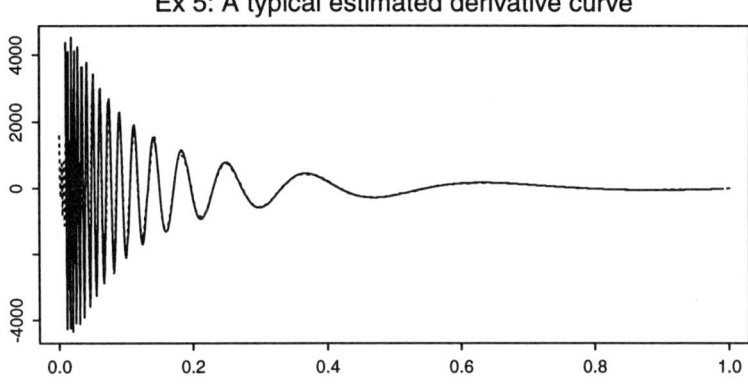

Figure 4.9. *Estimation of the derivative curve $m'(\cdot)$ of Example 5, with the true derivative curve (solid line) and a typical estimated derivative curve (dotted line). Adapted from Fan and Gijbels (1995a).*

example. This section is devoted to the application of smoothing techniques, and in particular the proposed procedures, to various data sets reflecting different situations encountered in practice. For concreteness, the bandwidth used in this section is always selected by (4.21), although other procedures can also be applied. Whenever a density estimator is referred to, we always use (2.39) with bandwidth (2.42).

Depending on the situation, we can be interested in estimating a regression curve or its derivative (rate of growth). These curves can be estimated directly using nonparametric techniques. Illustrations of this are scattered throughout this book. On other occasions, one combines parametric and nonparametric techniques to give sensible analyses of data. Here we use three examples as illustrations. Other applications of nonparametric techniques can be found in Chapters 5 and 6.

4.8.1 Example 1: modelling and model diagnostics

As the quality of life improves, an increasing concern is the percentage of Fat Free Weight (FFW), particularly for individuals participating in fitness center programs. But the FFW cannot be measured easily. The percentage of FFW can be computed from an underwater weighing: a client has to be undressed and sunk

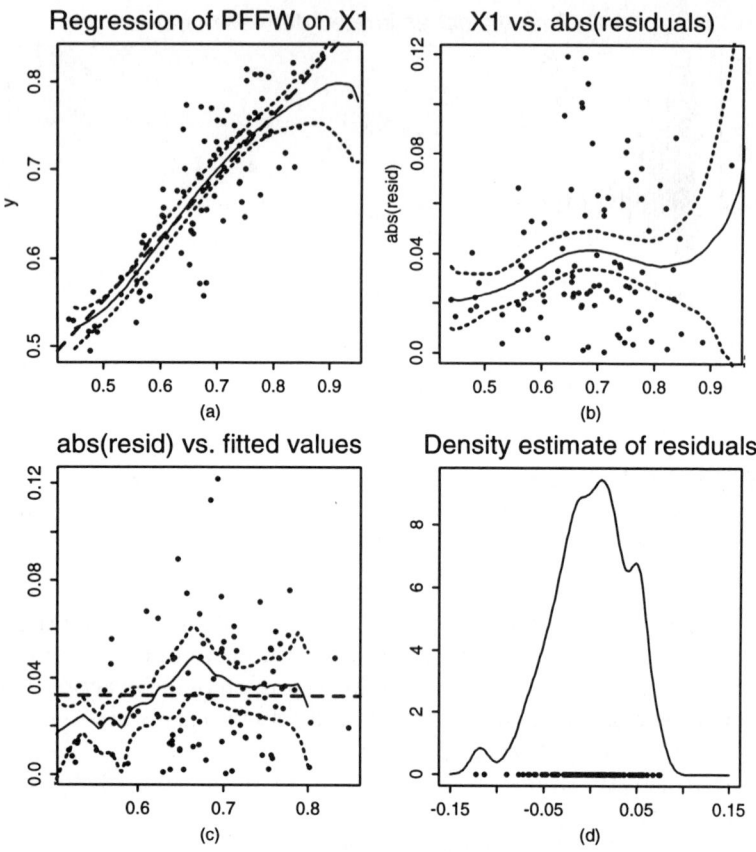

Figure 4.10. *Predicting the percentage of fat free body weight. (a) Regression of percentage of fat free body weight on variable X_1, the dashed curves are pointwise approximate 95% confidence limits; (b) regression of the absolute residuals on X_1; (c) regression of absolute residuals on the fitted values of the multiple regression; (d) the kernel density estimate of the residuals from the multiple regression.*

completely into a tank of water in order to compute the increase of volume. Then the body density is computed as well as the percentage of FFW. This kind of measurement is considered as the golden standard, but is time consuming and complicated. The purpose of this study is to find the best prediction equation for the percentage of FFW using other variables such as bioelectrical impedance, height, weight, skinfolds, physical activity, etc. The data set an-

SMOOTHING TECHNIQUES IN USE 135

alyzed here is based on a group of 97 women aged 18–40, collected at the Department of Physical Education (courtesy of Dr. B. Ainsworth), University of North Carolina at Chapel Hill. To simplify the analysis, we only consider the bioelectrical impedance, weight and height covariates. The response variable is the percentage of FFW.

Regarding the body approximately as a cylinder, it can be shown under ideal assumptions that FFW is linear in squared height divided by bioelectrical impedance (see e.g. Kushner (1992)). So we first study the relation between the percentage of FFW and

$$X_1 = \frac{\text{height}^2}{(\text{bioelectrical impedance}) \times (\text{weight})}.$$

Figure 4.10 (a) shows the scatter plot of the data (with correlation coefficient 0.839), the nonparametric estimated curve (solid curve), its pointwise approximate 95% confidence limits (short-dashed curves), and the linear regression line (long-dashed line). The figure reveals the fact that the linearity holds reasonably. In Figure 4.10 (b) the absolute value of residuals after the linear regression is plotted against X_1. The picture also shows the smoothed curve (solid curve) and the pointwise approximate 95% confidence limits (dashed curves). One can see some heteroscedasticity in the plot: the variation at the middle of the plot tends to be larger. Ignoring the right-hand tails (caused by sparsity of design points), one can reasonably model the variance as a symmetric quadratic function around 0.7 and use this to improve further the estimated coefficients from the linear regression.

We next regress linearly the percentage of FFW on the variables X_1, height, bioelectrical impedance and weight. Since the intercept term is not significant, it is not included in the regression model. The resulting prediction equation is

$$Y = 0.9349X_1 - 0.0026X_2 + 0.0006X_3 + 0.002X_4,$$

where Y = percentage of FFW, X_2 = height in centimeters, X_3 = bioelectrical impedance in Ohms, and X_4 = weight in kilograms. To diagnose the linear fit, we give both the absolute residuals plotted versus fitted values (Figure 4.10 (c)) and the residual plot (Figure 4.10 (d)). The nonparametric smoothed curve (solid curve), 95% pointwise confidence limits (short-dashed curves), and the mean of the absolute residuals (long-dashed line) are presented in Figure 4.10 (c). This figure does not show strong heteroscedasticity. Figure 4.10 (d) presents the kernel density estimate of the residuals. Given the sample size 97, it does not seem to depart seriously from normality. Hence, it is reasonable to employ the Gaussian theory for constructing confidence intervals for parametric linear regression for this data set.

4.8.2 Example 2: comparing two treatments

It is often of interest to compare the outcomes of two treatments, after adjusting some confounding factors. For example, one might be interested in comparing the salary of two populations after adjusting confounding factors such as experience and education levels. Another example is to compare the side effect (such as blood pressure) of a drug between males and females after adjusting age effects. While the commonly used technique is the linear regression technique, nonparametric regression can provide a more objective comparison with a more flexible model, particularly when the adjustments are nonlinear.

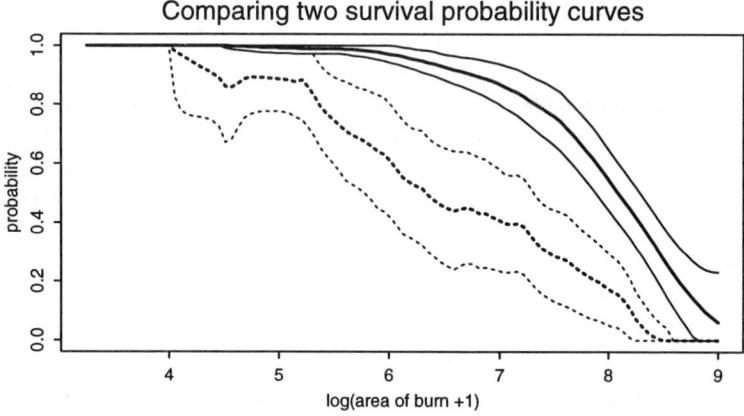

Figure 4.11. *Survival probability curves given the area of burn (on a logarithm scale) for burn victims aged 17–50 and 51–85. Thicker curves are estimated conditional probability curves and thin curves are confidence limits. Solid curves – patients aged 17–50; dashed curves – patients aged 51–85. It is clear that given the area of burn, the younger group has better probability of surviving.*

As an illustration, we use the burn data set introduced in Section 2.1 of Chapter 2. We divide the data into two groups according to the age of patients: age 17–50 and age 51–85. The younger group consists of 284 patients and the older group has 151 patients. The survival probability of the two groups is estimated and plotted against the area of burn (in log-scale). See Figure 4.11. It is clear that the older group is far more vulnerable when the area of the third degree burn is large enough. The confidence limits are also depicted in Figure 4.11. The biases are ignored in the confidence limits, since the method described in Section 4.4 does not apply well to this kind of binary data set. A better technique for analyzing binary data will be introduced in Section 5.4.

4.8.3 Example 3: analyzing a longitudinal data set

The following example concerns the correlation of two measures arising from a longitudinal study. The data were provided by the Neurotrauma Center of the University of Miami School of Medicine and made available to us by Dr. Frank W. Stitt and Dr. Ying Lu. People who are injured were transported to a hospital. The main question arising is how serious the injury is, so that a doctor can decide whether or not the patient needs emergency care, or hospital care or just home care after a certain number of days of treatment. Two possible head injury scores, APACH and TISS, are used to judge this, but can lead to a quite different decision. Our task here is to study the correlation between the two injury scores and the strength of individual effects. Data here are based on the daily measurements of APACH and TISS for 166 patients, among which 90 of them were treated for at least 5 days.

Firstly, all TISS scores were plotted against the APACH scores, and the mean regression curve between the scores was estimated nonparametrically. The scatter plot and the estimated regression curve are depicted in Figure 4.12 (a). There is evidence for departure from linearity, but the departure is practically negligible. Thus, the correlation coefficient is appropriate for measuring the association between the two scores. The correlation coefficient between all TISS and APACH scores is 0.594. A question arising now is 'What is the individual effect on the relationship between the two measures depicted above ?'. To examine this individual effect we computed for each patient with at least five observations, the correlation coefficient between TISS and APACH. A plot of these correlation coefficients is provided in Figure 4.12 (b), showing quite a few negative values and a large variability of the correlation coefficients among patients. This indicates a strong individual effect. In an attempt to account for the different lengths of observations, we

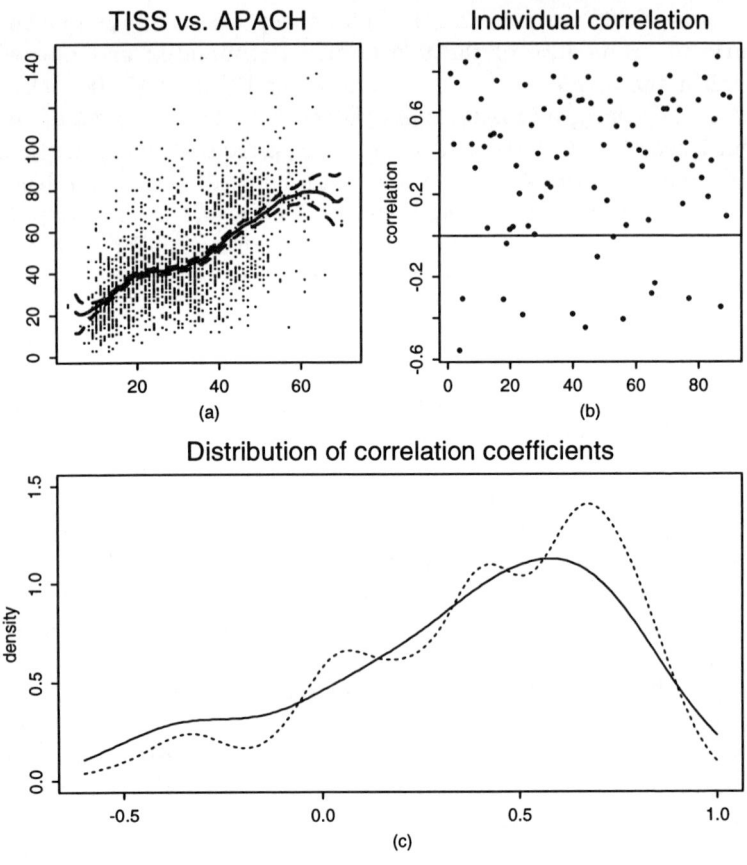

Figure 4.12. (a). Scatter plot of APACH and TISS, with estimated nonparametric regression curve and pointwise confidence limits; (b) scatter plot of individual correlation coefficients between APACH and TISS; (c) distribution of the individual correlation coefficients; solid curve – unweighted; dashed curve – weighted.

assign a weight, proportional to the number of observations for the patient, to each computed correlation coefficient. The weight here is not intended to improve on the value of the overall correlation coefficient, but is rather meant to make it report more reliably and honestly on the strength of the relationship (recall that the correlation coefficient for any two points is one and is high for any set of three points). The resulting overall weighted correlation coefficient, namely 0.415, is somewhat lower than the overall unweighted correlation coefficient 0.594. However, this overall weighted correlation coefficient is not informative and misleading since the distribution of the weighted correlation coefficients is highly skewed (see Figure 4.12 (c)). The distribution of the individual correlation coefficients, unweighted and weighted among patients, is investigated via a kernel density estimate, and depicted in Figure 4.12 (c). Recall that the bandwidths for both curves are computed from the normal reference rule (2.42). When using unweighted correlation coefficients, the resulting density function (solid curve) is qualitatively the same as the density function for the weighted correlation coefficients (dashed curve), except that for the latter one the largest mode is shifted a bit to the right. From this analysis we can conclude that the correlation coefficient between APACH and TISS is about 0.6–0.7 and that there is a strong individual effect. Further analyses should look at the correlation coefficients in a more homogeneous group with respect to injury types, age and gender.

To gain further insight into the variability for each patient a linear regression model, with TISS as covariate and APACH as response, is fitted, i.e.

$$Y_{i,j} = \alpha_i + \beta_i X_{i,j} + \varepsilon_{i,j} \qquad (4.24)$$

where $X_{i,j}$ and $Y_{i,j}$, $j = 1, \cdots, n_i$ are the TISS and the APACH scores of the ith patient, and $\varepsilon_{i,j} \sim N(0, \sigma_i^2)$. Under the above model we are interested in the distributions of the parameters $\{\alpha_i\}$ and $\{\beta_i\}$ and the parameters

$$\alpha = N^{-1} \sum_{i=1}^{N} \alpha_i \quad \text{and} \quad \beta = N^{-1} \sum_{i=1}^{N} \beta_i,$$

with N the number of patients. Note that β is closely related to the population correlation coefficient γ after adjusting the individual effect. Indeed, the relation between β and the correlation coefficient γ can be made explicit. Nonparametric kernel density estimates are used as a tool to summarize the variability (distribution) of the intercepts $\{\alpha_i\}$ and the slopes $\{\beta_i\}$ across the population. Figure 4.13 (a), respectively (b), shows the kernel density estimate for the intercepts, respectively the slopes. The distribution of the normalized residual sum of squares, i.e. $\sqrt{RSS/(n-2)}$, is shown in Figure 4.13 (c). Strictly speaking, the daily measurements are dependent and some adjustments are needed. However, for the purposes of simple graphical exploration, we ignore the dependence structure. For each linear regression model the t-value

140   AUTOMATIC DETERMINATION OF MODEL COMPLEXITY

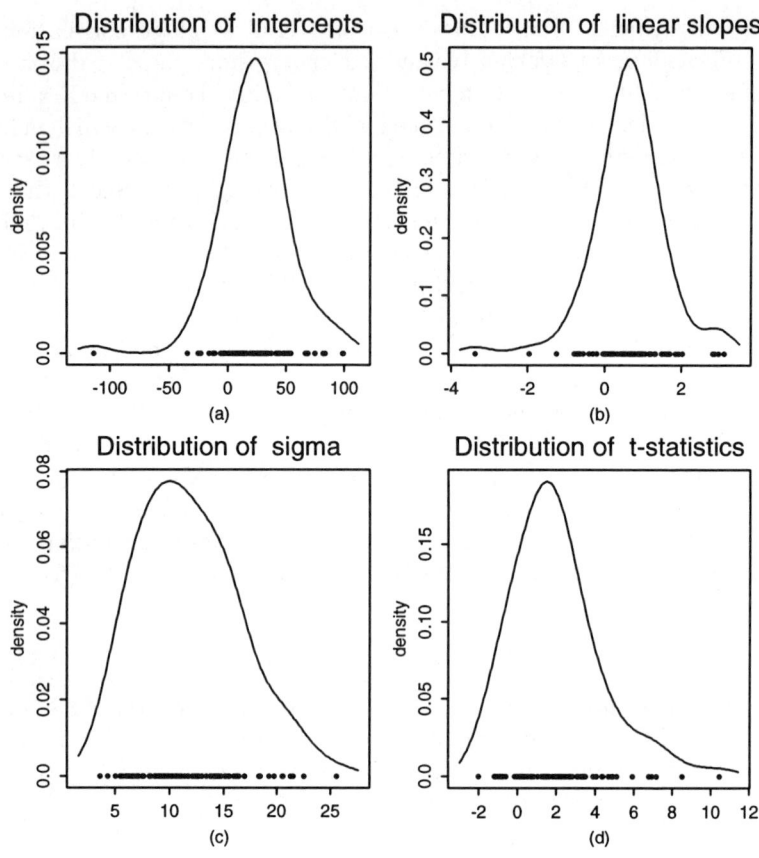

Figure 4.13. *Results of the linear fits: distribution of (a) the intercepts, (b) the linear slopes, (c) the normalized residual sum of squares, and (d) the t-values testing whether the linear coefficient is zero.*

for testing whether the slope is zero was calculated. The distribution of these t-values is depicted in Figure 4.13 (d). It indicates that for a large fraction of individuals in the population, there is a significant linear relationship between APACH and TISS. Note first of all that the intercepts and the slopes are nearly normally distributed and that the distribution of the normalized residual sum of squares is like a χ^2-distribution. The variability of these distributions is quite large, reflecting the individual effect. From the structural modelling point of view we can regard α_i, β_i and σ_i in model (4.24) as i.i.d. from some population. Then the issue is to find the posterior distribution of α, β and σ^2. Figure 4.13 can be used as a device for modelling, revealing that

$$\alpha \sim N(\mu_\alpha, v_\alpha) \qquad \beta \sim N(\mu_\beta, v_\beta) \qquad \text{and} \qquad \sigma \sim \Gamma(\mu_\sigma, v_\sigma)$$

would be a reasonable assumption. The parameters of interest are μ_α, v_α, μ_β, v_β, μ_σ, v_σ, as well as the correlation coefficient γ.

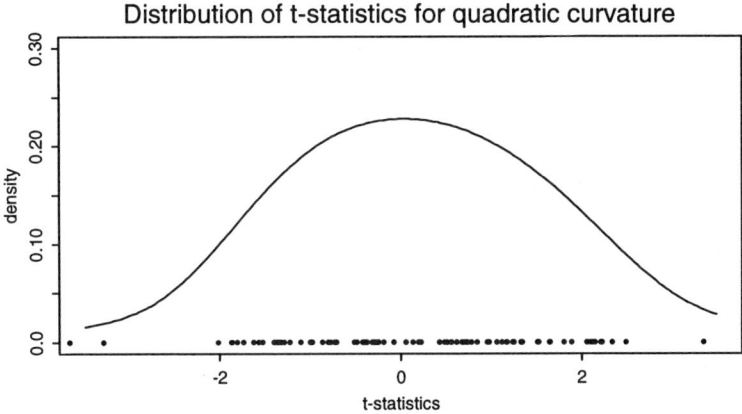

Figure 4.14. *Distribution of t-statistics for testing whether the quadratic coefficient is zero.*

Next, quadratic fits were tried out, and similar pictures were produced. We only present the distribution of the t-statistics for testing whether there is a significant quadratic relationship. Figure 4.14 shows that there is a weak effect on the quadratic term: few individuals have a large $|t|$-statistic.

In conclusion, the two measures APACH and TISS are approximately linearly related and individual effects are strong, and therefore a more homogeneous group with respect to type of injury and age should be focused.

4.9 A blueprint for local modelling

The basic idea of the refined bandwidth selection procedures developed in Section 4.6 was to get a good grip on the bias and variance, without relying heavily on asymptotics. The estimates of bias and variance required a pilot estimation step, which was accomplished by fitting locally a polynomial of higher order. Although these ideas were developed for the locally weighted least squares regression method, they can in fact be extended to likelihood-based methods, as will be shown in this section.

To fix ideas we start by outlining the maximum likelihood estimation (MLE) for a parametric model, and motivate from there a local modelling approach when dealing with the nonparametric situation. Suppose that for the ith observation (X_i, Y_i) the contribution to the conditional log-likelihood is $\ell\{g(X_i), Y_i\}$, with g being a function to be estimated, yielding the conditional log-likelihood of the n observations $(X_1, Y_1), \cdots, (X_n, Y_n)$:

$$\sum_{i=1}^{n} \ell\{g(X_i), Y_i\}.$$

Suppose further that g is parametrized as $g(x) = g_\theta(x)$, with θ an unknown parameter. Then the usual MLE method is to maximize

$$\sum_{i=1}^{n} \ell\{g_\theta(X_i), Y_i\} \equiv \ell(\theta; X_1, Y_1, \cdots, X_n, Y_n) \quad (4.25)$$

with respect to θ. We assume that the log-likelihood $\ell(\cdot; x_1, y_1, \cdots, x_n, y_n)$ satisfies the Bartlett identities of order 1 and 2, namely

$$E_\theta \left(\frac{\partial \ell}{\partial \theta}\right) = 0$$

$$E_\theta \left(\frac{\partial^2 \ell}{\partial \theta^2}\right) + \text{Var}_\theta \left(\frac{\partial \ell}{\partial \theta}\right) = 0. \quad (4.26)$$

Here, E_θ and Var_θ denote the conditional mean and variance given $X_1, \cdots, X_n$. Examples of log-likelihood functions which satisfy the above Bartlett identities are the log-likelihood functions for the generalized linear models and the partial-likelihood function applied in Cox's proportional hazards model. In Chapter 5 we will discuss briefly some estimation aspects in both model settings.

Suppose now that the function $g(\cdot)$ is completely unknown, and hence not parametrized. We formulate the *local modelling* approach to estimate $g(x_0)$. For a data point X_i in a neighborhood around x_0 we approximate $g(X_i)$, via a Taylor expansion, by a polynomial of degree p:

$$g(X_i) \approx g(x_0) + g'(x_0)(X_i - x_0) + \cdots + \frac{g^{(p)}(x_0)}{p!}(X_i - x_0)^p$$

$$\equiv \mathbf{X}_i^T \beta,$$

where $\mathbf{X}_i = (1, X_i - x_0, \cdots, (X_i - x_0)^p)^T$ and $\beta = (\beta_0, \cdots, \beta_p)^T$, with $\beta_\nu = g^{(\nu)}(x_0)/\nu!$, $\nu = 0, 1, \cdots, p$. The conditional *local log-*

A BLUEPRINT FOR LOCAL MODELLING 143

likelihood weighs the contribution of each datum point (X_i, Y_i) by a smooth function $K_h(X_i - x_0)$, leading to

$$\ell_p(\beta; h, x_0) = \sum_{i=1}^{n} \ell(\mathbf{X}_i^T \beta, Y_i) K_h(X_i - x_0). \tag{4.27}$$

The appearance of the subscript p and the dependence on h are clear. Maximization of this local log-likelihood with respect to β gives estimators for $g^{(\nu)}(x_0)$, $\nu = 0, 1, \cdots, p$.

To illustrate the concept of local log-likelihood let us consider the normal regression model

$$Y = g(X) + \varepsilon,$$

with $\varepsilon \sim N(0, \sigma^2)$, and X and ε independent. The conditional log-likelihood function under this normal model is

$$-n \log(\sqrt{2\pi}\sigma) - \frac{1}{2\sigma^2} \sum_{i=1}^{n} \{Y_i - g(X_i)\}^2.$$

When the form of $g(\cdot)$ is completely unknown and one has no idea about an appropriate parametric form, then the local modelling approach offers a way out. Following this approach we end up by considering the conditional local log-likelihood

$$-\log(\sqrt{2\pi}\sigma) \sum_{i=1}^{n} K_h(X_i - x_0)$$

$$-\frac{1}{2\sigma^2} \sum_{i=1}^{n} \left\{ Y_i - \sum_{j=0}^{p} \beta_j (X_i - x_0)^j \right\}^2 K_h(X_i - x_0),$$

which has to be maximized with respect to $\beta = (\beta_0, \cdots, \beta_p)^T$. This is precisely the weighted least squares regression problem considered in Chapter 3 for a more general regression model with general error term.

In the above example, the unknown function was a regression function. In other situations $g(\cdot)$ will be another unknown function of interest, such as for example the covariate effect function in a proportional hazards model encountered in survival analysis, or the transformed mean regression function in a generalized linear model. The latter examples will be discussed in Sections 5.3 and 5.4 respectively.

Denote the estimator resulting from the local log-likelihood in (4.27) by $\widehat{\beta} = \widehat{\beta}(x_0) = (\widehat{\beta}_0, \cdots, \widehat{\beta}_p)^T$. Then an estimator for $g^{(\nu)}(x_0)$ is $\widehat{g}_\nu(x_0) \equiv \widehat{\beta}_\nu \nu!$, $\nu = 0, 1, \cdots, p$. The important question now is how to get access to the bias and variance of the estimator $\widehat{\beta}$ of β. Let us first look at the bias. The bias of the estimate depends on the approximation error made by replacing the unknown target function $g(X_i)$ locally by a polynomial of degree p, namely $\sum_{j=0}^{p} g^{(j)}(x_0)(X_i - x_0)^j/j!$. Let $r(X_i)$ be this approximation error. Then,

$$r(X_i) \approx \beta_{p+1}(X_i - x_0)^{p+1} + \cdots + \beta_{p+a}(X_i - x_0)^{p+a} \equiv r_i, \quad (4.28)$$

where a discussion on the approximation order a, similar to that in Section 4.3, can be given. Suppose for a moment that the quantities r_i are known. Then, a more precise local log-likelihood is

$$\ell_p^*(\beta; h, x_0) = \sum_{i=1}^{n} \ell(\mathbf{X}_i^T \beta + r_i, Y_i) K_h(X_i - x_0).$$

Let $\widehat{\beta}^* = \widehat{\beta}^*(x_0)$ be the maximizer of the local log-likelihood $\ell_p^*(\beta; h, x_0)$. The bias of $\widehat{\beta}(x_0)$ can then be estimated by

$$\widehat{b}_p(x_0) = \widehat{\beta}(x_0) - \widehat{\beta}^*(x_0). \quad (4.29)$$

The computation of $\widehat{\beta}^* = \widehat{\beta}^*(x_0)$ can be avoided, as follows. Since $\widehat{\beta}^*(x_0)$ is the maximizer of $\ell_p^*(\beta; h, x_0)$, a Taylor expansion yields an approximation of (4.29):

$$\begin{aligned}
0 &= \frac{\partial}{\partial \beta} \ell_p^*(\widehat{\beta}^*; h, x_0) \\
&\approx \frac{\partial}{\partial \beta} \ell_p^*(\widehat{\beta}; h, x_0) + \frac{\partial^2}{\partial \beta^2} \ell_p^*(\widehat{\beta}; h, x_0)\{\widehat{\beta}^*(x_0) - \widehat{\beta}(x_0)\},
\end{aligned}$$

resulting in

$$\widehat{b}_p(x_0) \approx \left\{\frac{\partial^2}{\partial \beta^2} \ell_p^*(\widehat{\beta}; h, x_0)\right\}^{-1} \frac{\partial}{\partial \beta} \ell_p^*(\widehat{\beta}; h, x_0). \quad (4.30)$$

Hence, in this way one saves the computation of $\widehat{\beta}^*(x_0)$.

What does (4.30) give in the case of least squares local log-likelihood? There, except for a constant factor $-1/(2\sigma^2)$,

$$\ell_p^*(\beta; h, x_0) = (\mathbf{y}^* - \mathbf{X}\beta)^T \mathbf{W}(\mathbf{y}^* - \mathbf{X}\beta),$$

where $\mathbf{y}^* = \mathbf{y} - \mathbf{r}$, with $\mathbf{y} = (Y_1, \cdots, Y_n)^T$ and $\mathbf{r} = (r_1, \cdots, r_n)^T$. Now,

$$\left.\frac{\partial}{\partial \beta} \ell_p^*(\beta; h, x_0)\right|_{\beta = \widehat{\beta}} = -2\mathbf{X}^T \mathbf{W} \mathbf{y}^* + 2\mathbf{X}^T \mathbf{W} \mathbf{X} \widehat{\beta} = 2\mathbf{X}^T \mathbf{W} \mathbf{r}$$

and
$$\left.\frac{\partial^2}{\partial\beta^2}\ell_p^*(\beta;h,x_0)\right|_{\beta=\widehat{\beta}} = 2\mathbf{X}^T\mathbf{W}\mathbf{X}. \tag{4.31}$$

Therefore
$$\widehat{b}_p(x_0) \approx (\mathbf{X}^T\mathbf{W}\mathbf{X})^{-1}\mathbf{X}^T\mathbf{W}\mathbf{r}.$$

Note that this is exactly the approximation for the bias which was proposed in (4.4).

To get a grip on the variance, first note that

$$\begin{aligned}
0 &= \left.\frac{\partial}{\partial\beta}\ell_p(\beta;h,x_0)\right|_{\beta=\widehat{\beta}} \\
&\approx \frac{\partial}{\partial\beta}\ell_p(\beta;h,x_0) + \frac{\partial^2}{\partial\beta^2}\ell_p(\beta;h,x_0)(\widehat{\beta}-\beta),
\end{aligned}$$

with $\beta = (\beta_0,\cdots,\beta_p)^T$ being the true local parameters, leading to

$$\widehat{\beta} - \beta \approx -\left\{\frac{\partial^2}{\partial\beta^2}\ell_p(\beta;h,x_0)\right\}^{-1}\frac{\partial}{\partial\beta}\ell_p(\beta;h,x_0).$$

Hence, an approximation for the conditional variance is

$$\begin{aligned}
\mathrm{Var}(\widehat{\beta}|\mathbf{X}) &\approx \left\{\frac{\partial^2}{\partial\beta^2}\ell_p(\beta;h,x_0)\right\}^{-1}\mathrm{Var}\left\{\frac{\partial}{\partial\beta}\ell_p(\beta;h,x_0)|\mathbf{X}\right\} \\
&\quad \times \left\{\frac{\partial^2}{\partial\beta^2}\ell_p(\beta;h,x_0)\right\}^{-1}. \tag{4.32}
\end{aligned}$$

The unknown quantity $\frac{\partial^2}{\partial\beta^2}\ell_p(\beta;h,x_0)$ can be estimated by $\frac{\partial^2}{\partial\beta^2}\ell_p(\widehat{\beta};h,x_0)$ and for the conditional variance on the right-hand side of the above expression we can proceed as follows. From (4.27) we obtain

$$\begin{aligned}
&\mathrm{Var}\left\{\frac{\partial}{\partial\beta}\ell_p(\beta;h,x_0)|\mathbf{X}\right\} \\
&= \sum_{i=1}^n \mathrm{Var}\left\{\frac{\partial}{\partial\beta}\ell(\mathbf{X}_i^T\beta,Y_i)|\mathbf{X}\right\} K_h^2(X_i-x_0) \\
&= \sum_{i=1}^n \mathrm{Var}\left\{\ell'(\mathbf{X}_i^T\beta,Y_i)|\mathbf{X}\right\} \mathbf{X}_i\mathbf{X}_i^T K_h^2(X_i-x_0) \\
&\approx \mathrm{Var}\left[\ell'\{g(x_0),Y\}|X=x_0\right]\sum_{i=1}^n \mathbf{X}_i\mathbf{X}_i^T K_h^2(X_i-x_0) \\
&\equiv \mathrm{Var}\left[\ell'\{g(x_0),Y\}|X=x_0\right]\overline{S}_n, \tag{4.33}
\end{aligned}$$

where we denoted $\ell'(u,y) = \frac{\partial}{\partial u}\ell(u,y)$. From (4.32) and (4.33) we obtain the following approximation of the conditional variance of $\widehat{\beta}$:

$$\mathrm{Var}(\widehat{\beta}|\mathbb{X}) \approx \mathrm{Var}\left[\ell'\{g(x_0),Y\}|X=x_0\right]\left\{\ell''_p(\widehat{\beta};h,x_0)\right\}^{-1}\overline{S}_n$$
$$\times \left\{\ell''_p(\widehat{\beta};h,x_0)\right\}^{-1}, \qquad (4.34)$$

where we have introduced the notation $\ell''_p(u;h,y) = \frac{\partial^2}{\partial u^2}\ell_p(u;h,y)$. The only unknown quantity in the above expression is $\mathrm{Var}\left[\ell'\{g(x_0),Y\} \mid X=x_0\right]$, which has to be replaced by an estimator. Two cases can occur. The first case is that $\mathrm{Var}\left[\ell'\{g(x_0),Y\} \mid X=x_0\right] = V\{g(x_0)\}$ for some known function $V(\cdot)$ such as for the Bernoulli, Poisson and exponential distributions in the context of generalized linear models. In this case, we estimate the conditional variance by $V\{\widehat{g}(x_0)\}$. The second case is that we do not have such a form, as in the case of the normal likelihood model. In this case, we proceed with the variance estimation as follows. Suppose that a $(p+a)^{th}$ order polynomial is fitted locally, resulting in the pilot estimator $\widehat{\beta}^{(p+a)} = (\widehat{\beta}_0,\cdots,\widehat{\beta}_{p+a})^T$, as in the bias estimation. Then a possible estimator for $\mathrm{Var}\left[\ell'\{g(x_0),Y\}|X=x_0\right]$ is given by

$$\frac{\sum_{i=1}^n\left\{\ell'(\mathbf{X}_i^T\widehat{\beta}^{(p+a)},Y_i)\right\}^2 K_h(X_i-x_0)}{\sum_{i=1}^n K_h(X_i-x_0)}. \qquad (4.35)$$

To get a better insight into the above approximation of the conditional variance let us consider once again the special case of the least squares local log-likelihood, for which

$$\ell(x,y) = -(y-x)^2/2,$$

which implies

$$\begin{aligned}\mathrm{Var}\left[\ell'\{g(x_0),Y\}|X=x_0\right] &= \mathrm{Var}\{Y-g(x_0)|X=x_0\} \\ &= \sigma^2(x_0).\end{aligned}$$

Hence for this special least squares local log-likelihood

$$\mathrm{Var}(\widehat{\beta}|\mathbb{X}) \approx \sigma^2(x_0)S_n^{-1}\overline{S}_nS_n^{-1},$$

with $S_n = \mathbf{X}^T\mathbf{W}\mathbf{X}$. Note that this is exactly the expression in (4.7). Further it is easily seen that the estimator (4.35) is asymptotically the same as (4.8).

To summarize, this section gives a general principle for estimating an unknown function in a likelihood-based model. The estimation procedure consists of maximizing the local log-likelihood (4.27). To determine the size of the neighborhood which controls the model complexity, we first fit a $(p+a)^{th}$ order polynomial locally with a pilot bandwidth h^*, and then compute the approximated bias in (4.30) and the approximated variance in (4.34) with the unknown parameters replaced by the pilot estimates. This estimated bias and variance lead to an estimator for the mean squared error denoted by $\widehat{\mathrm{MSE}}_{\nu,p}(x_0;h)$. Then one can select either a constant or variable bandwidth as in Section 4.6. We will refer to the resulting bandwidth selector as the refined bandwidth selector. The pilot bandwidth can be selected via the following procedure, which is a natural extension of the RSC procedure discussed in Section 4.5.

As in Section 4.5.1, our interest focuses on selecting the bandwidth $h_{\nu,p}$ for estimating $g^{(\nu)}(x_0)$ based on the local log-likelihood (4.27). Note first of all that the local log-likelihood problem (4.27) can be regarded as an iterative local least squares problem: given the current value β_c of β, update β_c via the local p^{th} order polynomial regression of the working variable

$$Z_i = \mathbf{X}_i^T \beta_c - \frac{\ell'(\mathbf{X}_i^T \beta_c, Y_i)}{E[\ell''\{g(x_0), Y\}|X = x_0]} \qquad (4.36)$$

on X_i, where the conditional expectation is computed using the parameter β_c. Thus, at the last step of the iteration, we can regard the local log-likelihood problem as a local polynomial regression problem, and apply the residual squares criterion

$$\mathrm{ERSC}(x_0;h) = \widehat{\sigma_*}^2(x_0)\{1 + (p+1)V\}, \qquad (4.37)$$

to this least squares problem, where $\widehat{\sigma_*}^2(x_0)$ is the normalized residual sum of squares using the working variable Z_i. Compare with (4.14). We will refer to criterion (4.37) as the *Extended Residual Squares Criterion* (ERSC). This criterion reduces to the RSC when we are in the least squares setting. We can now select the pilot bandwidth as follows. Let

$$\widehat{h}^*_{\nu,p} = \arg\min_h \int \mathrm{ERSC}(y;h)w(y)dy,$$

for some given weight function w. Then, define the *ERSC bandwidth selector* as

$$\widehat{h}^{\mathrm{ERSC}}_{\nu,p} = \mathrm{adj}_{\nu,p} \widehat{h}^*_{\nu,p}. \qquad (4.38)$$

We do not intend to provide any mathematical justification of the above procedure. The only aim of this section is to reveal that the basic ideas established in the previous sections can be extended to likelihood-based methods, resulting in a *nearly universal principle*.

4.10 Other existing methods

A vast amount of literature has been devoted to bandwidth selection problems. Various techniques for selecting smoothing parameters have been proposed during the last decades in different setups, mainly in kernel density estimation and specifically in kernel regression. A good survey of recent developments in bandwidth selection procedures for kernel density estimation can be found in Jones, Marron and Sheather (1992, 1995). Commonly used methods for bandwidth selection in nonparametric estimation are cross-validation techniques, the normal reference method, plug-in methods and nearest neighbor bandwidths, among others. Some of these techniques have been developed particularly for the density estimation setup, whereas others are devoted originally to the regression estimation setup. It is not always clear whether methods which were established for a certain estimation context can be extended to other contexts.

The aim of this section is to give readers an impression of the general ideas behind techniques for choosing a bandwidth. We try to accomplish this by discussing briefly some of the available bandwidth selection methods. We do not intend to give an overview of all existing methods, nor to give any recommendations about when to use which of them. We present the methods in the setup, density or regression estimation, in which they were originally proposed. So, for simplicity we say that the target function in this section is either a density function $f(\cdot)$ or a regression function $m(\cdot)$.

Possible measures for investigating the performance of an estimator are the mean squared error (for local performance) and the mean integrated squared error (for global performance). These quantities are unknown and hence they have to be estimated in one way or another, or one can focus directly on their asymptotic expressions. In (2.40) in Section 2.7.1 an asymptotic expression for the MSE of a kernel density estimator was given. Minimization of the asymptotic MISE led to the theoretical optimal constant bandwidth in (2.41). Asymptotic expressions for the conditional MSE and MISE for local polynomial regression estimators were provided in Chapter 3. From these asymptotic expressions theoretical optimal variable and constant bandwidths (see (3.20) and

OTHER EXISTING METHODS 149

(3.21) respectively) were derived. The theoretical optimal bandwidths depend on unknown quantities and are of no direct practical use. But they are a starting point for deriving practical bandwidth selectors as will be seen in the following sections.

4.10.1 Normal reference method

A *rule of thumb* type of bandwidth selector for a kernel density estimator of f is obtained by making reference to a normal density, i.e. by considering f to be a Gaussian density with standard deviation σ. See (2.42). This idea was used by Bickel and Doksum (1977) for the uniform kernel density estimator and by Deheuvels (1977) for histogram type estimators. It was further exploited for kernel density estimators by Silverman (1986). The unknown standard deviation σ can be substituted by the sample standard deviation s_n, or by another robust estimator such as the interquartile range divided by 1.3490.

The above normal reference method is particularly designed for the density estimation setup. Transferring the idea to the regression estimation problem is not straightforward since in that setup one has to deal with a more complex model, involving an unknown regression function, a design density and an error distribution. The rule of thumb presented in Section 4.2 can be viewed as an attempt to extend the normal reference idea to regression estimation. Indeed, in order to obtain the rule of thumb in (4.3) we made reference to a parametric polynomial regression function.

4.10.2 Cross-validation

The simplest way to understand cross-validation techniques is in the regression setup. Let $\widehat{m}_h(\cdot)$ denote any estimate, involving a smoothing parameter h, of the regression function $m(\cdot)$. For each given i, we use data $\{(X_j, Y_j), j \neq i\}$ to build a regression function $\widehat{m}_{h,-i}(\cdot)$ and then validate the model by examining the prediction error $Y_i - \widehat{m}_{h,-i}(X_i)$. The *least squares cross-validation technique* uses the weighted average of squared errors

$$n^{-1} \sum_{i=1}^{n} \{Y_i - \widehat{m}_{h,-i}(X_i)\}^2 w(X_i), \qquad (4.39)$$

as an overall measure of the effectiveness of the estimation scheme $\widehat{m}_h(\cdot)$. A version of this was presented in (2.37) for smoothing splines and its principle is applicable to all other estimation procedures $\widehat{m}_h(\cdot)$. Later on in Chapter 5 we apply cross-validation to select a smoothing parameter in a censored regression problem. The least squares criterion (4.39) estimates

$$n^{-1}\sum_{i=1}^{n}\{m(X_i) - \widehat{m}_h(X_i)\}^2 w(X_i) + n^{-1}\sum_{i=1}^{n}\sigma^2(X_i).$$

Note that the first term is a discrete approximation of the weighted integrated squared error, with weight function $w(x)f(x)$, and that the second term is independent of h. The *least squares cross-validation bandwidth selector* is the one that minimizes (4.39).

The above cross-validation idea also carries over to the estimation of derivative curves, such as $m'(\cdot)$. Details on how to do this can be found in Härdle (1990).

Cross-validation techniques are also applicable to the density estimation setting. There, the basic idea of least squares cross-validation is as follows. Suppose one aims at approximating the minimizer of the mean integrated squared error of the kernel density estimate $\widehat{f}_h(\cdot)$ in (2.39). Consider the integrated squared error which equals

$$\begin{aligned}\text{ISE}(h) &= \int\{\widehat{f}_h(x) - f(x)\}^2 dx \\ &= \int \widehat{f}_h^2(x)dx - 2\int \widehat{f}_h(x)f(x)dx + \int f^2(x)dx.\end{aligned}$$

The last term does not depend on h. Minimization of this integrated squared error reduces to the minimization of

$$\int \widehat{f}_h^2(x)dx - 2\int \widehat{f}_h(x)f(x)dx. \qquad (4.40)$$

An estimator for this quantity is provided by

$$\text{CV}(h) = \int \widehat{f}_h^2(x)dx - \frac{2}{n}\sum_{i=1}^{n}\widehat{f}_{h,-i}(X_i), \qquad (4.41)$$

where $\widehat{f}_{h,-i}(\cdot)$ is the density estimate based on the sequence of observations $X_1,\cdots,X_{i-1},X_{i+1},\cdots,X_n$, obtained by leaving out the ith observation X_i. This *leave-one-out* method validates the ability to predict $f(X_i)$ across the subsample $X_1,\cdots,X_{i-1},X_{i+1},\cdots,X_n$, resulting in the above *cross-validation* quantity. It is easy to check that

$$E\{\text{CV}(h)\} = \text{MISE}(h) - \int f^2(x)dx.$$

OTHER EXISTING METHODS 151

Hence the minimizer of the cross-validation function CV(h), namely

$$\widehat{h}_{\mathrm{CV}} = \arg\min_h \mathrm{CV}(h), \qquad (4.42)$$

provides an estimator for the minimizer of the MISE. Here, the performance of the estimator is measured via the mean integrated squared error, resulting in the least squares cross-validation bandwidth selector (4.42). However, the above cross-validation idea can also be applied when the *Kullback-Leibler distance* between the true density and the estimated density is used as a measure of performance. This results in the *likelihood cross-validation* method which was introduced in the literature even before least squares cross-validation. The latter cross-validation technique gained more popularity, since it was shown that the former one can have some undesirable practical behavior. We do not elaborate further on this point.

4.10.3 Nearest neighbor bandwidth

In this section we discuss an adaptive bandwidth selector which is very simple in nature and is therefore quite appealing to use when the design points are not uniform.

Suppose that we have observations $X_1, \cdots, X_n$. These can be single observations as in the density estimation setup or the covariates from the bivariate observations $(X_1, Y_1), \cdots, (X_n, Y_n)$ in the regression context. The ordered observations are denoted by $X_{(1)} \leq X_{(2)} \leq \cdots \leq X_{(n)}$. Consider the following adaptive variable bandwidth

$$\widehat{h}_k(x) = \{X_{(\ell+k)} - X_{(\ell-k)}\}/2, \qquad (4.43)$$

where $X_{(\ell)}$ is the design point closest to the point x, and k is a positive integer determining the number of local data points. Note that this bandwidth is a kind of hybrid between a variable bandwidth and the nearest neighbor method. The use of a bandwidth as in (4.43) results in practice in using a small (respectively large) bandwidth in a dense (respectively sparse) design region. The justification of this interpretation of the bandwidth selector in (4.43) is given in the next theorem, which was established in Fan and Gijbels (1994). The quantity $f(\cdot)$ in the theorem can be regarded as the density function f in the density estimation problem or the design density in the regression problem. In Section 4.11 we outline the proof of this result.

Theorem 4.2 *Suppose that $f(\cdot)$ is positive and continuous on a compact interval $[a,b]$, and $k_n \to \infty$ such that $k_n/n \to 0$ and $k_n/\log n \to \infty$. Then,*

$$\widehat{h}_{k_n}(x) = \frac{k_n}{nf(x)}\{1 + o_P(1)\}$$

uniformly in $x \in [a,b]$.

The above theorem reveals that the adaptive variable bandwidth in (4.43) behaves asymptotically as $k/\{nf(x)\}$. This proportionality to $f(x)$ means that a large bandwidth will be used in a sparse design region, whereas a small bandwidth is used in a dense design region. Moreover, one can easily change the degree of dependence on the density $f(x)$, by introducing a power α ($0 \leq \alpha \leq 1$) and considering a bandwidth $\widehat{h}_k^\alpha(x)$. The choice $\widehat{h}_k^{1/4}(x)$ would for example correspond to a variable bandwidth which is asymptotically proportional to $f^{1/4}(x)$, and the resulting estimator for the regression function corresponds asymptotically to a smoothing spline estimator. See e.g. Silverman (1984). In Section 5.2.3 we use the bandwidth $\widehat{h}_k(x)$ in a simulation study on censored regression. Such a model is already complex and a simple variable bandwidth as in (4.43) turned out to be quite useful.

4.10.4 Plug-in ideas

The idea of a *direct plug-in* method was introduced by Woodroofe (1970) in density estimation. An asymptotic expression for the mean integrated squared error of the density estimate (2.39) is given by

$$\frac{1}{4}\{\int u^2 K(u)du\}^2 R(f'')h^4 + R(K)\frac{1}{nh}, \qquad (4.44)$$

with $R(g) = \int g^2(x)dx$ for any function g (see also (2.40)). Minimization of this asymptotic MISE leads to the theoretical optimal bandwidth given in (2.41), which depends on the unknown quantity $R(f'')$. The direct plug-in idea consists of estimating the density roughness $R(f'')$ from an initial density estimate, and then substituting this estimate into (4.44). This brings up the choice of an initial bandwidth, say h_0. An attempt to avoid the choice of an initial bandwidth is to introduce an iterative procedure of which a simple form reads as follows. Starting with a large value for h_0

find subsequent values $h_1, h_2, \cdots$, satisfying (see also (2.41))

$$h_i = \alpha(K) \left\{ R(\widehat{f}''_{h_{i-1}}) \right\}^{-1/5} n^{-1/5},$$

and continue the iteration until convergence. This is the iteration procedure as proposed by Scott, Tapia and Thompson (1977). We elaborate further on this iterative procedure in Section 4.10.5.

We conclude this section by mentioning that similar plug-in techniques are applied in the regression estimation setup. See for example Gasser, Kneip and Köhler (1991) and Ruppert, Sheather and Wand (1995).

4.10.5 Sheather and Jones' bandwidth selector

In the previous section we described briefly the early ideas about plug-in techniques. As an example we now present a more sophisticated plug-in procedure which was developed recently by Sheather and Jones (1991). Other plug-in bandwidth selectors have been proposed, but it is beyond the scope of this book to discuss all these. See the survey papers by Jones, Marron and Sheather (1992, 1995). We restrict to one particular example to highlight the idea.

In the plug-in iterative procedure, the bandwidth for estimating $R(f'')$ is taken to be the same as that for estimating the density function. An appropriate bandwidth for estimating $R(f'')$, say g, should differ from a bandwidth appropriate for estimating the density function f itself, say h. Jones and Sheather (1991) showed that when

$$R(\widehat{f}''_g) = n^{-2} g^{-5} \sum_{i=1}^n \sum_{j=1}^n (K * K)^{(4)} \left(\frac{X_i - X_j}{g} \right), \qquad (4.45)$$

where $K * K$ denotes the convolution of K with itself, is used to estimate the unknown quantity $R(f'')$ then the theoretical optimal bandwidth for this estimation task is of the form

$$g_{\text{opt}} = \alpha_1(K) D_1(f) n^{-1/7}, \qquad (4.46)$$

where $\alpha_1(K)$ is a constant depending only on K, and $D_1(f)$ is a function of f. The explicit expressions for these functions are known but are not important for the present discussion. From (4.46) and (2.41) it follows that g_{opt} is an order of magnitude larger than the optimal bandwidth h_{opt}. One can write

$$g_{\text{opt}} = \alpha_2(K) D_2(f) h_{\text{opt}}^{5/7} \equiv g_1(h_{\text{opt}}), \qquad (4.47)$$

for appropriate functions $\alpha_2(\cdot)$ and $D_2(\cdot)$. The above considerations now lead to the bandwidth selector which is defined as the value of H which solves the equation

$$H = \left[\frac{R(K)}{\{\int u^2 K(u) du\}^2 R\{\widehat{f}''_{g_1(H)}\}} \right]^{1/5} n^{-1/5}. \tag{4.48}$$

This is the bandwidth selector proposed by Sheather and Jones (1991). The function $D_2(f)$ appearing in (4.47) is a function of $R(f'')/R(f''')$ and Sheather and Jones (1991) consider kernel estimates for this function where the bandwidth in this initial step is chosen by making reference to a normal density with unknown standard deviation (see the idea presented in Section 4.10.1).

4.11 Complements

Derivation of the conditional expectation of RSC in Theorem 4.1

The proof of the result in Theorem 4.1 is quite simple. Recall that in the definition of RSC, $\widehat{\sigma}^2(x_0)$ denotes the normalized weighted residual sum of squares resulting from a local polynomial fit of order p, based on a bandwidth h (see also (4.8)). With $d_n = \operatorname{tr}\{\mathbf{W} - \mathbf{W}\mathbf{X}(\mathbf{X}^T\mathbf{W}\mathbf{X})^{-1}\mathbf{X}^T\mathbf{W}\}$ and $W_i = K_h(X_i - x_0)$ we have

$$\begin{aligned}
\widehat{\sigma}^2(x_0) &= d_n^{-1} \sum_{i=1}^n (Y_i - \widehat{Y}_i)^2 W_i \\
&= d_n^{-1} (\mathbf{y} - \mathbf{X}\widehat{\beta})^T \mathbf{W} (\mathbf{y} - \mathbf{X}\widehat{\beta}) \\
&= d_n^{-1} \mathbf{y}^T \{\mathbf{W} - \mathbf{W}\mathbf{X}(\mathbf{X}^T\mathbf{W}\mathbf{X})^{-1}\mathbf{X}^T\mathbf{W}\} \mathbf{y}.
\end{aligned}$$

Using the local homoscedasticity we obtain

$$\begin{aligned}
\mathrm{E}\{\widehat{\sigma}^2(x_0) | \mathbb{X}\} &= d_n^{-1} \mathbf{m}^T \{\mathbf{W} - \mathbf{W}\mathbf{X}(\mathbf{X}^T\mathbf{W}\mathbf{X})^{-1}\mathbf{X}^T\mathbf{W}\} \mathbf{m} + \sigma^2(x_0) \\
&= d_n^{-1} \mathbf{r}^T \{\mathbf{W} - \mathbf{W}\mathbf{X}(\mathbf{X}^T\mathbf{W}\mathbf{X})^{-1}\mathbf{X}^T\mathbf{W}\} \mathbf{r} + \sigma^2(x_0),
\end{aligned} \tag{4.49}$$

since $\mathbf{r} = \mathbf{m} - \mathbf{X}\beta$. The ith element of this residual vector $\mathbf{r}$ can be approximated by

$$r_i = m(X_i) - \sum_{j=0}^p \beta_j (X_i - x_0)^j = \beta_{p+1} (X_i - x_0)^{p+1} + O_P(h^{p+2}),$$

so that the first term on the right-hand side of (4.49) can be written

as
$$d_n^{-1}(S_{n,2p+2} - c_n^T S_n^{-1} c_n)\beta_{p+1}^2\{1 + O_P(h)\}, \qquad (4.50)$$
where the notation for S_n and c_n was adopted from Section 3.7. Further, it is easily seen from (3.61) and (3.56) that
$$\begin{aligned}
d_n &= \operatorname{tr}\{\mathbf{W} - \mathbf{W}\mathbf{X}(\mathbf{X}^T\mathbf{W}\mathbf{X})^{-1}\mathbf{X}^T\mathbf{W}\} \\
&= S_{n,0} - \operatorname{tr}\{(\mathbf{X}^T\mathbf{W}\mathbf{X})^{-1}\mathbf{X}^T\mathbf{W}^2\mathbf{X}\} \\
&= f(x_0)n + O_P(h^{-1}). \qquad (4.51)
\end{aligned}$$
From (4.49)–(4.51) and the asymptotic behavior of $S_{n,j}$ and c_n, as established in Section 3.7, it follows that
$$\begin{aligned}
E\{\hat{\sigma}^2(x_0)|\mathbf{X}\} &= d_n^{-1}(S_{n,2p+2} - c_n^T S_n^{-1} c_n)\beta_{p+1}^2 + \sigma^2(x_0) \\
&\quad + o_P(h^{2p+2}) \\
&= C_p \beta_{p+1}^2 h^{2p+2} + \sigma^2(x_0) + o_P(h^{2p+2}), \qquad (4.52)
\end{aligned}$$
with C_p as in the formulation of Theorem 4.1. Finally, by using (3.55) and (3.57),
$$\begin{aligned}
V &= e_1^T S_n^{-1}(\mathbf{X}^T\mathbf{W}^2\mathbf{X})S_n^{-1} e_1 \\
&= \frac{1}{nhf(x_0)} e_1^T S^{-1} S^* S^{-1} e_1 + o_P\{(nh)^{-1}\} \\
&= \frac{a_0}{nhf(x_0)} + o_P\{(nh)^{-1}\}. \qquad (4.53)
\end{aligned}$$
Combination of (4.52) and (4.53) leads to
$$\begin{aligned}
E\{\operatorname{RSC}(x_0; h)|\mathbf{X}\} &= E\{\hat{\sigma}^2(x_0)|\mathbf{X}\}\{1 + (p+1)V\} \\
&= \sigma^2(x_0) + C_p \beta_{p+1}^2 h^{2p+2} + (p+1)a_0 \frac{\sigma^2(x_0)}{nhf(x_0)} \\
&\quad + o_P\{h^{2p+2} + (nh)^{-1}\}
\end{aligned}$$
which is exactly the statement in Theorem 4.1.

Behavior of the nearest neighbor type bandwidth selector (4.43)

Let $F_X(\cdot)$ denote the (marginal) distribution function of X. For simplicity in what follows, we suppress the dependence on n in k_n. Suppose it is shown that
$$\max_{k < c \leq n-k} \left|\frac{F_X(X_{(c+k)}) - F_X(X_{(c-k)})}{2k/n} - 1\right| = o_P(1). \qquad (4.54)$$
Then by the mean value theorem we obtain
$$F_X(X_{(\ell+k)}) - F_X(X_{(\ell-k)}) = 2\hat{h}_k(x) f\{\xi_n(x)\}, \qquad (4.55)$$

with $\xi_n(x) \in [X_{(\ell-k)}, X_{(\ell+k)}]$. Now, from (4.54) it follows that $\sup_{x \in [a,b]} |\xi_n(x) - x| = o_P(1)$, and hence, one can conclude from (4.55) that

$$\widehat{h}_k(x) = \frac{F_X(X_{(\ell+k)}) - F_X(X_{(\ell-k)})}{2f(x)}\{1 + o_P(1)\},$$

uniformly in $x \in [a, b]$. Thus the result stated in Theorem 4.2 follows from (4.54), which we will prove now.

Note that from the representation of uniform order statistics (see e.g. Pyke (1965)), we have

$$F_X(X_{(c+k)}) - F_X(X_{(c-k)}) = \sum_{j=c-k+1}^{c+k} E_j \Big/ \sum_{j=1}^{n+1} E_j,$$

where $\{E_j : 1 \leq j \leq n+1\}$ is a sequence of i.i.d. standard exponentially distributed random variables. Hence, (4.54) is equivalent to

$$\max_{k < c \leq n-k} \Big| \sum_{j=c-k+1}^{c+k} E_j/(2k) - 1 \Big| = o_P(1), \qquad (4.56)$$

since $\sum_{j=1}^{n+1} E_j/n = 1 + o_P(1)$ by the weak law of large numbers. In order to prove (4.56), we need to evaluate

$$P\Big\{\Big| \sum_{j=c-k+1}^{c+k} E_j/(2k) - 1 \Big| \geq \varepsilon\Big\}$$
$$= P\Big\{\sum_{j=c-k+1}^{c+k} E_j \leq (1-\varepsilon)2k\Big\} + P\Big\{\sum_{j=c-k+1}^{c+k} E_j \geq (1+\varepsilon)2k\Big\}$$

for any $\varepsilon > 0$. Now $\sum_{j=c-k+1}^{c+k} E_j$ is Gamma($2k$)-distributed, from which it is easy to see that

$$P\Big\{\sum_{j=c-k+1}^{c+k} E_j \leq (1-\varepsilon)2k\Big\} = \sum_{i=2k}^{\infty} \frac{\{2k(1-\varepsilon)\}^i}{i!} \exp\{-2k(1-\varepsilon)\}.$$
(4.57)

Note that for $i \geq 2k$, the sequence $\{2k(1-\varepsilon)\}^i/(i-2)!$ is decreasing. Therefore, (4.57) can be bounded by

$$\frac{\{2k(1-\varepsilon)\}^{2k}}{(2k-2)!} \sum_{i=2k}^{\infty} \frac{1}{i(i-1)} \exp\{-2k(1-\varepsilon)\}$$
$$\leq \frac{1}{\sqrt{2\pi}e} \sqrt{2k-1} \, (1-1/2k)^{-2k} \exp\Big[2k\{\varepsilon + \log(1-\varepsilon)\}\Big]$$

$$= o\Big\{\exp(-k\varepsilon^2/2)\Big\}, \tag{4.58}$$

where Stirling's formula and the fact that $\log(1-\varepsilon) < -\varepsilon - \varepsilon^2/2$ were used. Using similar arguments, it is shown that $P\{\sum_{j=c-k+1}^{c+k} E_j \geq (1+\varepsilon)2k\} = o\Big\{\exp(-k\varepsilon^2/2)\Big\}$, which together with (4.58) yields $P\{|\sum_{j=c-k+1}^{c+k} E_j/(2k) - 1| \geq \varepsilon\} = o\Big\{\exp(-k\varepsilon^2/2)\Big\}$. Hence, (4.56) follows from

$$P\Big\{\max_{k<c\leq n-k}\Big|\sum_{j=c-k+1}^{c+k} E_j/(2k) - 1\Big| \geq \varepsilon\Big\} = o\Big\{2n\exp(-k\varepsilon^2/2)\Big\}.$$

4.12 Bibliographic notes

The idea of local likelihood was discussed by Tibshirani and Hastie (1987). Recent papers exploiting the idea of local likelihood are Fan, Heckman and Wand (1995), Loader (1995) and Hjort and Jones (1996a), among others. The idea of estimating the bias and variance appears to be new.

Eubank and Speckman (1993) study a method for constructing confidence bands in nonparametric regression.

Likelihood cross-validation was proposed by Habbema, Hermans and van der Broeck (1974) and by Duin (1976). Hall (1987) studied in more detail the use of the Kullback-Leibler loss function in density estimation. For a comparative study with other selection methods see Marron (1989) and Cao, Cuevas and González-Manteiga (1994).

Least squares cross-validation is due to Rudemo (1982) and Bowman (1984). Theoretical properties of the least squares cross-validation bandwidth were established by Scott and Terrell (1987) and Hall and Marron (1988), among others. Vieu (1991) uses local cross-validation to select local bandwidths. Hart and Vieu (1990) study the cross-validation criterion for dependent data. A nice overview of the cross-validation methods is given by Hall and Johnstone (1992). Least squares cross-validation suffers from a lot of sample variation and searching the minimum of the CV function (assuming that one exists) can cause problems. Some improved versions of least squares cross-validation have been proposed in the literature. Scott and Terrell (1987) introduced the biased cross-validation method, and Hall, Marron and Park (1992) use presmoothing on the differences of the observations $X_i - X_j$ involved in the cross-validation quantity. Fan, Hall, Martin and Patil (1996) apply smoothed cross-validation to choose a variable bandwidth function.

Chiu (1991), Hall, Sheather, Jones and Marron (1991), and Marron, Park and Kim (1994) have all proposed root-n consistent bandwidth selectors in the density estimation setting. The best possible efficient lower bound for a bandwidth selector was derived by Fan and Marron (1992). In the homoscedastic regression setting, Huang (1995) shows that the refined bandwidth selector (4.21) with $a = 3$ is root-n consistent when the local linear fit is used.

CHAPTER 5

Applications of local polynomial modelling

5.1 Introduction

The key idea of local modelling is explained in the context of least squares regression models. The simplicity, interpretability and its good statistical properties indicate that the local modelling approach can also be very useful in a wide array of statistical problems. The theme of this chapter is to apply this approach in survival analysis, generalized linear models, and quantile regression. Readers are encouraged to read Section 4.9 to get a general idea of local modelling.

Important developments in statistics in the past three decades include analyzing health sciences related data, in which the interest frequently centers on understanding how risk factors contribute to the survival times of a group of individuals. The survival times are often not fully observable due to, for example, the termination of a study or to the deaths of certain individuals even before the beginning of a study. There are many existing statistical approaches designated for analyzing these types of incomplete data. They can basically be classified as regression based approaches and hazard risk based methods. To utilize these approaches, a parametric form is popularly imposed. However, this form might not be satisfied for a given data set or needs at least to be validated. Sections 5.2 and 5.3 describe how these two tasks can be accomplished via the local modelling approach.

A family of useful models for data analyses is the class of generalized linear models. These models include many commonly used distributions such as the Gaussian, binomial, Poisson and gamma distributions, and can be used to analyze both discrete and continuous types of data. An important question is to determine a structural relationship between the response variable and the covariates. This issue is discussed in Section 5.4 using the local modelling approach.

In many applications, the variance of the stochastic components can be large. This seriously affects the quality of the estimated functional relationship between the response variable and its associated explanatory variables. In that case, a robustified version of the estimator is necessary in order to draw a meaningful conclusion. Section 5.5 describes how to robustify the least squares approaches described in Chapters 3 and 4. As a result, we also demonstrate how the local modelling approach can be used to construct predictive intervals for a future response.

The key technical device used in developing the asymptotic theory is given in Section 5.6. It attempts to give an overview of how to establish the asymptotic normality when an estimator does not admit an explicit form.

5.2 Censored regression

5.2.1 Preliminaries

Here we introduce some related background materials on survival analysis. For a more thorough introduction, readers can consult the books by Kalbfleisch and Prentice (1980), Miller (1981), Cox and Oakes (1984), Fleming and Harrington (1991) and Andersen, Borgan, Gill and Keiding (1993) among others.

In this brief introduction, we use the famous Stanford heart transplant data, analyzed by Miller and Halpern (1982) and Doksum and Yandell (1982) and reproduced in Table 8.1 of Cox and Oakes (1984). Between October 1967 and February 1980, 184 patients out of the 249 patients admitted to the program, received heart transplantation. We only focus on two variables, with survival time after transplantation as the response and age as the covariate. Patients alive beyond February 1980 were considered as censored. To understand better the *censoring mechanism*, let S_i and E_i denote respectively the calendar time of the surgery and of the death for the ith patient. Some patients died before February 1980 and their survival times were recorded. Others died after February 1980 and hence their survival times were not observable. What is observed for each of the patients is the censoring indicator (1 – death; 0 – surviving) and the minimum of the survival time T_i and the censoring time C_i, where

$$T_i = E_i - S_i; \quad C_i = \text{Termination date of the study} - S_i.$$

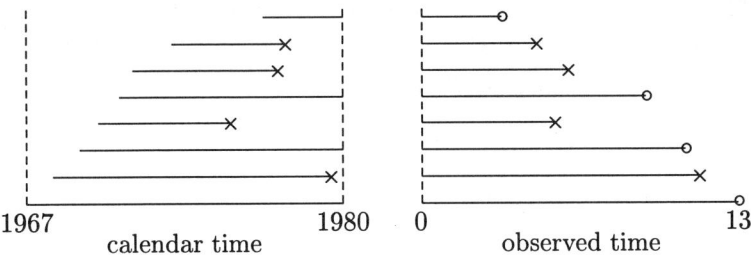

Figure 5.1. *Calculation of survival times and censoring times. The symbol × represents death and ○ indicates censoring. (a) Calendar time. (b) Time from entry into the study.*

Figure 5.1 illustrates this calculation. Only the 157 patients for which complete information on the tissue type was available are included in the analysis. For those 157 patients the logarithm of the observed times $\min(T_i, C_i)$ are plotted against their age at the time of transplantation. There were 55 cases of censoring. See Figure 5.2.

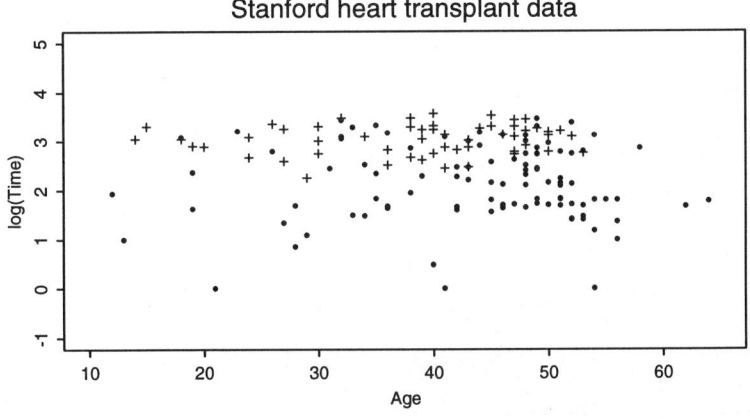

Figure 5.2. *Stanford heart transplant data – log-survival time plotted against age. The character '+' indicates the censored observations, and the uncensored observations are represented by '.'.*

Let T, C and **X** be respectively the survival time, the censoring time and their associated covariates. Correspondingly, let $Z = \min(T, C)$ be the observed time and $\delta = I(T \leq C)$ be the censoring indicator. To analyze the association between T and $\mathbf{X}$ in a statistical framework, it is customary to regard the observed data

$$\{(\mathbf{X}_i, Z_i, \delta_i) : i = 1, \cdots, n\} \tag{5.1}$$

as an i.i.d. random sample from a certain population $(\mathbf{X}, Z, \delta)$. Assume further that T and C are conditionally independent given $\mathbf{X}$. We will refer to this assumption as *independent censoring*. Further, the censoring mechanism is called *noninformative censoring*, if the conditional distribution of C given $\mathbf{X}$ does not involve the parameters of interest.

One way to study the association between the response and the covariate variables is via the regression model:

$$Y \equiv g(T) = m(\mathbf{X}) + \sigma(\mathbf{X})\varepsilon, \tag{5.2}$$

for a given transformation g, often a logarithmic function. This approach attempts to assess the contributions of risk factors via a mean response function $m(\cdot)$.

Another approach is to study the contributions of the risk factors via the conditional *hazard rate* function. For a given covariate $\mathbf{X}$, this hazard rate at a given time t is

$$\lambda(t|\mathbf{X}) = \lim_{\Delta \downarrow 0} \frac{P(t \leq T < t + \Delta | T \geq t, \mathbf{X})}{\Delta}, \tag{5.3}$$

representing the risk that an individual fails immediately after time t given survival at the time t. There are two popular models based on the hazard rate: the *accelerated lifetime model* and the *proportional hazards model*. The former assumes that the hazard rate has the form

$$\lambda(t|\mathbf{x}) = \lambda_0\{t\psi(\mathbf{x})\}\psi(\mathbf{x}), \tag{5.4}$$

and the latter describes the hazard rate via

$$\lambda(t|\mathbf{x}) = \lambda_0(t)\exp\{\psi(\mathbf{x})\}, \tag{5.5}$$

where $\lambda_0(\cdot)$ is the *baseline hazard function*, representing the hazard risk at the covariate $\mathbf{x} = 0$, and $\psi(\mathbf{x})$ is the function that depicts the contribution of covariates $\mathbf{x}$.

Deviations from model (5.5) are sometimes necessary. For example, when comparing two levels of treatment (0 and 1, say), model (5.5) implies

$$\frac{\lambda(t|x=1)}{\lambda(t|x=0)} = \exp\{\psi(1) - \psi(0)\},$$

for *any* $t > 0$. In other words, the hazard risk changes instantaneously upon the introduction of the treatment, which is implausible in many applications (the treatment effect is typically only observed after a certain period). One possible deviation from model (5.5) is to allow interaction with a binary variable when comparing two risk groups (or treatments): each group satisfies model (5.5) with possibly different baseline hazard functions and covariate effects $\psi(\cdot)$. Details of this kind of deviation can be found in Dabrowska, Doksum, Feduska, Husing and Neville (1992).

We now interpret the above hazard-based models. Assume for simplicity that the covariates $\mathbf{X}$ are not time varying and that the random variable T is absolutely continuous. First note that

$$\lambda(t|\mathbf{x}) = f(t|\mathbf{x})/\bar{F}(t|\mathbf{x}), \quad \text{with} \quad \bar{F}(t|\mathbf{x}) = 1 - F(t|\mathbf{x}) \qquad (5.6)$$

being the conditional *survivor function*, where $f(t|\mathbf{x})$ and $F(t|\mathbf{x})$ are respectively the conditional density and distribution function of T given $\mathbf{X} = \mathbf{x}$. Thus, from (5.6), the conditional survivor function can be represented as

$$\bar{F}(t|\mathbf{x}) = \exp\{-\int_0^t \lambda(u|\mathbf{x})du\}, \qquad (5.7)$$

for a nonnegative random variable T. Hence, modelling the hazard rate function is equivalent to assuming a specific form for the conditional distribution. Further, it can easily be shown that the conditional hazard function for $Y = g(T)$ is given by

$$\lambda_Y(y|\mathbf{x}) = \lambda_T\{g^{-1}(y)|\mathbf{x}\}/g'\{g^{-1}(y)\}. \qquad (5.8)$$

We are now ready to interpret model (5.4). Let T_0 be a random variable whose hazard function is $\lambda_0(\cdot)$, independent of $\mathbf{x}$. By (5.8), the hazard risk function of $T = T_0/\psi(\mathbf{X})$ is given by (5.4). In other words, the covariates $\mathbf{X}$ reduce the lifetime of an individual by a factor of $\psi(\mathbf{X})$. Thus, the random variable T admits the following regression form:

$$\log T = -\log\{\psi(\mathbf{X})\} + \varepsilon, \quad \text{with} \quad \varepsilon = \log(T_0),$$

that is, T satisfies the regression model (5.2). On the other hand, when T satisfies model (5.2) with $\sigma(\mathbf{X})$ independent of $\mathbf{X}$ and $g(t) = \log(t)$, it can easily be shown that T satisfies the accelerated lifetime model (5.4). Hence model (5.2), with $g(t) = \log t$ has an interpretation as an accelerated lifetime model.

Model (5.5) has a direct interpretation in terms of the hazard risk. The covariates $\mathbf{X}$ increase the hazard risk of an individual by a factor of $\exp\{\psi(\mathbf{X})\}$. In terms of conditional distribution functions, model (5.5) can be written as

$$\bar{F}(t|\mathbf{x}) = \{\bar{F}_0(t)\}^{\exp\{\psi(\mathbf{X})\}},$$

where $\bar{F}_0(\cdot)$ denotes the survivor function at the covariate $\mathbf{x} = 0$.

In many situations, one is also interested in estimating the marginal distribution of T based on the censored data $\{(Z_i, \delta_i), i = 1, \cdots, n\}$. The marginal survivor function of T is popularly estimated by the Kaplan-Meier (1958) *product-limit* estimator:

$$\begin{aligned}
1 - \widehat{F}_n(t) &= \prod_{j: Z_j \leq t} \left(1 - \frac{1}{\#\{i : Z_i \geq Z_j\}}\right)^{\delta_j} \\
&= \prod_{j: T_j^o \leq t} \left(1 - \frac{1}{\#\{i : Z_i \geq T_j^o\}}\right),
\end{aligned} \quad (5.9)$$

where $\{T_j^o\}$ denote the ordered uncensored survival times. Note

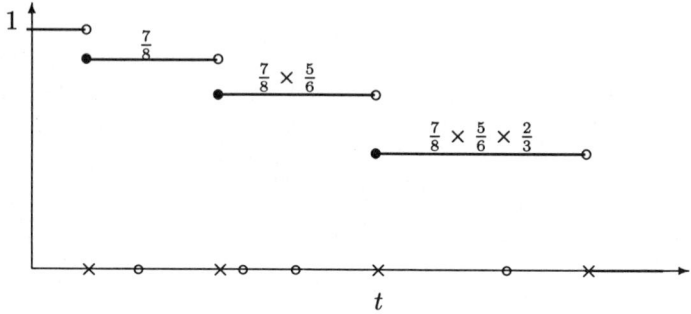

Figure 5.3. *Construction of a Kaplan-Meier estimator. The survivor function* $1 - \widehat{F}_n(t)$ *is plotted against time* t.

that $\#\{i : Z_i \geq T_j^o\}$ represents the number of subjects at risk just before time T_j^o and the factor $(1 - \frac{1}{\#\{i: Z_i \geq T_j^o\}})$ can be explained as the chance of surviving the risk period T_j^o at which an individual failed. Estimator (5.9) has an intuitive explanation: the chance that an individual survives at least time t is the product of the chances that he survives each risk period T_j^o before time t. Figure 5.3 illustrates the calculation of the Kaplan-Meier product-limit estimator for eight hypothetical observations, four censored and four uncensored, represented on the x-axis. Evidently, when the largest observation is censored, $\widehat{F}_n(\cdot)$ will not be a proper distribution function. A popular convention in this case is to treat the largest observation as an uncensored one. This convention is also adopted in this section. When all data are uncensored the Kaplan-Meier estimator reduces to the usual empirical cumulative distribution function.

5.2.2 Censoring unbiased transformation

To establish the regression model (5.2), we need to transform the censored data by the function g. For convenience of notation, we continue to denote the resulting observations by $(\mathbf{X}_i, Z_i, \delta_i)$. So, with a slight abuse of notation, $Z_i = \min\{g(T_i), g(C_i)\}$, where $g(C_i)$ is denoted again by C_i.

How does one estimate the relationship between $Y = g(T)$ and $\mathbf{X}$ in the case of censored data? The usual statistical tools, such as for example scatter plots, are not directly applicable. The basic idea is then to adjust for the censoring effect by transforming the data in an unbiased way. We now explain this idea. Let $\phi_1(\cdot, \cdot)$ and $\phi_2(\cdot, \cdot)$ be the transformation functions acting respectively on the uncensored and the censored observations. More precisely, the datum point $(\mathbf{X}, Z, \delta)$ will be replaced by the transformed datum point $(\mathbf{X}, Y^*)$ according to

$$\begin{aligned} Y^* &= \begin{cases} \phi_1(\mathbf{X}, Y) & \text{if uncensored} \\ \phi_2(\mathbf{X}, C) & \text{if censored} \end{cases} \\ &= \delta \phi_1(\mathbf{X}, Z) + (1-\delta)\phi_2(\mathbf{X}, Z). \end{aligned} \tag{5.10}$$

Since statistical inferences for $m(\cdot)$ will be based on $(\mathbf{X}, Y^*)$, the basic requirement is that $E(Y^*|\mathbf{X}) = m(\mathbf{X}) = E(Y|\mathbf{X})$, ensuring that we estimate the right object. Note that

$$\begin{aligned} E\{\delta \phi_1(\mathbf{X}, Z)|\mathbf{X}\} &= E[E\{\delta \phi_1(\mathbf{X}, Y)|\mathbf{X}, Y\}|\mathbf{X}] \\ &= E\{\phi_1(\mathbf{X}, Y)\bar{G}(Y|\mathbf{X})|\mathbf{X}\}, \end{aligned}$$

and analogously $E\{(1-\delta)\phi_2(\mathbf{X}, Z)|\mathbf{X}\} = E\{\phi_2(\mathbf{X}, C)\bar{F}(C|\mathbf{X})|\mathbf{X}\}$, where $\bar{F}(\cdot|\mathbf{x})$ and $\bar{G}(\cdot|\mathbf{x})$ are respectively the conditional survivor function of the random variables Y and C given $\mathbf{X} = \mathbf{x}$. Using these equalities, the basic requirement is equivalent to

$$E\left[\left\{\phi_1(\mathbf{X}, Y)\bar{G}(Y|\mathbf{X}) + \phi_2(\mathbf{X}, C)\bar{F}(C|\mathbf{X})\right\}\Big|\mathbf{X}\right] = E(Y|\mathbf{X}).$$

Any pair of transformations (ϕ_1, ϕ_2) satisfying this requirement is called *censoring unbiased transformation*. To avoid technicalities we assume that $\bar{G}(\cdot|\mathbf{x})$ is continuous. The above equation can be written as

$$\int_0^{+\infty} \left[\phi_1(\mathbf{x},y)\bar{G}(y|\mathbf{x}) - \int_0^y \phi_2(\mathbf{x},c)d\bar{G}(c|\mathbf{x}) - y\right] d\bar{F}(y|\mathbf{x}) = 0, \tag{5.11}$$

for all $\mathbf{x}$. Since least squares techniques are employed, the quality of the transformation is reflected in $\text{Var}(Y^*|\mathbf{X}=\mathbf{x})$.

An intuitive and appealing transformation is given by Buckley and James (1979):

$$\phi_1(\mathbf{x},y) = y \quad \text{and} \quad \phi_2(\mathbf{x},y) = E(Y|Y>y, \mathbf{X}=\mathbf{x}), \tag{5.12}$$

which leaves the uncensored survival times Y unchanged and imputes the censored observations by the best restoration. This transformation can be written as $Y_0^* = E(Y|\delta, Z, \mathbf{X})$. It is closest to the original response in the sense that $E(Y - Y_0^*)^2 \leq E(Y - Y^*)^2$ for any censoring unbiased transformation Y^*. Thus, it can be regarded as the 'best restoration'. However, this transformation depends on the unknown regression function, and can lead to an iterative procedure. The computation for such a procedure would be formidable, especially when using nonparametric techniques. Besides, computing the conditional expectation in (5.12) requires strong model assumptions (for example $m(x)$ is a linear function), limiting the applicability. A fast and dynamical transformation requires estimates with an explicit form. Here we offer two approaches based on the local average estimate and the local linear estimate.

Suppose that the point $(\mathbf{X}_i, Z_i, \delta_i)$ is a censored observation (i.e. $\delta_i = 0$). Here, for simplicity, assume $\mathbf{X}_i$ is univariate. The multivariate extension is simple but not very useful due to the 'curse of dimensionality' which will be explained in Chapter 7. Useful multivariate extensions require certain 'dimensionality reduction principles' discussed in Chapter 7. Consider a neighborhood of X_i, and replace the censored observation Z_i by the weighted average of all uncensored responses in that neighborhood which are larger than Z_i. Formally, let k be an integer which determines the neighborhood of X_i, and denote by K a nonnegative weight function. We assume that all observations $\{(X_i, Z_i, \delta_i) : i = 1, \cdots, n\}$ have been ordered according to the X_i' s. Define

$$Y_i^* = \frac{\sum_{j:Z_j>Z_i} Z_j K\left\{\frac{X_j-X_i}{(X_{i+k}-X_{i-k})/2}\right\}\delta_j}{\sum_{j:Z_j>Z_i} K\left\{\frac{X_j-X_i}{(X_{i+k}-X_{i-k})/2}\right\}\delta_j}, \tag{5.13}$$

with the convention of truncating the neighborhood when the indices $i+k$ or $i-k$ are out of the index range. The above transformation is a nonparametric estimate of the conditional expectation in (5.12). Uncensored data points remain unchanged, i.e. $Y_i^* = Z_i$, if $\delta_i = 1$. The transformed data are $\{(X_i, Y_i^*) : i = 1, \cdots, n\}$. We refer to (5.13) as the *local average unbiased transformation*, based on the *nearest neighbor* bandwidth $h_k(X_i) = (X_{i+k} - X_{i-k})/2$ and the kernel function K. The *local linear unbiased transformation* is defined by

$$Y_i^* = \sum_{j:Z_j>Z_i} w_j Z_j / \sum_{j:Z_j>Z_i} w_j \qquad (5.14)$$

with

$$w_j = \delta_j K\left\{\frac{X_j - X_i}{h_k(X_i)}\right\} \{S_{n,2} - (X_j - X_i)S_{n,1}\},$$

where $S_{n,\ell} = \sum_{j:Z_j>Z_i} \delta_j K\left\{\frac{X_j - X_i}{h_k(X_i)}\right\}(X_j - X_i)^\ell$. See Section 2.3.1. This transformation corresponds to estimating ϕ_2 in (5.12) by a local linear fit. Since transformation (5.12) is the 'best restoration', the estimated transformations (5.13) and (5.14) are favorable and intuitively appealing. However, the scope of this application is somewhat limited: the dimensionality of X is usually small.

Another useful approach is to choose transformations which do not explicitly depend on $F(\cdot|\mathbf{x})$, according to the following procedure. Consideration of (5.11) leads to the requirement that

$$\phi_1(\mathbf{x},y)\bar{G}(y|\mathbf{x}) - \int_0^y \phi_2(\mathbf{x},c)d\bar{G}(c|\mathbf{x}) - y = 0, \qquad (5.15)$$

for all x and y. The major strength of transformation (5.15) is that it can easily be applied to multivariate covariates $\mathbf{X}$ when the conditional censoring distribution is independent of the covariates. However, a price, in terms of variance increase, has to be paid, as will be illustrated in Figure 5.4. In the following we discuss some simple transformations obtained via this approach.

Put $\phi_2(\mathbf{x},y) = \phi_1(\mathbf{x},y) + d(\mathbf{x},y)$. Then, the solution to (5.15) can be expressed as

$$\begin{cases} \phi_1(\mathbf{x},y) = \int_0^y \{\bar{G}(t|\mathbf{x})\}^{-1}dt + \int_0^y d(\mathbf{x},t)\{\bar{G}(t|\mathbf{x})\}^{-1}d\bar{G}(t|\mathbf{x}), \\ \phi_2(\mathbf{x},y) = \phi_1(\mathbf{x},y) + d(\mathbf{x},y), \end{cases}$$

(5.16)

which is given in Zheng (1987) with a different motivation. Different choices of the function $d(\cdot,\cdot)$ result in a different degree of variability. The exact quantification of this can be found in Fan and Gijbels (1994).

A specific subclass of transformation (5.16) is that with $d(\mathbf{x}, y) = \alpha y / \bar{G}(y|\mathbf{x})$, leading to the class of transformations

$$\begin{cases} \phi_1(\mathbf{x}, y) = (1+\alpha) \int_0^y \{\bar{G}(t|\mathbf{x})\}^{-1} dt - \alpha y \{\bar{G}(y|\mathbf{x})\}^{-1}, \\ \phi_2(\mathbf{x}, y) = (1+\alpha) \int_0^y \{\bar{G}(t|\mathbf{x})\}^{-1} dt. \end{cases} \quad (5.17)$$

We remark that the Koul, Susarla and Van Ryzin (1981) transformation corresponds to $\alpha = -1$ which is inadmissible and is dominated by (has a larger variance than) Leurgans' (1987) transformation with $\alpha = 0$. For details see Fan and Gijbels (1994). The tuning parameter α in the class of *distribution-based unbiased transformations* (5.17) creates an opportunity for improvement. The choice $\alpha > 0$ transforms more on the censored observations than on the uncensored observations, which is more intuitive than Leurgans' equal choice ($\alpha = 0$). On the other hand, when α is large, $\phi_1(\mathbf{x}, y) < y$ causing more variability. This leads us to take the largest α such that $\phi_1(X_i, Y_i) \geq Y_i$ for the uncensored response:

$$\widehat{\alpha} = \min_{\{i:\delta_i=1\}} \frac{\int_0^{Y_i} \{\bar{G}(t|\mathbf{X}_i)\}^{-1} dt - Y_i}{Y_i \{\bar{G}(Y_i|\mathbf{X}_i)\}^{-1} - \int_0^{Y_i} \{\bar{G}(t|\mathbf{X}_i)\}^{-1} dt}. \quad (5.18)$$

This choice of α, proposed by Fan and Gijbels (1994), reduces the variability and will be used in our implementations.

In many applications, censoring is caused by the termination of a study. Then, it is reasonable to assume that $G(c|\mathbf{x}) \equiv G(c)$, which can be estimated by the Kaplan-Meier product-limit estimator. Regarding C_i as survival time and Y_i as censoring time we find, from (5.9),

$$\widehat{\bar{G}}(c) = \prod_{j: Z_j \leq c} \left(1 - \frac{1}{\#\{i : Z_i \geq Z_j\}}\right)^{1-\delta_j}.$$

This leads to the transformation

$$\begin{cases} \phi_1(y) = (1+\alpha) \int_0^y \{\widehat{\bar{G}}(t)\}^{-1} dt - \alpha y \{\widehat{\bar{G}}(y)\}^{-1}, \\ \phi_2(y) = (1+\alpha) \int_0^y \{\widehat{\bar{G}}(t)\}^{-1} dt. \end{cases} \quad (5.19)$$

The key strength of transformation (5.19) is its simplicity: after transformation of the data such that it does not involve covariates, one reduces the censored regression to the usual regression even in the multivariate setting.

When C and $\mathbf{X}$ are not independent, transformation (5.17) is more complicated. We illustrate the idea when $\mathbf{X}$ is univariate. In this case, one can use Beran's (1981) local product-limit estimator for $\bar{G}(c|x)$, defined as

$$\widehat{\bar{G}}_n(c|x) = \prod_{j \in I_k(x): Z_j \leq c} \left\{1 - \widehat{\lambda}_j(x)\right\}^{1-\delta_j}, \qquad (5.20)$$

where $\widehat{\lambda}_j(x) = 1/\#\{i: Z_i \geq Z_j, i \in I_k(x)\}$, with $I_k(x)$ the set of indices of the $(2k+1)$ nearest neighbors of x. The estimator for $\bar{G}(c|x)$ is inaccurate at the tail, and to avoid this inaccuracy, we do not transform a particular datum point when $\widehat{\bar{G}}_n(Z_j|X_j) \leq 1/k$, say.

We illustrate the effectiveness of the above censoring unbiased transformations via a simulation study. We simulated 200 data points from the following model:

$$\begin{cases} Y_i = 4.5 - 64X_i^2(1-X_i)^2 - 16(X_i - 0.5)^2 + 0.25\varepsilon_i, \\ X_i \sim_{\text{i.i.d.}} \text{Uniform}[0,1], \quad \varepsilon_i \sim_{\text{i.i.d.}} N(0,1), \end{cases}$$

where $\{X_i\}$ and $\{\varepsilon_i\}$ are independent. Given X_i, the censoring time C_i is conditionally independent of the survival time Y_i and is distributed as $(C_i|X_i = x) \sim \text{Exponential}\{c(x)\}$, where $c(x)$ is the mean conditional censoring time which is given by

$$c(x) = \begin{cases} 3(1.25 - |4x - 1|), & \text{if } 0 \leq x \leq 0.5, \\ 3(1.25 - |4x - 3|), & \text{if } 0.5 < x \leq 1. \end{cases}$$

Further, the censoring times $C_1, \cdots, C_n$ are independent and all have the same conditional distribution Exponential$\{c(x)\}$. In this example, about 40% of the 200 observations are censored. The resulting transformed data are presented in Figure 5.4. The choice $k = 7$ was selected by a cross-validation criterion that will be explained in Section 5.2.3. Clearly, the variability of the data induced by (5.13) is smaller than the variability of those induced by (5.17). Both procedures are very effective since the estimated curves using the transformed data are almost indistinguishable from those using the unobserved data.

In summary, we recommend using transform (5.13) for a univariate covariate and transformation (5.19) for a multivariate setup with α given by (5.18), namely

$$\widehat{\alpha} = \min_{\{i:\delta_i=1\}} \frac{\int_0^{Y_i} \{\bar{G}(t)\}^{-1} dt - Y_i}{Y_i\{\bar{G}(Y_i)\}^{-1} - \int_0^{Y_i} \{\bar{G}(t)\}^{-1} dt}.$$

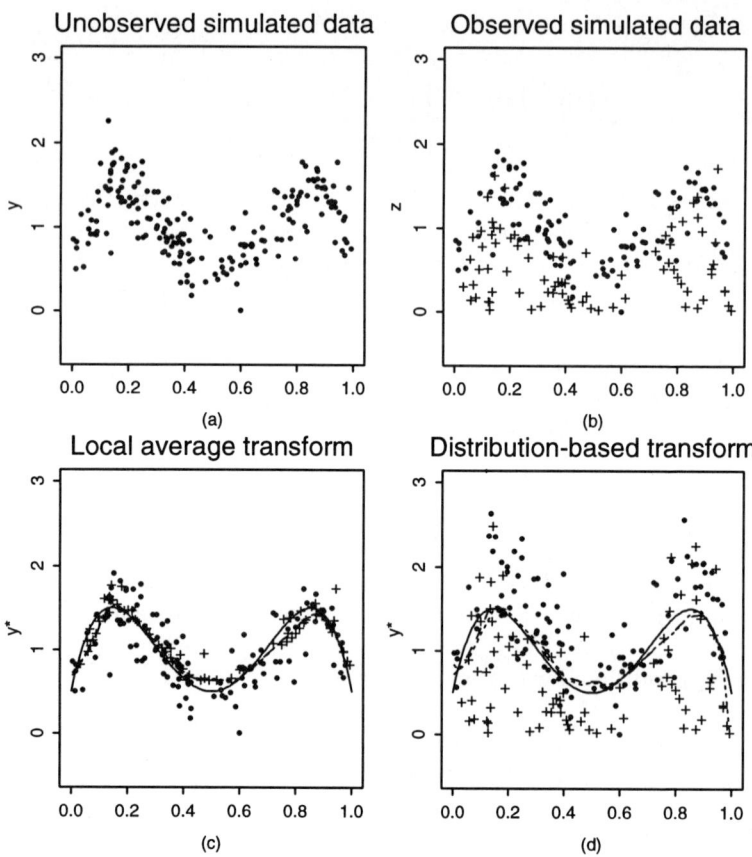

Figure 5.4. *A simulation study. The same characters as in Figure 5.2 are used. (a) Unobserved simulated data; (b) observed simulated data; (c) data transformed by local average using $k = 7$; (d) data transformed by distribution-based unbiased transformation with α given by (5.18) and $\bar{G}(\cdot|x)$ estimated by (5.20) with $k = 7$. For (c) and (d), we also superimpose the true curve (solid) and the local linear regression curves by using the transformed data (dotted) and the unobservable data (dashed). Adapted from Fan and Gijbels (1994).*

5.2.3 Local polynomial regression

The transformed data $\{(X_i, Y_i^*) : i = 1, \cdots, n\}$ can be treated as uncensored data and standard regression techniques can be applied. While the Y_i^* variables are not independent given X, they are nearly independent and produce consistent estimates. See Section 5.2.4 for more detailed discussions. Among standard regression techniques, one can use either global modelling approaches (parametric) or local modelling approaches (nonparametric). In a situation involving a few covariates, one can transform the data via (5.19). After transformation of the data one can apply either the traditional parametric regression techniques or semiparametric regression techniques such as partially linear models. We omit the details of these techniques. See Bickel, Klaassen, Ritov and Wellner (1993), Green and Silverman (1994) and Chapter 7 for the related semiparametric techniques.

We now focus on the univariate local polynomial regression approach. Suppose that the primary interest is in estimating $m^{(\nu)}(\cdot)$. Then, for a given bandwidth, one can run a local polynomial regression (2.8) of degree $p = \nu+1$ or $p = \nu+3$. The bandwidth h can either be determined subjectively by data analysts or objectively by any of the bandwidth selection rules in Chapter 4.

A special case is to estimate the regression function using a local linear fit. Transformation (5.13) in conjunction with a local linear fit leads to a simple and easily implemented procedure. We now describe a method for selecting the smoothing parameter k in the transformation step and in the estimation step. This method, proposed in Fan and Gijbels (1994), is designed to be simple so that we use the same smoothing parameter in the transformation and the estimation step. The method consists of the following steps:

Transformation of the data. For a given integer k and a nonnegative weight function K, replace the observed data $\{(X_i, Z_i, \delta_i)\}$ by $\{(X_i, Y_i^*)\}$, where Y_i^* is given by (5.13) for censored data and equals Y_i for uncensored data.

Application of the local linear regression technique. Starting from the transformed data $\{(X_i, Y_i^*)\}$, do a local linear regression using the smoothing parameter k. Namely, compute

$$\widehat{m}(x) = \sum_{i=1}^{n} w_i(x) Y_i^* / \sum_{i=1}^{n} w_i(x), \qquad (5.21)$$

where with $s_{n,\ell} = \sum_{i=1}^{n} K\left\{\frac{X_i - x}{\widehat{h}_k(x)}\right\} (X_i - x)^{\ell}, \ell = 0, 1, 2,$

$$w_i(x) \equiv K\left\{\frac{X_i - x}{\widehat{h}_k(x)}\right\} \{s_{n,2} - (X_i - x)s_{n,1}\}.$$

Here we use the bandwidth

$$\widehat{h}_k(x) = (X_{\ell+k} - X_{\ell-k})/2, \qquad (5.22)$$

where ℓ is the index of the design point X_ℓ closest to x. This bandwidth is similar to the one used in LOWESS (see Cleveland (1979)).

Selection of the smoothing parameter k by cross-validation. For a given k, do the following. Transform the data as in step 1, then disregard the ith transformed datum point (X_i, Y_i^*), and denote by $\widehat{m}_{-i}(x)$ the estimator calculated from (5.21) without using the ith observation. Compute $CV(k) = \sum_{i=1}^{n} \{Y_i^* - \widehat{m}_{-i}(X_i)\}^2$. Denote by $\widehat{k}$ the cross-validation smoothing parameter, i.e. the parameter which minimizes $\{CV(k) : k = 1, \cdots, [(n-1)/2]\}$, where $[a]$ denotes the greatest integer part of a.

Note that another possible approach is to delete the ith observation before carrying through the transformation step. However, this would result in an extra loop operation, which would increase the computational cost by a factor of nearly n. Hence, this approach is not recommended.

Calculation of the local linear regression estimator. With $\widehat{k}$ as a smoothing parameter, transform the observed data as in step 1 and estimate the regression function as in step 2.

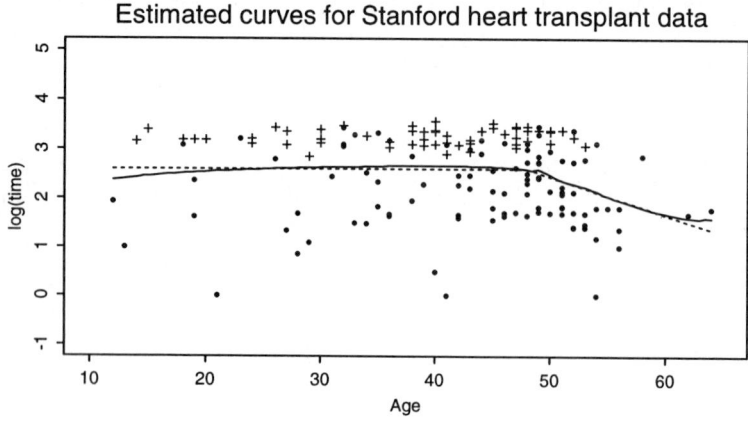

Figure 5.5. *Transformed data by local average and their estimated curve for the Stanford heart transplant study. Solid curve – the locally linearly estimated curve. Dotted curve – the suggested relationship* $2.74 - 0.078(x-48)_+$. *The figure is adapted from Fan and Gijbels (1994).*

CENSORED REGRESSION

We now analyze the Stanford heart transplant data presented in Figure 5.2. The cross-validation criterion leads to $\widehat{k} = 23$. With this $\widehat{k}$ we transform the data via (5.13) and then estimate the curve via (5.21). The resulting transformed data and its local linear regression estimator can be found in Figure 5.5. Note that the estimated curve is flat for ages ranging from 10 to 48. From age 48 on, the curve decreases linearly with a slope of approximately -0.078 and with intercept 2.74 at age 48. This reflects the fact that for earlier age, the log-survival time is nearly independent of age, and at later age it decreases linearly with age. In terms of hazard rates, as presented in (5.8), this means that the transplantation risk is nearly the same for patients aged less than 48 and increases from age 48 on according to

$$\lambda(t|x) = \lambda_0\{t\psi(x)\}\psi(x) \text{ with } \psi(x) = \exp\{-2.74+0.078(x-48)_+\},$$

where $\lambda_0(\cdot)$ is the baseline hazard function. The result supports our intuition, but moreover gives a deep insight into the risk of heart-transplantation at various ages. Should one be interested in accessing the variability of the estimated parameters, then one can parametrize the function $\psi(x)$ by

$$\psi(x) = \exp\{-\theta_1 + \theta_2(x - \theta_3)_+\},$$

and then apply a parametric approach. This piecewise log-linear model is also called a *thresholding model* or *change point model*.

The above example illustrates two key points: the nonparametric method is effective in extracting information from data and can be used to suggest a parametric model and to check an existing parametric model which is a part of *sensitivity analysis*. Since nonparametric methods impose fewer modelling assumptions, one typically gets somewhat cruder conclusions and needs fewer model diagnostics. In contrast, parametric modelling imposes more stringent assumptions and needs more sensitivity analyses.

5.2.4 An asymptotic result

We now outline a theorem that supports the use of the standard nonparametric regression method. For simplicity of presentation, we consider the case that $\nu = 0$ and $p = 1$. Let $\widehat{m}(x; \phi_1, \phi_2)$ be the local linear regression estimator introduced in Section 2.3.1, using the transformed data (5.10). The asymptotic properties of $\widehat{m}(x; \phi_1, \phi_2)$ follow directly from the standard theory presented in Chapter 3. A basic requirement in the consistency result for

$\widehat{m}(x; \widehat{\phi}_1, \widehat{\phi}_2)$ is that the estimators $\widehat{\phi}_1(t, z)$ and $\widehat{\phi}_2(t, z)$ are consistent, uniformly for t in a neighborhood of x and for z in an interval which is chosen such that the instability of $\widehat{G}_n$ in the tail can be dealt with. Formally, assume that

$$\beta_n(x) = \max_{j=1,2} \left\{ \sup_{z \in (0, \tau_n), t \in (x \pm \gamma)} |\widehat{\phi}_j(t, z) - \phi_j(t, z)| \right\} = o_P(1), \tag{5.23}$$

with $\tau_n > 0$ and $\gamma > 0$. Then, extend the definition of $\widehat{\phi}_j$ $(j = 1, 2)$ as follows:

$$\widehat{\phi}_j(t, z) = \begin{cases} \widehat{\phi}_j(t, z), & \text{if } z \leq \tau_n \\ z, & \text{elsewhere.} \end{cases} \tag{5.24}$$

In other words, do not transform the data in the region of instability of the estimated $\widehat{\phi}_j$. This approach is justified as long as the contribution from the tail is negligible:

$$\kappa_n(x) = \max_{j=1,2} \left\{ \sup_{t \in (x \pm \gamma)} E\left(1_{[Z > \tau_n]} |Z - \phi_j(t, Z)| \,\Big|\, X = t \right) \right\} = o(1).$$

Theorem 5.1 *Let K be a nonnegative function with compact support. Then*

$$\widehat{m}(x; \widehat{\phi}_1, \widehat{\phi}_2) - \widehat{m}(x; \phi_1, \phi_2) = O_P\{\beta_n(x) + \kappa_n(x)\},$$

provided that $h \to 0$ and $nh \to \infty$, and the design density is positive and continuous at the point x.

As a consequence of Theorems 3.1 and 5.1, we obtain that $\widehat{m}(x; \widehat{\phi}_1, \widehat{\phi}_2)$ is a consistent estimator of $m(x)$ for *any* consistent estimators $\widehat{\phi}_1$ and $\widehat{\phi}_2$. Let $\tau_F(x) = \sup_y \{y : F(y|x) < 1\}$ and define $\tau_G(x)$ similarly. If $\tau_F(u) < \tau_G(u)$ for u in a neighborhood of x, then $\kappa_n(x) = 0$ by taking $\tau_n \in (\tau_F(u), \tau_G(u))$. With the optimal choice of $h \sim n^{-1/5}$, the local average estimator (5.13) is of uniform rate $(\log n/n)^{2/5}$, i.e., $\beta_n(x) = O_P\{(\log n/n)^{2/5}\}$. Thus, the resulting estimator $\widehat{m}(x; \widehat{\phi}_1, \widehat{\phi}_2)$ has the same rate.

Next, consider transformation (5.19). If the censoring time C is independent of the covariate X, then $\beta_n(x) = O_P\{(\log n/n)^{1/2}\}$. This rate is fast enough for most nonparametric applications.

The above result deals only with a constant bandwidth h. For the nearest neighbor type bandwidth $\widehat{h}_k(x)$ in (5.22), the above result also holds. See Fan and Gijbels (1994).

5.3 Proportional hazards model

One of the most popular models in survival analysis is the proportional hazards model (5.5). There are several important reasons for this popularity. Firstly, many data collected in various scientific disciplines do not seem to depart significantly from the model. Secondly, the model itself has good physical interpretation. Thirdly, the complicated structure of censored data can easily be incorporated via the *partial likelihood* method. In this section we summarize the key ingredients of the parametric partial likelihood method and show how the local modelling idea can be used to estimate the function $\psi(x)$. For more detailed accounts for parametric partial likelihood inferences, see the books by Kalbfleisch and Prentice (1980), Cox and Oakes (1984), Fleming and Harrington (1991) and Andersen, Borgan, Gill and Keiding (1993), among others.

5.3.1 Partial likelihood

The partial likelihood, proposed by Cox (1975), is an illuminating idea for handling complicated data structures where the full likelihood function is hard to obtain. It is particularly useful for deleting nuisance parameters, leading to a simplified likelihood equation for the parameters of interest. In this subsection, we first outline the original idea of Cox (1975) and then apply it to the parametric proportional hazards model (5.5), with the function ψ being parametrized as

$$\psi(\mathbf{x}) = \mathbf{x}^T \beta, \qquad (5.25)$$

for a vector of covariates $\mathbf{x}$ and a vector of regression coefficients β. Readers wishing to see more details are recommended to read Cox's original paper.

Suppose that the observed data $\mathbf{Y}$ can be regarded as a random vector having a density $f_\mathbf{Y}(\mathbf{y}; \theta)$. Here, θ is a vector of parameters (β, γ) and the primary interest is on β with γ being nuisance parameters. In the proportional hazards model (5.5), $\beta = \psi(\cdot)$ and $\gamma = \lambda_0(\cdot)$. If further information contained in $\mathbf{Y}$ can be decomposed into two components $\mathbf{V}$ and $\mathbf{W}$ such that

$$f_\mathbf{Y}(\mathbf{y}; \theta) = f_{\mathbf{W}|\mathbf{V}}(\mathbf{w}|\mathbf{v}; \theta) f_\mathbf{V}(\mathbf{v}; \theta), \qquad (5.26)$$

with $\mathbf{v}$ and $\mathbf{w}$ being functions of $\mathbf{y}$ (ignoring the determinant of the Jacobian matrix which is independent of θ), then inferences based on the first factor (*conditional inferences*) or the second factor (*marginal inferences*) can gain in simplicity and computational efficiency. These inferences are particularly useful when one of the factors does not depend on the nuisance parameters γ – it can be used directly for the inferences about β. However, this often results in a loss of statistical efficiency since the other factor typically also contains information about β, i.e. is a function of β. Of course, one tries to decompose (5.26) such that the loss of efficiency is minimal. But only very few general guidelines are available on this aspect. One rather relies on intuition and heuristics in factorizing (5.26).

In a complicated data structure, neither factors of (5.26) are independent of the parameters γ. The partial likelihood extends the idea of the marginal and conditional inferences by permitting more flexible decompositions. Assume that the data vector $\mathbf{Y}$ is transformed into a sequence of random variables $\mathbf{V}_1, \mathbf{W}_1, \cdots, \mathbf{V}_N, \mathbf{W}_N$ with N possibly depending on $\mathbf{Y}$. Then, write the full likelihood as

$$\begin{aligned}
f_{\mathbf{Y}}(\mathbf{y};\theta) &= f_{\mathbf{V}_1,\mathbf{W}_1,\cdots,\mathbf{V}_N,\mathbf{W}_N}(\mathbf{v}_1,\mathbf{w}_1,\cdots,\mathbf{v}_N,\mathbf{w}_N;\theta) \\
&= \prod_{j=1}^N f_{\mathbf{W}_j|\mathcal{T}_j}(\mathbf{w}_j|\mathbf{v}_1,\mathbf{w}_1,\cdots,\mathbf{v}_{j-1},\mathbf{w}_{j-1},\mathbf{v}_j;\theta) \\
&\quad \times f_{\mathbf{V}_j|\mathcal{S}_j}(\mathbf{v}_j|\mathbf{v}_1,\mathbf{w}_1,\cdots,\mathbf{v}_{j-1},\mathbf{w}_{j-1};\theta) \\
&= \prod_{j=1}^N f_{\mathbf{W}_j|\mathcal{T}_j}(\mathbf{w}_j|\mathcal{T}_j;\theta) \times \prod_{j=1}^N f_{\mathbf{V}_j|\mathcal{S}_j}(\mathbf{v}_j|\mathcal{S}_j;\theta), \quad (5.27)
\end{aligned}$$

where

$$\mathcal{S}_j = (\mathbf{V}_1, \mathbf{W}_1, \cdots, \mathbf{V}_{j-1}, \mathbf{W}_{j-1}), \quad \mathcal{T}_j = (\mathcal{S}_j, \mathbf{V}_j).$$

With this flexibility of decomposition, the first factor can depend only on the parameters β. The partial likelihood uses the first part

$$\prod_{j=1}^N f_{\mathbf{W}_j|\mathcal{T}_j}(\mathbf{w}_j|\mathcal{T}_j;\theta) \qquad (5.28)$$

of the full likelihood for statistical inferences about β. Cox called this term (5.28) the *partial likelihood* on β based on $\mathbf{W}$ in the sequence $\mathbf{V}_1, \mathbf{W}_1, \cdots, \mathbf{V}_N, \mathbf{W}_N$.

A number of questions arise about the use of a partial likelihood. Firstly, decomposition (5.27) may not be unique or there may not exist a natural decomposition as in (5.27). Secondly, since the partial likelihood is not a standard likelihood, one cannot expect that the estimators based on the partial likelihood would always be consistent or asymptotically normal. Wong (1986) has shown that under certain regularity conditions, this is generally true. Thirdly, the amount of information about β contained in the second factor of (5.27) is generally unknown. One does not always know how much efficiency gets lost in partial likelihood inferences. Nevertheless, partial likelihood inferences have been very successful in dealing with the proportional hazards model (5.5).

We now derive a partial likelihood for the proportional hazards model. For simplicity, assume that we have an absolutely continuous failure time with non-time varying covariates. A natural decomposition of (5.27) is to observe events as time evolves. To this end, let $t_1^o < \cdots < t_N^o$ denote the ordered observed failure times. Let (j) provide the label for the item failing at t_j^o so that the covariates associated with the N failures are $\mathbf{X}_{(1)}, \cdots, \mathbf{X}_{(N)}$. Since the form of $\lambda_0(\cdot)$ in (5.5) is not specified, the values of the failure times provide little information about $\psi(x)$. As an extreme example, $\lambda_0(\cdot)$ can be identically zero except at a small neighborhood around t_j^o. This means that there is no risk at times other than $\{t_j^o, j = 1, \cdots, N\}$ and these times are completely determined by the function $\lambda_0(\cdot)$. Therefore, little information about $\psi(\cdot)$ is contained in the observed times t_j^o when $\lambda_0(\cdot)$ is not specified. At the time of failure however, the label (j) tells us which item fails among the remaining subjects at risk with different covariates. Hence, it provides useful information on the covariate effects, namely, on the relative risk $\exp\{\psi(x)\}$ among surviving subjects. Thus, intuitively, it is reasonable to base the inferences of $\psi(x)$ on these labels.

To write down the partial likelihood formally, let $W_j = (j)$ and

$\mathbf{V}_j = \{$History from the $(j-1)^{th}$ failure to the jth failure, and $t_j^o\}$.

More precisely, $\mathbf{V}_j$ consists of all events occurring between the time right after t_{j-1}^o and just before t_j^o, and the random variable t_j^o itself. Then, $\mathcal{T}_j$ is the history (events) from time 0 to t_j^o- and t_j^o, where t_j^o- denotes the time instantaneously before t_j^o. Let R_j denote the risk set at time t_j^o-:

$$R_j = \{i : Z_i \geq t_j^o\}. \tag{5.29}$$

The notation and the idea of the partial likelihood are illustrated in Figure 5.6.

For $\ell \in R_j$, we have

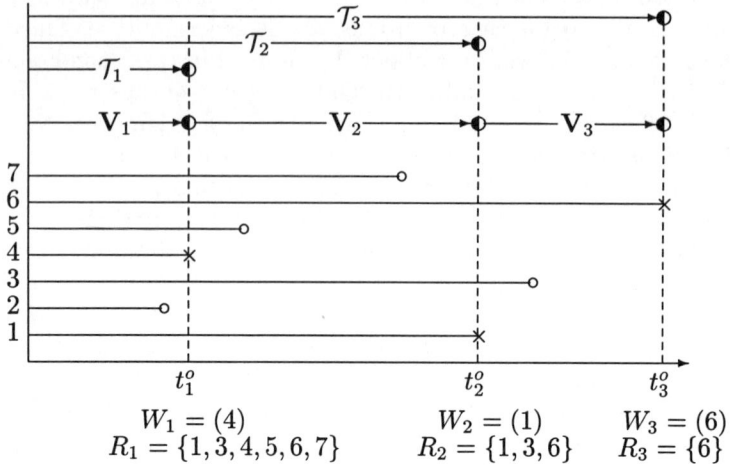

Figure 5.6. *Identifying risk sets for censored data and construction of a partial likelihood.*

$$P\{W_j = (\ell)|\mathcal{T}_j\} = \lim_{dt \downarrow 0}$$

$$\frac{P\{T_{(\ell)} \in [t_j^o, t_j^o + dt)|\mathcal{T}_j\} \prod_{i \in R_j \setminus \{(\ell)\}} P\{T_i \notin [t_j^o, t_j^o + dt)|\mathcal{T}_j\}}{\sum_{k \in R_j} P\{T_k \in [t_j^o, t_j^o + dt)|\mathcal{T}_j\} \prod_{i \in R_j \setminus \{k\}} P\{T_i \notin [t_j^o, t_j^o + dt)|\mathcal{T}_j\}},$$

where $R_j \setminus \{k\}$ is the risk set R_j with the element k deleted. Now, as $dt \to 0$, by independence and (5.3),

$$\begin{aligned} P\{T_k \in [t_j^o, t_j^o + dt)|\mathcal{T}_j\} &= P\{T_k \in [t_j^o, t_j^o + dt)|T_k \geq t_j^o, \mathbf{X}_k\} \\ &\sim \lambda(t_j^o|\mathbf{X}_k) dt \end{aligned}$$

and similarly $\lim_{dt \downarrow 0} P\{T_i \notin [t_j^o, t_j^o + dt)|\mathcal{T}_j\} = 1$. From these considerations and model (5.5), we obtain

$$\begin{aligned} P\{W_j = (\ell)|\mathcal{T}_j\} &= \frac{\lambda(t_j^o|\mathbf{X}_{(\ell)})}{\sum_{k \in R_j} \lambda(t_j^o|\mathbf{X}_k)} \\ &= \frac{\exp\{\psi(\mathbf{X}_{(\ell)})\}}{\sum_{k \in R_j} \exp\{\psi(\mathbf{X}_k)\}}. \end{aligned}$$

Consequently, the partial likelihood (5.28) for the proportional hazards model is given by

$$\prod_{j=1}^{N} \frac{\exp\{\psi(\mathbf{X}_{(j)})\}}{\sum_{k \in R_j} \exp\{\psi(\mathbf{X}_k)\}}. \tag{5.30}$$

In particular, when ψ is parametrized as in (5.25), this partial likelihood equals

$$\prod_{j=1}^{N} \frac{\exp(\mathbf{X}_{(j)}^T \beta)}{\sum_{k \in R_j} \exp(\mathbf{X}_k^T \beta)}. \tag{5.31}$$

Statistical inferences about β are based on (5.31). Maximizing the partial likelihood (5.31) results in the *maximum partial likelihood estimator*. Asymptotically, the covariance matrix of this estimator is approximately the inverse of the Hessian matrix of (5.31). With this approximated covariance structure, one can easily carry out statistical inferences such as constructing confidence intervals and testing hypotheses of parameters. Details about the parametric inferences can be found in the books by Fleming and Harrington (1991) and Andersen, Borgan, Gill and Keiding (1993).

5.3.2 Local partial likelihood

The likelihood (5.31) relies strongly on the parametric assumption (5.25), which is not always satisfied or at least needs to be diagnosed. In this section, we relax this parametric assumption, and estimate $\psi(\mathbf{x})$ directly using the local approximation method. For simplicity, we treat the univariate covariate case. The multivariate extension is straightforward, but some dimensionality reduction principles need to be incorporated in order to yield a practical method.

The univariate *hazard regression* arises when one wishes to single out a certain covariate and examine its effect on the survival risk. It also provides useful tools for model diagnostics and sensitivity analysis. Let X^0 be the variable that is singled out and denote by $\mathbf{X}^1$ the vector of remaining covariates. If the proportional hazards model (5.5) holds, and the risk contribution is multiplicative so that $\psi(\mathbf{x}) = \psi_0(x^0) + \psi_1(\mathbf{x}^1)$, then

$$\lambda(t|\mathbf{x}) = \lambda_0(t) \exp\{\psi_1(\mathbf{x}^1)\} \exp\{\psi_0(x^0)\}.$$

This can be regarded as a proportional hazards model for the univariate covariate X^0, if X^0 is not highly correlated with the remaining variables. Thus, via univariate hazard regression, one can suggest a specific form for $\psi_0(x^0)$ even in the original multivariate hazard regression problem.

Let K be a nonnegative symmetric probability density function and h be a bandwidth, controlling the size of the local neighborhood around a point x_0. Suppose that in this neighborhood, $\psi(x)$ can be approximated by

$$\psi(x) \approx \beta_0 + \cdots + \beta_p(x - x_0)^p. \tag{5.32}$$

Since no parametric form for $\psi(\cdot)$ is specified, a remote datum point contains no information about $\psi(x_0)$. Thus, it suffices to consider the data points in the neighborhood around x_0. From the statistical modelling point of view, $\psi(\cdot)$ is modelled *locally* by (5.32). From (5.30) the following *local partial log-likelihood* is obtained:

$$\sum_{j=1}^{N} I\left(\frac{|X_{(j)} - x_0|}{h} \leq 1\right)\left(\psi(X_{(j)})\right.$$
$$\left. - \log\left[\sum_{k \in R_j} \exp\{\psi(X_k)\} I\left(\frac{|X_k - x_0|}{h} \leq 1\right)\right]\right).$$

The above local log-likelihood assumes that each datum point around the point x_0 is equally influential. An improved idea is to weigh down smoothly the contributions, leading to

$$\sum_{j=1}^{N} K_h(X_{(j)} - x_0)\left(\psi(X_{(j)}) - \log\left[\sum_{k \in R_j} \exp\{\psi(X_k)\} K_h(X_k - x_0)\right]\right),$$

where $K_h(\cdot) = K(\cdot/h)/h$. Note that the function $\psi(x)$ is only identifiable up to a constant term: one can multiply with and divide by a constant factor in the proportional hazards model (5.5). Using the local model (5.32), we obtain the local log-likelihood

$$\sum_{j=1}^{N} K_h(X_{(j)} - x_0)\left[\mathbf{X}_{(j)}^T \beta - \log\left\{\sum_{k \in R_j} \exp(\mathbf{X}_k^T \beta) K_h(X_k - x_0)\right\}\right], \tag{5.33}$$

where

$$\beta = (\beta_1, \cdots, \beta_p)^T \text{ and } \mathbf{X}_j = \{(X_j - x_0), \cdots, (X_j - x_0)^p\}^T.$$

The function value $\psi(x_0)$ is not directly estimable (see also the remark above): (5.33) does not depend on β_0, the intercept. Let $\widehat{\beta}(x_0)$ be the maximum local log-likelihood estimate, which maximizes (5.33). The approximation (5.32) suggests that

$$\widehat{\psi}^{(\nu)}(x_0) = \nu! \widehat{\beta}_\nu(x_0), \tag{5.34}$$

i.e. we propose to estimate the νth derivative of $\psi(x)$ in the point x_0 by $\nu! \widehat{\beta}_\nu(x_0)$. See also Section 2.3.

To ensure identifiability of $\psi(x)$, we impose the condition $\psi(0) = 0$. With this extra constraint, the function $\psi(x) = \int_0^x \psi'(t)dt$ can be estimated by

$$\widehat{\psi}(x) = \int_0^x \widehat{\psi}'(t) dt. \tag{5.35}$$

In practical implementation, one often evaluates the function $\psi'(x)$ either at grid points or at the design points. Assume that the maximizer of the local log-likelihood (5.33) is computed at points $x_0, \cdots, x_m$, namely $\widehat{\psi}'(x_j) = \widehat{\beta}_1(x_j) \equiv \widehat{\beta}_{1,j}$. Then, $\widehat{\psi}(x_i)$ can be approximated by the trapezoidal rule

$$\widehat{\psi}(x_i) = \sum_{j=1}^{i} (x_j - x_{j-1})(\widehat{\beta}_{1,j} + \widehat{\beta}_{1,j-1})/2.$$

A natural question to ask is which value of p should be chosen in the local model (5.32). If the primary interest is to estimate $\psi'(x)$, then as explained in Section 3.3, p should be 2 or occasionally 4. If the primary interest is to estimate $\psi(x)$ via (5.35), we still recommend using $p = 2$. Should we use $p = 1$, this would create large bias (e.g. at the boundary) for derivative estimation. The integration in (5.35) would usually not reduce the order of the bias.

At this point, one may wonder why we do not estimate $\exp\{\psi(x)\}$ directly. There are many reasons supporting the log-transformed function. Firstly, the range of $\psi(x)$ is $(-\infty, +\infty)$. Thus, the estimate $\widehat{\psi}(\cdot)$ is range-preserving. Secondly, the local log-likelihood (5.33) is concave in β. Consequently, computing $\widehat{\beta}(x)$ is much easier, requiring only a few iterations, and we can always find the global maximizer. Thirdly, when h is large, the model reduces to the usual parametric model – this provides parsimonious models for the parametric polynomial models.

As an illustration of the local modelling idea, we revisit the Stanford heart transplant data presented in Figure 5.5. Depicted in Figure 5.7 is the local partial likelihood estimation using bandwidths $h = 10$, $h = 20$ and $h = 30$. Note that as h gets larger, the resulting estimated curve gets closer to a parametric quadratic fit (see

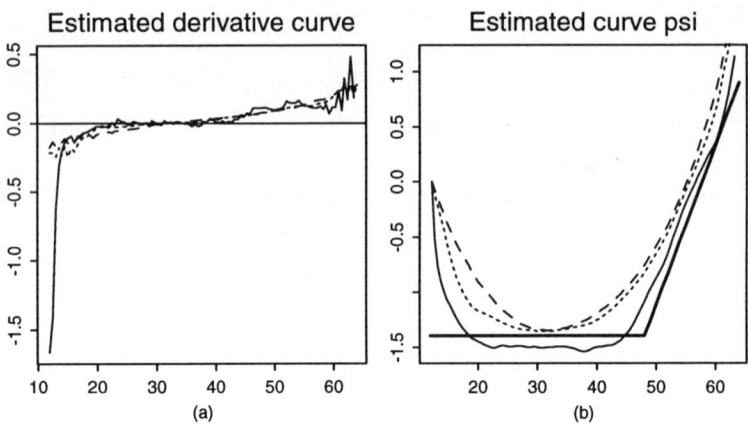

Figure 5.7. *Local partial likelihood estimation for Stanford heart transplant data. (a) Estimated derivative curves by using local partial likelihood with $p = 2$. (b) Estimated curve $\psi(\cdot)$; thin solid curve – bandwidth $h = 10$; short-dashed curve – bandwidth $h = 20$; long-dashed curve – bandwidth $h = 30$; thick solid curve – result from a parametric analysis inspired by the nonparametric analyses.*

Figure 5.7 (b)). Figure 5.7 (b) clearly indicates the possible nonlinearity of the estimated $\psi(\cdot)$. We tried to decrease the bandwidth to $h = 5$, but had problems with convergence of the local partial likelihood. The thin solid curve in Figure 5.7 suggests that we use the following thresholding model:

$$\psi(x) = \theta_1 (20 - x)_+ + \theta_2 (x - 48)_+.$$

Fitting this parametric model, we obtained the estimated parameters $\widehat{\theta}_1 = 0.0397$ and $\widehat{\theta}_2 = 0.1455$ with estimated standard errors respectively 0.1049 and 0.0306. Thus, the first term is not statistically significant and is deleted from the model. Refitting the parametric model after the deletion, we obtain the following model

$$\psi(x) = \theta(x - 48)_+ \quad \widehat{\theta} = 0.144, \quad \widehat{\text{SE}}(\widehat{\theta}) = 0.0305.$$

This parametric estimation (reduced by -1.4 for comparison purposes since $\psi(\cdot)$ is identifiable within a constant) is superimposed in Figure 5.7 (b) (thick solid curve). The conclusion obtained here is very analogous to that in Section 5.2.3: for earlier ages, the risk of heart transplant is nearly constant and at later ages it increases (exponentially) linearly with age.

This is another example where a combination of parametric and nonparametric methods can yield sensible statistical analyses. The role of nonparametric analysis here is to suggest a parametric model and to diagnose the parametric analysis.

5.3.3 Determining model complexity

As explained in Chapters 3 and 4, the bandwidth h controls effectively the model complexity of a nonparametric method. When $h = +\infty$, the local log-likelihood (5.33) becomes the global log-likelihood. Thus, the resulting estimator corresponds to a parametric polynomial fit. With a data-driven bandwidth the data themselves determine the complexity of the nonparametric model.

There are three possible methods for selecting a bandwidth in this local log-likelihood context: the cross-validation approach, the plug-in method, and the principles introduced in Section 4.9. The cross-validation method is simple, but is very slow to compute and has slow rate of convergence in terms of statistical efficiency. The plug-in idea is also simple and efficient, but relies on asymptotic formulas and pilot estimates. From our experiences, in particular those in the least squares case, it appears that the principles introduced in Section 4.9 are efficient. They can be applied directly to the current setup and only need some programming effort. Here we describe a simple rule of thumb for the current situation, which can be used directly in many applications.

It can be shown that under certain regularity conditions and for a symmetric kernel K, the estimator $\widehat{\psi}'(x)$ with $p = 2$ and $\nu = 1$ is asymptotically normally distributed:

$$\sqrt{nh^3}\left\{\widehat{\psi}'(x) - \psi'(x) - \frac{\int t^3 K_1^*(t)dt}{6}\psi^{(3)}(x)h^2\right\}$$
$$\xrightarrow{D} N\left\{0, \int K_1^{*2}(t)dt\ \sigma^2(x)/f(x)\right\}, \tag{5.36}$$

where $K_1^*(t) = tK(t)/\mu_2$ is the equivalent kernel for the derivative estimation as indicated in Table 3.1 and

$$\sigma^{-2}(x) = \exp\{\psi(x)\}\int_0^\infty \lambda_0(t)P\{Z \geq t|X = x\}dt. \tag{5.37}$$

See Fan, Gijbels and King (1995). The asymptotic bias and variance exhibit a similar form to that of the local polynomial fitting in Section 3.2 (see (3.18) and (3.19)). Thus, from (3.21) it is easily seen that the theoretical optimal bandwidth, minimizing the asymptotic weighted MISE, is given by

$$h_{opt} = C_{1,2}(K) \left[\frac{\int \sigma^2(x)w(x)/f(x)dx}{\int \{\psi^{(3)}(x)\}^2 w(x)dx} \right]^{1/7} n^{-1/7}, \qquad (5.38)$$

where $C_{1,2}(K)$ is as in (3.20) and is tabulated in Table 3.2.

The simplest way to implement (5.38) is to use the rule of thumb, introduced in Section 4.2. Let $[a,b]$ be the interval where $\widehat{\psi}'(x)$ is to be evaluated. We fit a global polynomial model (5.25) of order 5 and denote the resulting fit as

$$\check{\psi}(x) = \check{\alpha}_1 x + \cdots + \check{\alpha}_5 x^5.$$

Denote by $\check{\Lambda}_0(t)$ the Breslow (1972, 1974) estimator for the cumulative baseline function $\Lambda_0(t) = \int_0^t \lambda_0(s)ds$ defined as

$$\check{\Lambda}_0(t) = \sum_{j=1}^N I(t_j^0 \leq t)[\sum_{k \in R_j} \exp\{\check{\psi}(X_k)\}]^{-1}.$$

Regarding the conditional probability in (5.37) as the unconditional one, the rule of thumb bandwidth selector given in Section 4.2 translates into the current setting as follows:

$$\check{h}_{ROT} = C_{1,2}(K) \left[\frac{\int_a^b \exp\{\check{\psi}(x)\}dx \check{R}}{n \sum_{i=1}^n \{\check{m}^{(3)}(X_i)\}^2 I_{[a,b]}(X_i)} \right]^{1/7}, \qquad (5.39)$$

where

$$\check{R} = \int \sum_{i=1}^n I(Z_i \geq t) d\check{\Lambda}_0(t)$$

$$= \sum_{j=1}^N \#(R_j)[\sum_{k \in R_j} \exp\{\check{\psi}(X_k)\}]^{-1}.$$

Equation (5.39) gives an idea of how large the bandwidth should be when one is interested in estimating the derivative curve using a local quadratic fit. When the primary interest is in estimating $\psi(x)$, an undersmooth form of $\widehat{\psi}'(x)$ is needed since integration (5.35) makes $\widehat{\psi}(x)$ smoother. In other words, a smaller bandwidth than $\check{h}_{ROT}$ is more desirable.

We conclude this section with an analysis of the Primary Biliary Cirrhosis (PBC) data set. Between January 1974 and May 1984, the Mayo Clinic collected data on PBC, a rare but fatal chronic liver disease of unknown cause. Of the 424 registered patients, 312

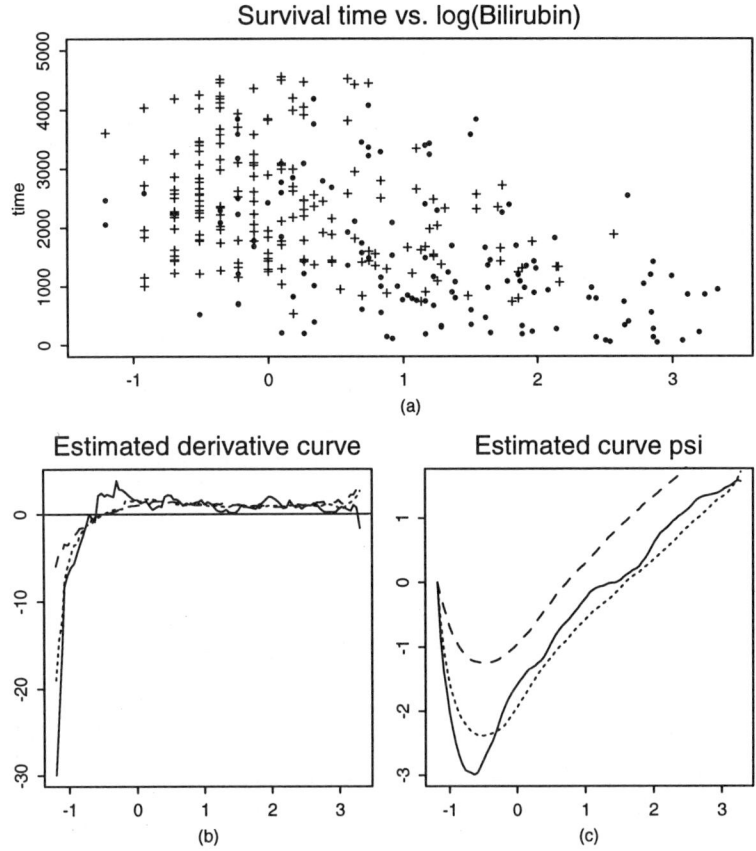

Figure 5.8. *The Cox regression of the survival time on log(Bilirubin) for the primary biliary cirrhosis data. (a) Observed times versus log(Bilirubin) with '+' indicating censored observations. (b) Estimated derivative curve $\psi'(x)$ by using local partial log-likelihood with $p = 2$. (c) Estimated curve $\psi(x)$; solid curve – bandwidth $h = 0.6$; short-dashed curve – bandwidth $h = 1.0$; long-dashed curve – bandwidth $h = 1.4$.*

participated in the randomized trial, and our analysis is based on those patients. Of the 312 patients, 187 cases were censored. For a detailed description and analysis of this data set, see page 2 and Chapter 4 of Fleming and Harrington (1991). In our analysis we take the time (in days) between registration and death, or liver transplantation or time of the study analysis (July 1986) as response and logarithm of Serum Bilirubin (in mg/dl) as the co-

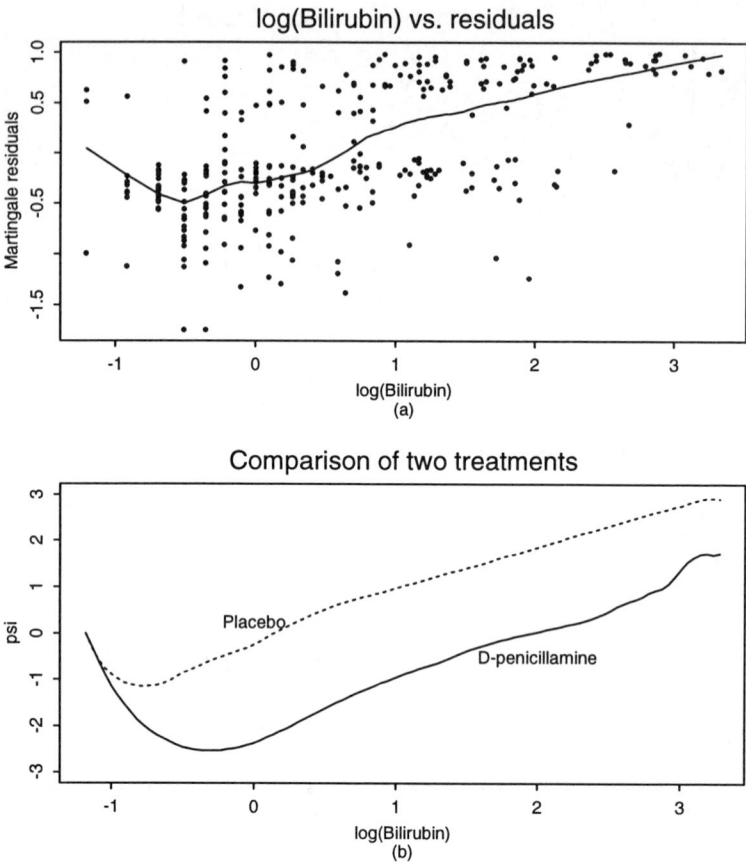

Figure 5.9. *A study for the primary biliary cirrhosis data. (a) Scatter plot of the log(Bilirubin) versus the martingale residuals with log(Bilirubin) omitted. (b) Comparisons of two treatments. The function $\psi(\cdot)$ is plotted against log(Bilirubin). Solid curve – D-penicillamine; dashed curve – Placebo.*

variate. We use the local partial log-likelihood method (5.33) with $p = 2$ for various bandwidths h. The Epanechnikov kernel is employed in all cases. Figure 5.8 depicts the results. Figures 5.8 (b) and (c) indicate nonlinearity when the Bilirubin level is small. Figure 5.8 (c) also reveals that a piecewise linear model or a quadratic model would be reasonable for this data set.

Another method for suggesting a functional form of a covariate in the Cox proportional hazards model (5.5) with $\psi(\mathbf{x}) = \mathbf{x}^T\beta$ is to omit first the covariate and to fit the resulting parametric model, and then to smooth the resulting *martingale residuals*. See Fleming and Harrington (1991), pages 165–168. To illustrate this alternative method, we use treatment codes (1 = D-penicillamine, 2 = placebo), age, gender, and logarithm of Bilirubin as covariates. The martingale residuals with the covariate log(Bilirubin) omitted are plotted against the logarithm of Bilirubin. The result is shown in Figure 5.9 (a). The smoothed curve is obtained by using the LOWESS smoother with d=1/4, which shows a similar pattern to Figure 5.8 (c).

To examine the effect of treatment, a reasonable attempt is to fit the parametric proportional hazards model. The treatment effect is, however, not statistically significant (P-value = 0.711). One possible reason for this is that the proportional hazards model (5.5) does not hold, as discussed after (5.5). To investigate the effect of the treatment, we divided the data into two groups according to the treatment code. For each group, we fitted the model (5.5) using log(Bilirubin) as a covariate and bandwidth $h = 1.2$. The resulting curves are depicted in Figure 5.9 (b). Recall that $\psi(\cdot)$ is only identifiable within a constant. The shift of the curves can possibly be due to this effect. However, since the trial was randomized, it is reasonable to assume that both groups have the same baseline hazards. Therefore, the treatment effect (reduces the risk) is real and cannot be explained as the identifiability problem of the function $\psi(\cdot)$.

5.3.4 Complete likelihood

The aim of this section is to derive the partial likelihood and the Breslow estimator from the complete likelihood perspective, under independent and noninformative censoring. This helps one to understand the partial likelihood and why it is efficient under the semiparametric model (5.5) with $\psi(\cdot)$ being parametrized.

The conditional likelihood for obtaining the ith datum point $(\mathbf{X}_i, Z_i, \delta_i)$ (see (5.1)) is

$$\lim_{\varepsilon \to 0} P\{Z \in Z_i \pm \varepsilon, \delta = \delta_i | \mathbf{X} = \mathbf{X}_i\}/(2\varepsilon).$$

When $\delta_i = 1$, the above conditional probability is given by

$$\lim_{\varepsilon \to 0} P\{T \in Z_i \pm \varepsilon, C \geq Z_i | \mathbf{X} = \mathbf{X}_i\}/(2\varepsilon).$$

Using the independent censoring assumption, the above limit is $f(Z_i|\mathbf{X}_i)\bar{G}(Z_i|\mathbf{X}_i)$, where $f(\cdot|\cdot)$ denotes the conditional probability of T given $\mathbf{X}$ and $\bar{G} = 1 - G$ with $G(\cdot|\cdot)$ being the conditional distribution of C given $\mathbf{X}$. Using the same argument, we can obtain the conditional likelihood for $\delta_i = 0$. Combining these two expressions, we find that the conditional likelihood for the ith datum point is

$$f(Z_i|\mathbf{X}_i)^{\delta_i}\bar{F}(Z_i|\mathbf{X}_i)^{1-\delta_i}g(Z_i|\mathbf{X}_i)^{1-\delta_i}\bar{G}(Z_i|\mathbf{X}_i)^{\delta_i}, \qquad (5.40)$$

where $g(\cdot|\cdot)$ is the conditional density of C given $\mathbf{X}$ and $\bar{F}$ is the conditional survivor function. Under the noninformative censoring assumption, the last two factors do not depend on the parameters of interest. Therefore, we can ignore these last two factors, and obtain the conditional likelihood

$$f(Z_i|\mathbf{X}_i)^{\delta_i}\bar{F}(Z_i|\mathbf{X}_i)^{1-\delta_i}. \qquad (5.41)$$

Under the proportional hazard models (5.5), this likelihood is

$$\begin{aligned} \text{lik} &= \prod_u f(Z_i|\mathbf{X}_i) \prod_c \bar{F}(Z_i|\mathbf{X}_i) \\ &= \prod_u \lambda(Z_i|\mathbf{X}_i) \prod_{i=1}^n \bar{F}(Z_i|\mathbf{X}_i) \\ &= \prod_{i=1}^N \lambda_0(Z_{(i)})\exp\{\psi(\mathbf{X}_{(i)})\} \prod_{i=1}^n \exp[-\Lambda_0(Z_i)\exp\{\psi(\mathbf{X}_i)\}], \end{aligned} \qquad (5.42)$$

where the subscripts u and c denote the product on the censored and uncensored data respectively, and $\Lambda_0(\cdot)$ is the cumulative baseline hazard function.

Now consider the 'least informative' (see Section 5.3.1) nonparametric modelling for $\Lambda_0(\cdot)$, i.e. $\Lambda_0(t)$ has a possible jump λ_j only at the observed failure time t_j^0. More precisely,

$$\Lambda_0(t) = \sum_{j=1}^N \lambda_j I(t_j^0 \le t). \qquad (5.43)$$

Then,

$$\Lambda_0(Z_i) = \sum_{j=1}^N \lambda_j I(i \in R_j).$$

Using this and (5.42), we have

$$\log(\text{lik}) = \sum_{j=1}^{N}\{\log(\lambda_j) + \psi(\mathbf{X}_{(j)})\}$$
$$- \sum_{i=1}^{n}[\sum_{j=1}^{N}\lambda_j I(i \in R_j)\exp\{\psi(\mathbf{X}_i)\}]. \quad (5.44)$$

Taking the derivative with respect to λ_j and setting it to zero, we obtain

$$\widehat{\lambda}_j = [\sum_{i \in R_j}\exp\{\psi(\mathbf{X}_i)\}]^{-1}. \quad (5.45)$$

Substituting this into (5.44), we obtain

$$\max_\lambda\{\log(\text{lik})\} = \sum_{j=1}^{N}\left\{\psi(\mathbf{X}_{(j)}) - \log[\sum_{i \in R_j}\exp\{\psi(\mathbf{X}_i)\}]\right\} - N. \quad (5.46)$$

Maximizing (5.46) with respect to the parameters in $\psi(\cdot)$ yields the nonparametric maximum likelihood estimate. Note that (5.46) corresponds to the logarithm of the partial likelihood function (5.30). Therefore, the partial likelihood arises naturally from this likelihood derivation. The Breslow estimator is also obtained by this likelihood method by combining (5.43) and (5.45).

5.4 Generalized linear models

In many applications, the response variable can be either discrete or continuous, and the conditional variance can in one application depend on the mean regression function while in another application it does not. Different situations require a different model for the stochastic component, i.e. for the conditional distribution. A popular family of distributions is the *exponential family*, which includes commonly used distributions such as Gaussian, binomial, Poisson, and gamma distributions. Under these conditional distributions, the mean regression function is often assumed to be linear via a monotonic transformation. These models are called *generalized linear models*. A comprehensive account of generalized linear models can be found in McCullagh and Nelder (1989). In this section, we show how the mean regression function can be estimated without the restrictive linear assumption. The form of the function will be determined by the data.

5.4.1 Exponential family models

Let $(\mathbf{X}_1, Y_1), \cdots, (\mathbf{X}_n, Y_n)$ be a random sample from a certain population, where Y_i is a scalar response variable and $\mathbf{X}_i$ is the associated covariate vector. Assume that the conditional density of Y given $\mathbf{X} = \mathbf{x}$ belongs to a *canonical exponential family*

$$f(y|\mathbf{x}) = \exp([\theta(\mathbf{x})y - b\{\theta(\mathbf{x})\}]/a(\phi) + c(y, \phi)), \qquad (5.47)$$

for some known functions $a(\cdot)$, $b(\cdot)$ and $c(\cdot, \cdot)$. The parameter $\theta(\cdot)$ is called the *canonical parameter* and ϕ is called the *dispersion parameter*. Under this model, using the Bartlett (1954) identities (4.26), it can be shown easily that

$$m(x) = E(Y|\mathbf{X} = \mathbf{x}) = b'\{\theta(\mathbf{x})\},$$

$$\text{Var}(Y|\mathbf{X} = \mathbf{x}) = a(\phi)b''\{\theta(\mathbf{x})\}. \qquad (5.48)$$

In parametric generalized linear models, the unknown regression function $m(\mathbf{x})$ is modelled linearly via a known link function $g(\cdot)$:

$$g\{m(\mathbf{x})\} = \mathbf{x}^T \beta. \qquad (5.49)$$

The function g links the regression function to a linear space of the covariates. If $g = (b')^{-1}$, then g is called the *canonical link function* since in that case $g\{m(\mathbf{x})\}$ is the canonical parameter in the exponential family (5.47). Expressions (5.48) and (5.49) characterize the generalized linear models. In these models interest centers around statistical inferences about the parameter vector β, based on the maximum likelihood method.

It is clear that the dispersion function $a(\phi)$ does not affect the maximum likelihood estimate of the regression function $m(\cdot)$, when ϕ is independent of $m(\cdot)$. However, it affects the accuracy (e.g. the standard error) of the estimated $m(\cdot)$. An analogous situation appears in the least squares linear regression model – the noise level does not alter the parameter estimation derived from the least squares criterion, but affects the standard errors of the estimated parameters.

In some situations, the linear relationship (5.49) is not granted. Trial and error are required in order to search for a reasonable parametric link function. Instead of following this approach, we estimate $m(\mathbf{x})$ directly from the data without assuming a specific form. Since estimating $m(\mathbf{x})$ is equivalent to estimating $\theta(x) = (b')^{-1}\{m(\mathbf{x})\}$, we will focus on estimating $\theta(\mathbf{x})$. The advantages of estimating $\theta(\mathbf{x})$ are similar to those indicated in Section 5.3.2 concerning the estimation of $\psi(x)$: the estimators are range-preserving;

GENERALIZED LINEAR MODELS

the local log-likelihood is concave; and the local polynomial models provide parsimonious models for the parametric generalized linear models. In the parametric generalized linear models, different link functions correspond to different function forms of the mean regression function so that the choice of the link function is important. In contrast, for the nonparametric approach the link functions do not alter the model of the regression function so that it is sufficient to consider the canonical link functions.

We now give a few examples to illustrate model (5.47).

Example 1 If the conditional distribution of Y given $\mathbf{X} = \mathbf{x}$ is $N\{m(\mathbf{x}), \sigma^2\}$, then by writing the normal density as

$$\exp\left\{\frac{m(\mathbf{x})y - m^2(\mathbf{x})/2}{\sigma^2} - \frac{y^2}{2\sigma^2} - \log(\sqrt{2\pi}\sigma)\right\},$$

we can easily see that

$$\phi = \sigma^2, \ a(\phi) = \phi, \ b(m) = m^2/2,$$

and

$$c(y, \phi) = -y^2/(2\sigma^2) - \log(\sqrt{2\pi}\sigma).$$

This model is useful for continuous response with homoscedastic errors. The canonical link function is the identity link $g(t) = t$.

Example 2 Suppose that the conditional distribution is binomial $\{n_0, p(\mathbf{x})\}$. Then, its conditional density (with respect to the counting measure) can be written as

$$\binom{n_0}{y} p(\mathbf{x})^y \{1 - p(\mathbf{x})\}^{n_0-y} = \binom{n_0}{y} \exp\left[y\theta(\mathbf{x}) - b\{\theta(\mathbf{x})\}\right],$$

where $\theta(\mathbf{x}) = \log\frac{p(\mathbf{x})}{1-p(\mathbf{x})}$ is the *logit transformation* of $p(\mathbf{x})$ and $b(\theta) = n_0 \log\{1 + \exp(\theta)\}$. The canonical link is the logit type of transformation $g(t) = \log\{t/(n_0 - t)\}$. This model is commonly used for situations with a binary dependent variable.

Example 3 If the conditional distribution is Poisson$\{\lambda(\mathbf{x})\}$, then the conditional density can be written as

$$\exp\left\{y \log \lambda(\mathbf{x}) - \lambda(\mathbf{x}) - \log(y!)\right\}.$$

In this case, the canonical parameter is $\theta(\mathbf{x}) = \log \lambda(\mathbf{x})$ and $b(\theta) = \exp(\theta)$. The canonical link function is $g(t) = \log(t)$. This model is useful for situations where the response variable is a counting variable (e.g. number of car accidents per month) with the mean and variance approximately the same. When θ is modelled linearly, the resulting model is called the *log-linear model*.

Example 4 Assume that the conditional distribution is the gamma distribution with mean $\mu(\mathbf{x})$ and shape parameter α. Then, its conditional density admits the form

$$\exp\left\{\alpha(-y/\mu - \log \mu) + (\alpha - 1)\log y + \alpha \log \alpha - \log \Gamma(\alpha)\right\}.$$

For this case, $\phi = \alpha$, $\theta(\mathbf{x}) = -1/\mu(\mathbf{x})$, $a(\phi) = 1/\alpha$, $b(\theta) = -\log(-\theta)$. This model is useful for continuous responses with constant *coefficient of variation*.

For the generalized linear models (5.47) and (5.49), statistical inferences are often based on the maximum likelihood approach. Given n independent data, the log-likelihood can be written as

$$\sum_{i=1}^{n}[\{\theta_i Y_i - b(\theta_i)\}/a(\phi) + c(Y_i, \phi)],$$

where $\theta_i = \theta(\mathbf{X}_i)$. Without any restriction on the form of the function $\theta(\cdot)$, we would obtain the exact fit – the mean of the ith datum point is the same as the observed y variate, namely, the likelihood is maximized at $\widetilde{\theta}_i = (b')^{-1}(Y_i)$ by (5.48). The log-likelihood under this full model is

$$\sum_{i=1}^{n}[\{Y_i\widetilde{\theta}_i - b(\widetilde{\theta}_i)\}/a(\phi) + c(Y_i, \phi)].$$

Let $\widehat{\theta}_i$ be the maximum likelihood estimate under model (5.49). A natural measure for the discrepancy of $\widehat{\theta}_i$ is the difference in the maximum log-likelihood between the full model and model (5.49), namely $D(\mathbf{y}; \widehat{\mathbf{m}})/\{2a(\phi)\}$ with

$$D(\mathbf{y}; \widehat{\mathbf{m}}) = 2\sum_{i=1}^{n}[Y_i(\widetilde{\theta}_i - \widehat{\theta}_i) - \{b(\widetilde{\theta}_i) - b(\widehat{\theta}_i)\}]. \qquad (5.50)$$

The quantity $D(\mathbf{y}; \widehat{\mathbf{m}})$ is called the *deviance*, and plays a similar role to the *residual sum of squares* in the least squares model. The deviance has already been normalized so that it coincides with the residual sum of squares under normal models.

In the case of homoscedastic regression models, we can express the dependent variate as $y_i = \widehat{m}_i + \widehat{r}_i$, i.e., observed value = fitted value + residual. Residuals can be used to explore the adequacy of fit of a model, and they can also indicate the presence of anomalous values (e.g. outliers) requiring further investigation. For generalized linear models, we can analogously define the residuals. The contribution of the ith observation to the deviance (5.50) is

$$d_i^2 = 2[Y_i(\widetilde{\theta}_i - \widehat{\theta}_i) - \{b(\widetilde{\theta}_i) - b(\widehat{\theta}_i)\}],$$

which is a kind of squared residual. The residual for the fitted value $\widehat{m}_i$ can be defined as

$$r_{D,i} = \text{sgn}(y_i - \widehat{m}_i)d_i, \qquad (5.51)$$

which is referred to as the *deviance residual*. Note that the deviance residual reduces to the usual residual under normal models. It can be shown that the deviance residual $r_{D,i}$ increases when $y_i - \widehat{m}_i$ increases so that it measures the departure of the fitted value from the observed value.

The deviance residuals play a similar role as the residuals under normal models. They provide useful tools for model diagnostics. For example, one can plot $\widehat{m}$ against r_D, or $|r_D|$ and examine whether there are any trends in the plots. The first plot is often used to assess whether the mean regression is reasonably modelled. The second plot is often used to check possible over-dispersion. For example, an increasing trend in the plot of $\widehat{m}$ against $|r_D|$ indicates that the conditional variance function $V(\mu) = \text{Var}(Y|\mathbf{X} = \mathbf{x})$ is not properly modelled and suggests using $V(\mu) = \mu^2$ or some other power if the current fit assumes $V(\mu) = \mu$.

While we use the maximum likelihood estimation for the parametric model (5.49) as a backdrop, the concepts of deviance and deviance residuals apply to any estimation scheme.

5.4.2 Quasi-likelihood and deviance residuals

In many applications, one has only an idea about the conditional variance function, which can be modelled reasonably as

$$\text{Var}(Y|\mathbf{X} = \mathbf{x}) = V\{m(\mathbf{x})\},$$

for some specific function $V(\cdot)$. In this case, one would naturally try to use the weighted least squares method, but this method can be inconsistent. Instead, one replaces the conditional log-likelihood by a quasi-likelihood function $Q\{m(\mathbf{x}), y\}$ which satisfies

$$\frac{\partial}{\partial \mu} Q(\mu, y) = \frac{y - \mu}{V(\mu)}. \qquad (5.52)$$

The quasi-score (5.52) possesses properties similar to those of the usual log-likelihood score function. That is, it satisfies Bartlett's identities of the first and second order stated in (4.26). Quasi-likelihood methods behave analogously to the usual likelihood methods and thus are reasonable substitutes when the likelihood function is not available.

The log-likelihood of the one-parameter exponential family is a special case of a quasi-likelihood function with $V(\cdot) = a(\phi)b'' \circ (b')^{-1}(\cdot)$. In other words, when the conditional variance is specified as that of model (5.47), the quasi-likelihood method coincides with the log-likelihood method of model (5.47). For example for the model with a homogeneous variance $V(\cdot) = \sigma^2$, $Q(\mu, y) = (y - \mu)^2/(2\sigma^2)$. In this case, the quasi-likelihood method is equivalent to the normal log-likelihood based approach – both are the least squares method. Thus, the normal log-likelihood approach is applicable to all conditional distributions with a constant variance function. Similarly, when $V(\mu) = (1-\mu)\mu$, the quasi-likelihood method coincides with that of the Bernoulli log-likelihood. This indicates that the Bernoulli log-likelihood approach is applicable to all distributions with conditional variance function $V(\mu) = (1-\mu)\mu$. In summary, the quasi-likelihood method extends the scope of the applicability of model (5.47).

5.4.3 Local quasi-likelihood

Let $(X_1, Y_1), \cdots, (X_n, Y_n)$ be an i.i.d. sample with mean function $m(\cdot)$ and variance function $V\{m(\cdot)\}$ for a given function $V(\cdot)$. Of interest is to estimate $\eta(\mathbf{x}) = g\{m(\mathbf{x})\}$ or its derivative $\eta^{(\nu)}(\cdot)$, for some link function g. For simplicity of notation, we focus only on the univariate covariate case. Approximate $\eta(\cdot)$ locally by a polynomial function

$$\eta(z) \approx \beta_0 + \cdots + \beta_p(z-x)^p,$$

where z is in a neighborhood of x. From a statistical modelling point of view, we locally model the unknown function η by a polynomial. Using the local data around x, we obtain the local quasi-likelihood

$$\sum_{i=1}^{n} Q[g^{-1}\{\beta_0 + \ldots + \beta_p(X_i - x)^p\}, Y_i] K_h(X_i - x). \quad (5.53)$$

GENERALIZED LINEAR MODELS 195

Maximizing (5.53) with respect to the parameters $\beta_0, \cdots, \beta_p$ leads to the *maximum local quasi-likelihood estimator*

$$\widehat{\eta}_\nu(x) = \nu! \widehat{\beta}_\nu(x), \qquad (5.54)$$

for $\eta^{(\nu)}(x), \nu = 0, \cdots, p$. As discussed in Section 3.3, $p - \nu$ should preferably be odd, and often one takes $p = \nu + 1$ or occasionally $p = \nu + 3$. The mean regression function can be estimated via the inverse of the link function: $\widehat{m}(x) = g^{-1}\{\widehat{\eta}(x)\}$.

For the generalized linear model (5.47), $V(\cdot)$ is given by (5.48). In this case, the local quasi-likelihood reduces to the local log-likelihood. As argued in Section 5.4.1, the canonical link function should be employed in the current nonparametric setting. The bandwidth h controls effectively the model complexity: a very large h results in virtually a global polynomial fit and a very small h produces essentially interpolation of the data.

The quasi-likelihood (5.52) does not admit an explicit solution to the maximization of (5.53) unless $p = 0$. The maximization is carried out via the *Newton-Raphson method* or its variation – *Fisher's scoring method*. A few iterations are typically enough since the local quasi-likelihood (5.53) is concave in the β's with an appropriate link function and p is typically small. For the case that $p = 0$, we have

$$\widehat{\eta}_0(x) = \widehat{\eta}(x) = g\{\sum_{i=1}^n K_h(X_i - x) Y_i / \sum_{i=1}^n K_h(X_i - x)\},$$

which is the inverse of the Nadaraya-Watson estimator (2.4). This estimator is somewhat crude, but provides an intuitive understanding of the local quasi-likelihood estimator.

We now illustrate the local quasi-likelihood method. The scatter plot in Figure 5.10 (right panel) presents the proportions of surviving babies for 21 different birth weights, rounded to the nearest half kilogram. The data were taken from Karn and Penrose (1951). The probability of survival for a given birth weight x is estimated by a local linear fit. The left panel of Figure 5.10 shows how this estimate was constructed. The dashed curves show how the estimate was obtained at two particular values of x. At $x = 6$ a straight line was fitted to the weighted quasi-likelihood, which is the binomial log-likelihood in this case. Here, the weights correspond to the relative heights of the kernel function which, in this case, is a scaled normal density function centered around $x = 6$ and is shown at the base of the plot. The estimate at this point is the height of the line above $x = 6$. For estimation at the second point, $x = 10$, the

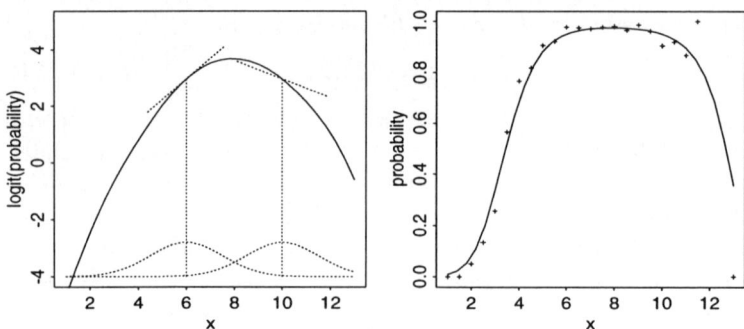

Figure 5.10. *Estimating surviving probability curve via a local linear fit. The left panel shows the estimated logit of the probability function and the right panel the estimated probability curve. Adapted from Fan, Heckman and Wand (1995).*

same principle is applied with the kernel function centered around $x = 10$. This process is repeated at all points x at which an estimator is required. The estimate of the probability itself is obtained by applying the inverse link function, in this case the inverse logit transformation, to the plot shown in the left panel of Figure 5.10.

5.4.4 Bias and variance

Further insight into the local quasi-likelihood method is provided by an asymptotic study. The following result establishes bias and variance of the local quasi-likelihood estimators. The proof of the result is outlined in Section 5.6. Let K be a symmetric and nonnegative function with compact support and let K_ν^* be its equivalent kernel function defined by (3.16). Denote by $f(x)$ the marginal density of X.

Theorem 5.2 *Assume that $p - \nu$ is odd. Suppose that Conditions (1)–(4) in Section 5.6 hold. Then, for x an interior point of the design density, we have*

$$\sigma_\nu^{-1}(x)\{\widehat{\eta}_\nu(x) - \eta^{(\nu)}(x) - b_\nu(x) + o(h^{p+1-\nu})\} \xrightarrow{D} N(0,1),$$

provided that $h \to 0$ and $nh \to \infty$, where with K_ν^* being the equivalent kernel defined by (3.16)

$$b_\nu(x) = \left\{\int t^{p+1}K_\nu^*(t)dt\right\}\frac{\nu!}{(p+1)!}\eta^{(p+1)}(x)h^{p+1-\nu}$$

GENERALIZED LINEAR MODELS 197

$$\sigma_\nu^2(x) = \int K_\nu^{*2}(t)dt \frac{\nu!^2[g'\{m(x)\}]^2 \, Var(Y|X=x)}{f(x)nh^{2\nu+1}}.$$

A similar result holds for estimating $\eta^{(\nu)}(\cdot)$ at a point near the boundary of the design density f. In other words, the local quasi-likelihood method adapts automatically to the boundary region of the design density. This and Theorem 5.2 are shown by Fan, Heckman and Wand (1995). The multivariate extension can also be found in that paper.

The bias in Theorem 5.2 does not depend on the quasi-likelihood function. This reflects the fact that the bias comes from the local approximation error. Theorem 5.2 holds regardless of whether the conditional variance V is correctly specified or not. This is due to the localization of the fitting. The cost due to misspecification of the conditional variance is only of a second order nature. In other words, the local quasi-likelihood is robust to the choice of the function $V(\cdot)$. This highlights another advantage of the local modelling approach.

The asymptotic distribution of $\widehat{m}(\cdot) = g^{-1}\{\widehat{\eta}(\cdot)\}$ is a direct corollary of Theorem 5.2.

Theorem 5.3 *Assume that the conditions of Theorem 5.2 hold, and suppose that $nh^3 \to \infty$. Then, the error $\widehat{m}(x) - m(x)$ has the same asymptotic behavior as $\widehat{\eta}(x) - \eta(x)$ given in Theorem 5.2 with $\nu = 0$ except that the asymptotic bias is divided by $g'\{m(x)\}$ and the asymptotic variance by $[g'\{m(x)\}]^2$.*

5.4.5 Bandwidth selection

The bandwidth parameter determines the complexity of the resulting estimate. It can be chosen either subjectively by users or objectively by the data. Section 4.9 describes two more sophisticated bandwidth selection methods: the deviance residual criterion and the refined bandwidth selector. They can be directly applied to the current setting. Here, we describe a simple rule of thumb for selecting the bandwidth. The rule can be used directly in applications.

A convenient criterion error for selecting the bandwidth h is the asymptotic weighted mean integrated squared error of $\widehat{\eta}_\nu(\cdot)$:

$$\int \{b_\nu^2(x) + \sigma_\nu^2(x)\}w(x)f(x)dx,$$

with $w(\cdot)$ a certain weight function. Minimization of the above

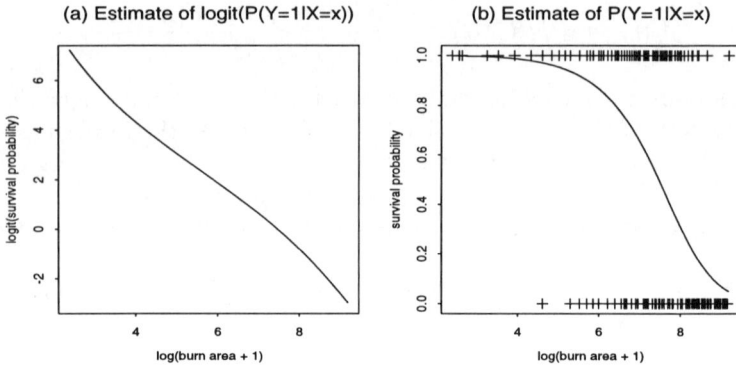

Figure 5.11. *Surviving probability of burn victims given the area of third degree burn injuries. (a) Estimation of logit of the conditional probability function. (b) Estimated conditional probability function. Results from Fan, Heckman and Wand (1995).*

expression leads to the theoretical optimal constant bandwidth

$$h_{\text{opt}} = C_{\nu,p}(K) \left(\frac{\int [g'\{m(x)\}]^2 V\{m(x)\}w(x)dx}{n \int \{\eta^{(p+1)}(x)\}^2 f(x)w(x)dx} \right)^{1/(2p+3)},$$
(5.55)

with $V\{m(x)\} = \text{Var}(Y|X = x)$ and where $C_{\nu,p}(K)$ is given in (3.20) and is tabulated in Table 3.1.

A simple way to implement the above formula is to consider a parametric pilot estimate. As in Section 4.2, fit a parametric polynomial model (5.49) of order $p + 3$. Let the resulting estimate of $\eta(\cdot)$ be

$$\widetilde{\eta}(x) = \widecheck{\alpha}_0 + \cdots + \widecheck{\alpha}_{p+3} x^{p+3}.$$

Correspondingly, let $\widetilde{m}(x) = g^{-1}\{\widetilde{\eta}(x)\}$. Substituting these pilot estimates into (5.55), we obtain

$$\widecheck{h}_{\text{ROT}} = C_{\nu,p}(K) \left(\frac{\int [g'\{\widetilde{m}(x)\}]^2 V\{\widetilde{m}(x)\}w(x)dx}{\sum_{i=1}^{n} \{\widetilde{\eta}^{(p+1)}(X_i)\}^2 w(X_i)} \right)^{1/(2p+3)}.$$
(5.56)

See (4.3) for a similar expression. In practical implementation, one often replaces the integration in (5.56) by a summation at grid points. Once $\widetilde{\eta}^{(p+1)}(\cdot)$ is computed at grid points then linear interpolation can be used for computing $\widetilde{\eta}^{(p+1)}(X_i)$.

The above bandwidth is very simple to implement. It works reasonably well for a variety of applications. When a refinement is needed, one can use the method proposed in Section 4.9.

We now illustrate this method using the burn data introduced in Section 2.1. Of interest is the estimation of the probability of surviving given the area of third degree burn injuries of victims. We employed the local linear fit using the Gaussian kernel function and the Bernoulli log-likelihood with the logit link. The bandwidth 1.242 was selected by Fan, Heckman and Wand (1995), using an idea similar to (5.56). We depict the results in Figure 5.11. Compare also with Figure 2.1 (a).

The estimated curves already describe the chance of surviving given the area of third degree burn injuries. When a parametric model is needed, Figure 5.11 (a) also suggests the use of a logit linear model. With this parametric linear model, one can easily construct confidence intervals for the estimated parameters.

5.5 Robust regression

Robust techniques are very useful when the variance of the error distribution is large. In this case, one often replaces the least squares criterion by robustified approaches. *Breakdown points* and *statistical efficiency* provide two important guidelines for selecting a robustified approach. Detailed discussions on robust statistics can be found in Huber (1981) and Hampel, Ronchetti, Rousseeuw and Stahel (1986), among others. In this section, we mainly focus on the robustification of the nonparametric method described in Chapter 3 and discuss how to perform a quantile regression.

5.5.1 Robust methods

Given the bivariate data $(X_1, Y_1), \cdots, (X_n, Y_n)$, one can replace the L_2-loss in (3.3) by

$$\sum_{i=1}^{n} \ell\{Y_i - \sum_{j=0}^{p} \beta_j (X_i - x_0)^j\} K_h(X_i - x_0), \qquad (5.57)$$

where $\ell(\cdot)$ is a loss function, usually more resistant to outliers than the L_2-loss function. Let the vector $\widehat{\beta}(x_0)$ be the minimizer of (5.57). Then, from Chapter 3, it is clear that

$$\widehat{m}_\ell^{(\nu)}(x_0) = \nu! \widehat{\beta}_\nu(x_0) \qquad (5.58)$$

is a good estimator for

$$m_\ell^{(\nu)}(x_0) \quad \text{with} \quad m_\ell(x) = \mathrm{argmin}_a E\{\ell(Y-a)|X=x\}. \qquad (5.59)$$

Note that the object function $m_\ell(x)$ depends on the loss function $\ell(\cdot)$. In particular, when $\ell(t) = t^2$, problem (5.57) reduces to the weighted least squares problem, estimating the conditional mean function. Other useful examples include:

Example 1 (Huber's ψ-estimator). Take

$$\psi_c(t) = \ell'(t) = \max\{-c, \min(c, t)\},$$

which leads to a more robust method than the least squares method. Different choices of c lead to different object functions (see (5.59)) unless the conditional distribution $G(\cdot|x)$ of Y given $X = x$ is symmetric around $m(x)$. Under the latter assumption $m_\ell(x) = m(x)$ for all c. If $c = +\infty$, the function $\ell(\cdot)$ becomes the L_2-loss, and as $c \to 0$, it results in the L_1-loss when suitably normalized. Huber (1981) recommends using $c = 1.35$. Another possibility is to replace the function $\psi_c(\cdot)$ by *Huber's bisquare function* $t\{1 - (t/c)^2\}_+^2$, which weighs down the tail contribution of t by a biweight function. In the parametric robustness literature the use of $c = 4.685$ is recommended.

Example 2 (Robustified likelihood). Huber's ψ function can be regarded as a robustification of the maximum local likelihood estimate under the Gaussian error. A further generalization is to use

$$\ell'(t) = \max[-c, \min\{c, -f'(t)/f(t)\}],$$

for a density function f. This can be regarded as a robustified version of the maximum local likelihood estimate with error density f.

Example 3 (Quantile regression). Let

$$\ell_\alpha(t) = |t| + (2\alpha - 1)t.$$

Then by (5.59), the object function is $\xi_\alpha(x) = G^{-1}(\alpha|x)$, the conditional quantile function, where $F(\cdot|x)$ is the conditional distribution of Y given $X = x$.

A useful robust procedure is LOWESS, introduced by Cleveland (1979), which reduces the influence of outliers by an iterative reweighted least squares scheme with weights proportional to the residuals from the previous iteration. This procedure also uses the nearest neighbor type of bandwidth (5.22), which enhances the design adaptation. Section 2.4 describes the relationship of LOWESS with the estimator (5.58) using Huber's bisquare loss.

The estimator (5.58), proposed and studied by Härdle and Gasser (1984) and Tsybakov (1986), and further investigated by Fan, Hu and Truong (1994) and Welsh (1994), inherits similar asymptotic properties from the local polynomial regression estimators. In particular, under certain regularity conditions, the estimator is asymptotically normal:

$$P\left\{\widehat{m}_\ell^{(\nu)}(x) - m_\ell^{(\nu)}(x) - b_{\nu,p}(x) \leq \sigma_{p,\nu}(x)t \big| \mathbb{X}\right\} \xrightarrow{P} \Phi(t), \quad (5.60)$$

for all t, where $b_{p,\nu}(x)$ is defined similarly to (3.18) with $m(\cdot)$ replaced by $m_\ell(\cdot)$ and $\sigma_{\nu,p}(x)$ is defined analogously to (3.19) replacing $\sigma^2(x)$ by

$$\sigma_\ell^2(x) = \mathrm{Var}[\ell'\{Y - m_\ell(x)\}|X = x]\{\phi''(0|x)\}^{-2}, \quad (5.61)$$

with $\phi(t|x) = E[\ell\{Y - m_\ell(x) + t\}|X = x]$. This estimator $\widehat{m}_\ell^{(\nu)}(\cdot)$ also adapts to the boundary of the design region: a property similar to Theorem 3.2 holds. Expression (5.60) exhibits an asymptotically optimal constant bandwidth

$$h_{\mathrm{opt}} = C_{\nu,p}(K)\left[\frac{\int \sigma_\ell^2(x)w(x)/f(x)dx}{\int \{m_\ell^{(p+1)}(x)\}^2 w(x)dx}\right]^{1/(2p+3)} n^{-1/(2p+3)}.$$

(5.62)

A simple use of (5.62) is to construct the rule of thumb. We do not pursue this further here, but will present the rule of thumb in the case of quantile regression. When Huber's ψ function is used with a large c, then one can apply a bandwidth $\widehat{h}_{\mathrm{LS}}$ developed for the least squares estimate, while when c is small, one can use a bandwidth $\widehat{h}_{\mathrm{LAD}}$ for the least absolute deviation estimate (quantile regression with $\alpha = 0.5$). For other values of c, one can interpolate the two bandwidths with for example the weight $(e^c - 1)/(e^c + 1)$ for $\widehat{h}_{\mathrm{LS}}$ and $2/(e^c + 1)$ for $\widehat{h}_{\mathrm{LAD}}$, so that when $c = 0$ one uses $\widehat{h}_{\mathrm{LAD}}$, and when $c = +\infty$ one uses $\widehat{h}_{\mathrm{LS}}$.

5.5.2 Quantile regression

Quantile regression is a special case of (5.57) with $\ell(\cdot)$ given in Example 3. Of particular interest is the median function $\xi_{0.5}(x)$, which is more explainable than the mean regression function for asymmetric conditional distributions. There are two important applications of quantile regression: constructing *predictive intervals* and detecting heteroscedasticity. A predictive interval is an interval that predicts, with certain coverage probability, the future value of the response variable Y for a given covariate $X = x$. More precisely, the $(1 - \alpha)100\%$ predictive interval is the interval $[\xi_{\alpha/2}(x), \xi_{1-\alpha/2}(x)]$. This interval can easily be constructed using quantile regression.

Suppose the observations can be modelled as

$$Y_i = m(X_i) + \sigma(X_i)\varepsilon_i. \tag{5.63}$$

Then the quantile function can be written as

$$\xi_\alpha(x) = m(x) + G_0^{-1}(\alpha)\sigma(x),$$

where $G_0(\cdot)$ denotes the distribution function of ε. In particular, a homoscedastic error, i.e. $\sigma(\cdot) = \sigma$, implies that the quantile curves are parallel. Therefore, an unparallel plot of the quantile functions suggests the presence of heteroscedasticity and proper care should be taken in the statistical analysis. A useful technique is to transform properly the variables to account for the heteroscedasticity. Carroll and Ruppert (1988) provide an excellent overview of this subject.

The asymptotic variance of $\widehat{\xi}_\alpha(x)$ is provided by (5.61) with

$$\sigma_\ell^2(x) = \alpha(1 - \alpha)[g\{\xi_\alpha(x)|x\}]^{-2} \equiv V_\alpha(x),$$

where $g(\cdot|x)$ is the conditional density of Y given $X = x$. In this case, the asymptotically optimal bandwidth (5.62) reads as

$$h_{opt} = C_{\nu,p}(K) \left[\frac{\int V_\alpha(x) w(x)/f(x) dx}{\int \{\xi_\alpha^{(p+1)}(x)\}^2 w(x) dx} \right]^{1/(2p+3)} n^{-1/(2p+3)}.$$

Note that for the location-scale model (5.63) with homoscedastic error, $g\{\xi_\alpha(x)|x\} = g_1\{G_1^{-1}(\alpha)\}$, where $G_1(\cdot)$ and $g_1(\cdot)$ are respectively the cumulative distribution and the density function of $\sigma\varepsilon_i$. When $\alpha = 0.5$ and $g_1(\cdot)$ is symmetric around zero, $G_1^{-1}(\alpha) = 0$. The rule of thumb is very simple to apply for this quantile regression setup. In the current setting we find

$$\check{h}_{ROT} = C_{\nu,p}(K) \left(\frac{\alpha(1-\alpha)[\check{g}_1\{\check{G}_1^{-1}(\alpha)\}]^{-2} \int w_0(x) dx}{\sum_{i=1}^n \{\check{\xi}_\alpha^{(p+1)}(X_i)\}^2 w_0(X_i)} \right)^{1/(2p+3)}, \tag{5.64}$$

using $w(x) = w_0(x) f(x)$ for a specific weight function $w_0(\cdot)$. Here, $\check{\xi}_\alpha(x) = \check{\alpha}_0 + \cdots + \check{\alpha}_{p+3} x^{p+3}$ is obtained from a global polynomial fit, $\check{g}_1(\cdot)$ can be obtained from a kernel density estimate of the residuals of the global polynomial fit, and $\check{G}_1^{-1}(\alpha)$ is the αth sample quantile of the residuals. See also Section 4.2 for a related technique.

We now illustrate the technique of constructing predictive intervals, using daily exchange rates (at closing time) between the British pounds and US dollars. The data consist of daily exchange rates at every business day between March 2, 1980 and April 3, 1993. The length of this time series is about 3,300. We analyze the data based on the exchange rates before March 2, 1992 and use the rest of the data set for validation. The entire series is plotted in Figure 5.12 (a). The linear regression of X_t against X_{t-1} explains about 99% of the variability of the data. To avoid such a trivial analysis, which does not help much in predicting future exchange rates, we consider a suitable difference of this time series. We searched for such a difference and found that the series $Y_t = X_t - X_{t-2}$ has non-trivial autocorrelation. In particular, the correlation between $Y_t = X_t - X_{t-2}$ and $Y_{t-1} = X_{t-1} - X_{t-3}$ is 0.50. This is not the case, for instance, for the series of the second order differencing $Y_t' = X_t - 2X_{t-1} + X_{t-2}$. The correlation $r(Y_t', Y_{t-1}') = -0.5$ is caused by the common factor X_{t-1} in the definition of Y_t' and Y_{t-1}'. Thus our analysis is now focused on the series Y_t, which is presented in Figure 5.12 (b). Figure 5.12 (c) depicts the 80% predictive intervals along with the median curve. The solid curves, corresponding to the conditional 10th, 50th and 90th percentile curves and obtained by using the Epanechnikov kernel with bandwidth 0.035 are fairly linear. The rule (5.64) for the median curve gives a bandwidth $\check{h}_{ROT} = 0.035$. The global linear fits, corresponding to bandwidth infinity, are also presented for comparison. We then use the exchange rates between March 2, 1992 to April 3, 1993 to validate the 80% predictive intervals. The scatter plot of these data are superimposed in Figure 5.12 (c): 77% of the data fall in the nonparametric predictive interval. Assuming the linear model for the median curve, we obtain the model

$$Y_t = 0.485 Y_{t-1} + \varepsilon_t.$$

Using this simple model, we predict the exchange rates with average error 0.015. Figure 5.12 (d) plots the actually observed series and the predicted series.

The analyses so far ignore the dependence structure. Indeed, the autocorrelation of ε_t suggests it can further be modelled by an AR(2) model:

$$\varepsilon_t = 0.261 \varepsilon_{t-1} - 0.386 \varepsilon_{t-2} + e_t$$

with $\text{Var}(e_t) \approx 0.000194$. This dependence can be used to improve

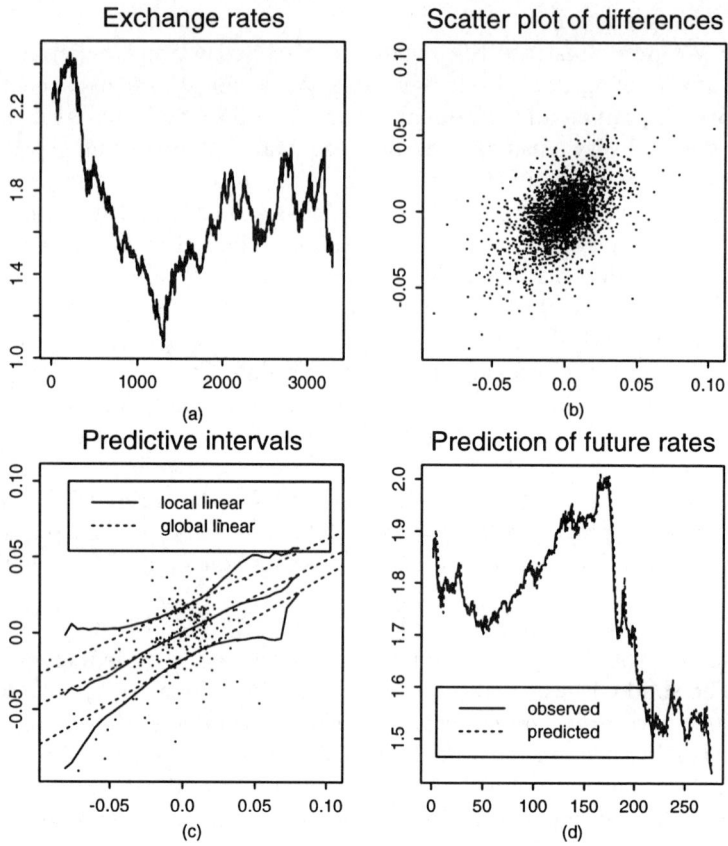

Figure 5.12. *An analysis of daily exchange rates between the British pound and US dollar. (a) Plot of time series; (b) scatter plot of lag 2 differences: $X_t - X_{t-2}$ against $X_{t-1} - X_{t-3}$; (c) 80% predictive intervals and validation of the predictive intervals; (d) prediction of future exchange rates using the simple equation $X_t - X_{t-2} = 0.485(X_{t-1} - X_{t-3})$.*

somewhat the predictive equation.

We now use quantile regression to detect heteroscedasticity and to build a statistical model. The esterase data set, introduced in Section 2.4, involves 115 assays for the concentration of the enzyme esterase. The response Y is the number of bindings counted by a radio-immunoassay (RIA) procedure and the covariate X is the observed esterase concentration. Here, we follow closely the analysis given in Welsh (1994) except that the bandwidths are

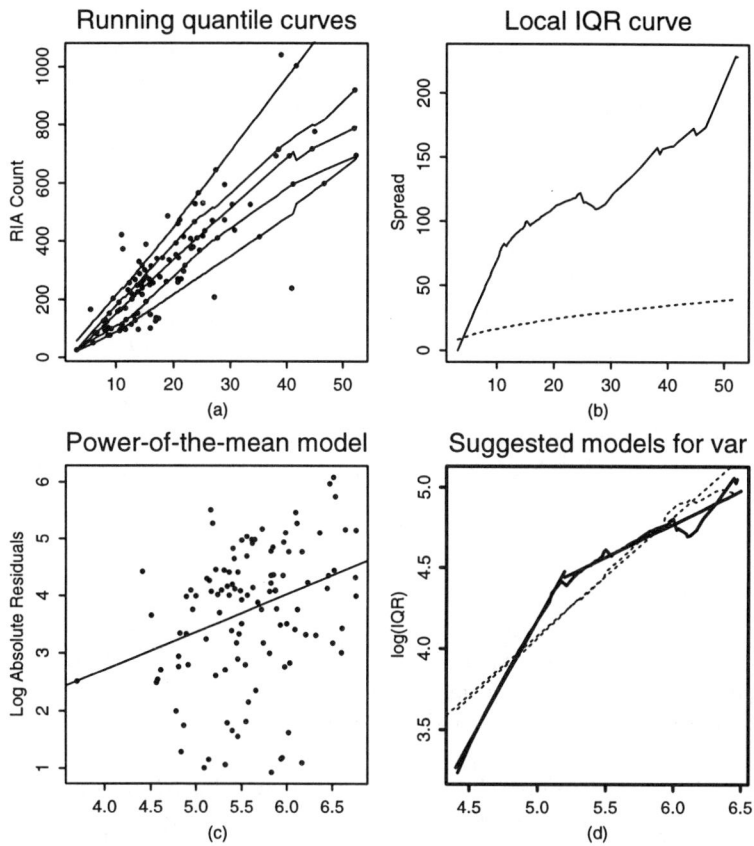

Figure 5.13. *An analysis of the esterase data. (a) Raw data and quantile functions. (b) Conditional interquartile range (IQR) function (solid curve) and the standard deviation function of the fitted Poisson distribution (dashed curve). (c) Scatter plot of log absolute Huber residuals against log-fitted values; solid line is the robust regression line. (d) Plot of the conditional log(IQR) against log-fitted values. Solid lines – piecewise linear model for the conditional log(IQR); dashed curve – conditional log(IQR) using Welsh's bandwidth; dashed line – possible linear model for the conditional log(IQR).*

chosen according to (5.64). The standard Gaussian kernel is employed in this analysis. Welsh (1994) uses the simple bandwidth $2 \times \min\{SD(X), MAD(X)\}$, a function of the Standard Deviation (SD) and the Mean Absolute Deviation (MAD) of the design points. This simple bandwidth depends only on the design points and is 16.9 for this data set, whereas $\check{h}_{\text{ROT}}$ depends also on the curvature and is about 7 for this data set when $\alpha = 0.25, 0.5$ and 0.75. The data are shown in Figure 5.13 (a) with the 10th, 25th, 50th, 75th and 90th percentile functions superimposed (the bandwidths for the 10th and the 90th quantile curve are chosen using Welsh's simple rule, since α or $1 - \alpha$ is too small). The quantile functions are nearly linear, but their fan shape indicates clearly heteroscedasticity. Figure 5.13 (b) plots the estimated *conditional interquartile range* (IQR) function $\widehat{\xi}_{.75}(x) - \widehat{\xi}_{.25}(x)$, which is increasing in x. This suggests that the variability is getting larger as x increases.

Since the response is a count, it makes sense, at least initially, to treat Y as having a Poisson distribution and to apply the techniques presented in Section 5.4. Figure 5.13 (a) reveals the fact that the relationship between Y and X is nearly linear. This suggests that the first two moments of Y can be modelled as

$$Y_i = \alpha + \beta X_i + \sigma(\alpha + \beta X_i)\varepsilon_i, \quad \text{with} \quad \sigma(t) = \sqrt{t},$$

appealing to the Poisson assumption. Comparing the Poisson dispersion $(\widehat{\alpha} + \widehat{\beta} x)^{1/2}$ (dashed curve in Figure 5.13 (b)) with the conditional IQR function in Figure 5.13 (b), we found evidence of over-dispersion. Possible models for the variance function include $\sigma(t) = \theta_0 t^{\theta_1}$ and $\sigma(t) = \theta_0 \exp(\theta_1 t)$.

A parametric approach to choosing a model for $\sigma(t)$, described in Carroll and Ruppert (1988, page 49), is to fit first a linear regression model, and then to plot the logarithm of the absolute residuals against different functions of the fitted values and to decide which of these plots exhibits a more linear relationship. Figure 5.13 (c) presents one of these plots, showing a reasonable linearity. An attractive alternative is to plot the logarithm of the estimated IQR function against a function of the fitted curve or the covariate and examine an appropriate function form for $\sigma(\cdot)$. This is done in Figure 5.13 (d), which reveals the piecewise linearity:

$$\sigma(t) = \begin{cases} \exp(-3.5)t^{1.54}, & \text{if } t \leq \exp(5.2) \\ \exp(2.26)t^{0.42}, & \text{if } t > \exp(5.2). \end{cases}$$

Another possible model for $\sigma(t)$, suggested by Carroll and Ruppert (1988) and Welsh (1994), can be obtained as follows. Compute the interquartile range using the bandwidth 16.9 suggested by Welsh (1994). This is shown as the dashed curve in Figure 5.13 (d), which looks fairly linear. Fitting a line (dashed line in Figure 5.13 (d)) through this curve, we obtain the power-of-the-mean model

$$\sigma(t) = \widehat{\theta}_1 t^{\widehat{\theta}_2}, \text{ with } \widehat{\theta}_1 = \exp(0.40) \text{ and } \widehat{\theta}_2 = 0.74.$$

The fact that $\widehat{\theta}_2 > 0.5$ indicates the over-dispersion relative to the Poisson model. Starting with this variance form, one can apply the quasi-likelihood method to complete the parametric analysis for α and β in the heteroscedastic linear model. Note that θ_1 has not been normalized yet since the IQR function differs from the conditional variance by a scalar factor under the location-scale model (5.63). Fortunately, in solving the quasi-likelihood equation, one does not need to know the value of θ_1.

5.5.3 Simultaneous estimation of location and scale functions

The conditional density for model (5.63) is given by

$$\sigma^{-1}(X_i) g_0 \left\{ \frac{Y_i - m(X_i)}{\sigma(X_i)} \right\},$$

where $g_0(\cdot)$ denotes the density of ε. If $m(\cdot)$ and $\log \sigma(\cdot)$ are modelled by a polynomial function, namely,

$$m(x) = \sum_{j=0}^{p} \beta_j (x - x_0)^j, \qquad \log \sigma(x) = \sum_{j=0}^{q} \theta_j (x - x_0)^j,$$

then the maximum conditional likelihood (given $X_1, \cdots, X_n$) estimators of the parameters satisfy the following system of equations

$$\sum_{i=1}^{n} \mathbf{X}_{i,p} \exp(-\mathbf{X}_{i,q}^T \theta) \psi \{ \exp(-\mathbf{X}_{i,q}^T \theta)(Y_i - \mathbf{X}_{i,p}^T \beta) \} = 0,$$

(5.65)

$$\sum_{i=1}^{n} \mathbf{X}_{i,q} \chi \{ \exp(-\mathbf{X}_{i,q}^T \theta)(Y_i - \mathbf{X}_{i,p}^T \beta) \} = 0,$$

where $\mathbf{X}_{i,p} = \{1, (X_i - x_0), \cdots, (X_i - x_0)^p\}^T$, $\mathbf{X}_{i,q}$ is defined similarly,

$$\psi(x) = -\frac{d \log g_0(x)}{dx} \qquad \text{and} \qquad \chi(x) = \psi(x) x - 1. \qquad (5.66)$$

Specifically, when g_0 is the standard normal density, $\psi(x) = x$ and $\chi(x) = x^2 - 1$. In general, neither ψ nor χ need to be determined by a probability density as is the case in (5.66). The assumption

can be relaxed to
$$E\psi(\varepsilon) = E\chi(\varepsilon) = 0.$$
This condition ensures that the estimation equations in (5.65) lead to consistent estimators of the parameters.

When the functions $m(\cdot)$ and $\sigma(\cdot)$ are not parametrized, we can use equations (5.65) locally, leading to the system of equations

$$\sum_{i=1}^{n} \mathbf{X}_{i,p} \exp(-\mathbf{X}_{i,q}^T \theta) \psi\{\exp(-\mathbf{X}_{i,q}^T \theta)(Y_i - \mathbf{X}_{i,p}^T \beta)\} \\ \times K_h(X_i - x_0) = 0,$$
$$\sum_{i=1}^{n} \mathbf{X}_{i,q} \chi\{\exp(-\mathbf{X}_{i,q}^T \theta)(Y_i - \mathbf{X}_{i,p}^T \beta)\} K_h(X_i - x_0) = 0. \tag{5.67}$$

Clearly then the functions $m(x)$ and $\sigma(x)$ as well as their derivatives can be estimated via

$$\widehat{m}_\nu(x_0) = \nu! \widehat{\beta}_\nu, \quad \widehat{\delta}_\nu(x_0) = \nu! \widehat{\theta}_\nu, \tag{5.68}$$

where $\delta_\nu(x) = \{\log \sigma(x)\}^{(\nu)}$.

The asymptotic properties of the above local polynomial estimators were studied by Welsh (1994), who showed that the robustified version of the estimators inherits the good properties of the local polynomial regression estimators discussed in Chapter 3.

In many applications, one takes $p = q = 1$ in (5.67), applying local linear fits. One popular choice of ψ and χ is *Huber's proposal 2* (Huber (1964)), namely

$$\psi(x) = \max\{-c, \min(c, x)\}, \quad \chi(x) = \psi^2(x) - \int \psi^2(t) d\Phi(t),$$

with $\Phi(t)$ the standard normal distribution function.

5.6 Complements

Proof of Theorem 5.1

Recall the transformed data (5.10) and denote the estimated transformed data by $\widehat{Y}^* = \delta \widehat{\phi}_1(X, Z) + (1 - \delta) \widehat{\phi}_2(X, Z)$, where $\widehat{\phi}_1$ and $\widehat{\phi}_2$ are given in (5.24). It follows from the definition of the local linear regression estimator that

$$\left| \widehat{m}(x; \widehat{\phi}_1, \widehat{\phi}_2) - \widehat{m}(x; \phi_1, \phi_2) \right| \leq \sum_{i=1}^{n} |w_i(x)| |\widehat{Y}_i^* - Y_i^*| / \sum_{i=1}^{n} w_i(x), \tag{5.69}$$

where $w_i(x) = K_h(X_i - x)\{S_{n,2} - (X_i - x)S_{n,1}\}$ with $S_{n,\ell} =$

$\sum_{i=1}^n K_h(X_i - x)(X_i - x)^\ell$. Note that

$$\sum_{i=1}^n |w_i(x)| \leq S_{n,0}S_{n,2} + \{\sum_{i=1}^n |X_i - x|K_h(X_i - x)\}^2 \leq 2S_{n,0}S_{n,2}.$$

Thus, by (3.61), we have

$$\frac{\sum_{i=1}^n |w_i(x)|}{\sum_{i=1}^n w_i(x)} \leq \frac{2S_{n,2}S_{n,0}}{S_{n,2}S_{n,0} - S_{n,1}^2} = \frac{2\mu_2\mu_0}{\mu_2\mu_0 - \mu_1^2} + o_P(1). \quad (5.70)$$

Since K has bounded support, the effective design-points are actually those local to the point x. Using this fact and combining (5.69) and (5.70), we find that

$$\left|\widehat{m}(x;\widehat{\phi}_1,\widehat{\phi}_2) - \widehat{m}(x;\phi_1,\phi_2)\right|$$

$$\leq \frac{1}{\sum_{i=1}^n w_i(x)} \sum_{i=1}^n \sum_{j=1}^2 |w_i(x)|I(Z_i > \tau_n)|Z_i - \phi_j(X_i, Z_i)|$$

$$+ 2\beta_n(x)\{1 + o_P(1)\}, \quad (5.71)$$

where $\beta_n(x)$ is given by (5.23). Now,

$$E\left\{\frac{1}{\sum_{i=1}^n w_i(x)} \sum_{i=1}^n \sum_{j=1}^2 |w_i(x)|I(Z_i > \tau_n)|Z_i - \phi_j(X_i, Z_i)|\,\Big|\,\mathbb{X}\right\}$$

$$\leq 2\kappa_n(x) \frac{\sum_{i=1}^n |w_i(x)|}{\sum_{i=1}^n w_i(x)},$$

and hence the result follows from (5.70) and (5.71).

Convexity lemma and quadratic approximation lemma

The target functions to be maximized or minimized in Sections 5.3–5.5 admit a similar structure: they are either concave or convex. Consistency and asymptotic normality of the resulting estimators can be proved in a similar way. The *convexity lemma* is a very useful technical device for establishing asymptotic normality. Here, we state a form of the lemma given in Pollard (1991). This will allow us to prove Theorem 5.2. The other results provided in Sections 5.2 and 5.4 can be proved analogously.

Convexity lemma *Let $\{\lambda_n(\theta) : \theta \in \Theta\}$ be a sequence of random convex functions defined on a convex, open subset Θ of $\mathbb{R}^d$. Suppose $\lambda(\cdot)$ is a real-valued function on Θ for which $\lambda_n(\theta) \to \lambda(\theta)$ in* *probability, for each θ in Θ. Then for each compact subset C of Θ,*

$$\sup_{\theta \in C} |\lambda_n(\theta) - \lambda(\theta)| \xrightarrow{P} 0.$$

Moreover, the function $\lambda(\cdot)$ is necessarily convex on Θ.

Establishing the uniform convergence of a random process is usually very tedious and cumbersome. With the convexity lemma, the task is reduced to proving the pointwise convergence without involving the supremum of the random process. In many statistical applications, the limiting process of a target function is typically of a quadratic form. Using the convexity lemma, we can easily establish asymptotic normality. We state a result of Fan, Heckman and Wand (1995), which allows us to specify second order terms in an asymptotic expansion. A similar version appears independently in Hjort and Pollard (1996).

Quadratic approximation lemma *Let $\{\lambda_n(\theta): \theta \in \Theta\}$ be a sequence of random concave functions defined on a convex open subset Θ of $I\!R^d$. Let $\mathbf{F}$ and $\mathbf{G}$ be non-random matrices, with $\mathbf{F}$ positive definite and $\mathbf{U}_n$ a stochastically bounded sequence of random vectors. Lastly, let α_n be a sequence of constants tending to zero. Write*

$$\lambda_n(\theta) = \mathbf{U}_n^T \theta - \frac{1}{2}\theta^T(\mathbf{F} + \alpha_n \mathbf{G})\theta + f_n(\theta).$$

If, for each $\theta \in \Theta$, $f_n(\theta) = o_P(1)$, then

$$\widehat{\theta}_n = \mathbf{F}^{-1}\mathbf{U}_n + o_P(1),$$

where $\widehat{\theta}_n$ (assumed to exist) maximizes $\lambda_n(\cdot)$. If, in addition, $f'(\theta) = o_P(\alpha_n)$ and $f_n''(\theta) = o_P(\alpha_n)$ uniformly in θ in a neighborhood of $\widehat{\theta}_n$, then

$$\widehat{\theta}_n = \mathbf{F}^{-1}\mathbf{U}_n - \alpha_n \mathbf{F}^{-1}\mathbf{G}\mathbf{F}^{-1}\mathbf{U}_n + o_P(\alpha_n).$$

Using the first part of the quadratic approximation lemma, we only need to show that the target function $\lambda_n(\theta)$ converges pointwise:

$$\lambda_n(\theta) = \mathbf{U}_n^T \theta - \frac{1}{2}\theta^T \mathbf{F} \theta + o_P(1).$$

The asymptotic normality of $\widehat{\theta}_n$ then follows directly from the fact that

$$\widehat{\theta}_n = \mathbf{F}^{-1}\mathbf{U}_n + o_P(1)$$

COMPLEMENTS

and the asymptotic normality of $\mathbf{U}_n$. We illustrate this idea in the next paragraphs.

Asymptotic normality for local quasi-likelihood estimators

We first impose some regularity conditions for Theorem 5.2. Let $q_\ell(x,y) = (\partial^\ell/\partial x^\ell)Q\{g^{-1}(x),y\}$. Note that q_ℓ is linear in y for fixed x and that

$$q_1\{\eta(x), m(x)\} = 0 \quad \text{and} \quad q_2\{\eta(x), m(x)\} = -\rho(x), \quad (5.72)$$

where $\rho(x) = ([g'\{m(x)\}]^2 V\{m(x)\})^{-1}$.

Conditions:
(1) The function $q_2(x,y) < 0$ for $x \in \mathbb{R}$ and y in the range of the response variable.
(2) The functions $f(\cdot)$, $\eta^{(p+1)}(\cdot)$, $\text{Var}(Y|X=\cdot)$, $V'(\cdot)$ and $g'''(\cdot)$ are continuous.
(3) Assume that $\rho(x) \neq 0$, $\text{Var}(Y|X=x) \neq 0$, and $g'\{m(x)\} \neq 0$.
(4) $E(Y^4|X=\cdot)$ is bounded in a neighborhood of x.

Recall that the vector $\widehat{\beta} = (\widehat{\beta}_0, \cdots, \widehat{\beta}_p)^T$ maximizes (5.53). Consider the normalized estimator

$$\widehat{\beta}^* = a_n^{-1}[\widehat{\beta}_0 - \eta(x), \cdots, h^p\{\widehat{\beta}_p - \eta^{(p)}(x)/p!\}]^T,$$

with $a_n = (nh)^{-1/2}$, so that each component has a nondegenerate asymptotic variance. Then, it can easily be seen that $\widehat{\beta}^*$ maximizes

$$\sum_{i=1}^n Q[g^{-1}\{\bar\eta(x,X_i) + a_n\beta^{*T}\mathbf{Z}_i\}, Y_i] K\{(X_i - x)/h\}$$

as a function of β^*, where

$$\bar\eta(x, X_i) = \eta(x) + \eta'(x)(X_i - x) + \cdots + \eta^{(p)}(x)(X_i - x)^p/p!$$

and

$$\mathbf{Z}_i = \left\{1, \frac{X_i - x}{h}, \cdots, \frac{(X_i - x)^p}{h^p}\right\}^T.$$

Equivalently, $\widehat{\beta}^*$ maximizes

$$\ell_n(\beta^*) = \sum_{i=1}^n \Big(Q[g^{-1}\{\bar\eta(x,X_i) + a_n\beta^{*T}\mathbf{Z}_i\}, Y_i]$$

$$- Q[g^{-1}\{\bar\eta(x,X_i)\}, Y_i]\Big) K\{(X_i - x)/h\}. \quad (5.73)$$

We remark that Condition (1) implies that ℓ_n is concave in β^*. Using a Taylor expansion of $Q\{g^{-1}(\cdot), Y_i\}$, we find

$$\ell_n(\beta^*) = \mathbf{W}_n^T \beta^{*T} + \frac{1}{2}\beta^{*T}\mathbf{A}_n\beta^*$$
$$+ \frac{a_n^3}{6}\sum_{i=1}^n q_3(\eta_i, Y_i)(\beta^{*T}\mathbf{Z}_i)^3 K\{(X_i - x)/h\}, \quad (5.74)$$

where η_i is between $\bar{\eta}(x, X_i)$ and $\bar{\eta}(x, X_i) + a_n\beta^{*T}\mathbf{Z}_i$,

$$\mathbf{W}_n = a_n \sum_{i=1}^n q_1\{\bar{\eta}(x, X_i), Y_i\}\mathbf{Z}_i K\{(X_i - x)/h\}$$

and

$$\mathbf{A}_n = a_n^2 \sum_{i=1}^n q_2\{\bar{\eta}(x, X_i), Y_i\} K\{(X_i - x)/h\} \mathbf{Z}_i \mathbf{Z}_i^T.$$

Next, we show that

$$\mathbf{A}_n = -\rho(x)f(x)\,(\mu_{i+j-2})_{1 \leq i,j \leq p+1} + o_P(1) \equiv \mathbf{A} + o_P(1). \quad (5.75)$$

This can easily be shown using the fact that

$$(\mathbf{A}_n)_{ij} = (E\mathbf{A}_n)_{ij} + O_P[\{\text{Var}(\mathbf{A}_n)_{ij}\}^{1/2}].$$

The mean in the above expression equals

$$E\mathbf{A}_n = h^{-1}E[q_2\{\bar{\eta}(x, X_1), m(X_1)\}K\{(X_1 - x)/h\}\mathbf{Z}_1\mathbf{Z}_1^T]$$

which tends to $\mathbf{A}$, and the variance term in the above expression is of order a_n. This verifies (5.75).

Note that $\mathbf{Z}_i$ in (5.74) is a bounded random vector since K has compact support. The expected value of the absolute value of the last term in (5.74) is bounded by

$$O\Big(na_n^3 E|q_3(\eta_1, Y_1)K\{(X_1 - x)/h\}|\Big) = O(a_n),$$

since q_3 is linear in Y_1 with $E(|Y_1|\,|X_1) < \infty$ and all other terms are bounded random variables. Therefore, the last term in (5.74) is of order $O_P(a_n)$. Combining (5.74) and (5.75) leads to

$$\ell_n(\beta^*) = \mathbf{W}_n^T \beta^{*T} + \frac{1}{2}\beta^{*T}\mathbf{A}\beta^* + o_P(1).$$

Using the quadratic approximation lemma, we obtain

$$\widehat{\beta}^* = \mathbf{A}^{-1}\mathbf{W}_n + o_P(1), \quad (5.76)$$

if $\mathbf{W}_n$ is a stochastically bounded sequence of random vectors. The asymptotic normality of $\widehat{\beta}^*$ follows from that of $\mathbf{W}_n$. Hence, it remains to establish the asymptotic normality of $\mathbf{W}_n$.

Since the random vector $\mathbf{W}_n$ is a sum of i.i.d. random vectors, its asymptotic normality follows by checking the Lyapounov condition. We first compute its first two moments. By Taylor's expansion,

$$q_1\{\bar{\eta}(x, x+hu), m(x+hu)\}$$
$$= \rho(x+hu)(hu)^{p+1}\frac{\eta^{(p+1)}(x)}{(p+1)!} + o(h^{p+1}).$$

Using this, we obtain

$$\begin{aligned}
E\mathbf{W}_n &= na_n Eq_1\{\bar{\eta}(x, X_1), m(X_1)\}\mathbf{Z}_1 K\{(X_1-x)/h\} \\
&= nha_n \int q_1\{\bar{\eta}(x, x+hu), m(x+hu)\}f(x+hu) \\
&\quad \mathbf{U}K(u)du + o(h^{p+1}) \\
&= a_n^{-1}\rho(x)h^{p+1}\frac{\eta^{(p+1)}(x)}{(p+1)!}\int u^{p+1}\mathbf{U}K(u)du\{1+o(1)\},
\end{aligned}$$
(5.77)

where $\mathbf{U} = (1, u, \cdots, u^p)^T$. Analogously,

$$\begin{aligned}
\text{Var}(\mathbf{W}_n) &= na_n^2 \text{Var}[q_1\{\bar{\eta}(x, X_1), Y_1\}\mathbf{Z}_1 K\{(X_1-x)/h\}] \\
&= h^{-1}[Eq_1^2\{\bar{\eta}(x, X_1), Y_1\}\mathbf{Z}_1\mathbf{Z}_1^T K^2\{(X_1-x)/h\} \\
&\quad + O(h^{2p+4})],
\end{aligned}$$

where (5.77) was used to compute the order of the mean function. Using the definition of q_1, we obtain that

$$\begin{aligned}
\text{Var}(\mathbf{W}_n) &= \frac{f(x)\text{Var}(Y|X=x)}{[V\{m(x)\}g'\{m(x)\}]^2}\int \mathbf{U}\mathbf{U}^T K^2(z)dz + o(1) \\
&\equiv \mathbf{B} + o(1).
\end{aligned}$$
(5.78)

To prove that

$$\{\text{Var}(\mathbf{W}_n)\}^{-1/2}(\mathbf{W}_n - E\mathbf{W}_n) \xrightarrow{D} N(0, I_{p+1}), \qquad (5.79)$$

we employ the Cramér-Wold device: for any unit vector $\mathbf{a}$,

$$\{\mathbf{a}^T\text{Var}(\mathbf{W}_n)\mathbf{a}\}^{-1/2}(\mathbf{a}^T\mathbf{W}_n - \mathbf{a}^T E\mathbf{W}_n) \xrightarrow{D} N(0,1),$$

which can easily be shown via checking Lyapounov's condition.

Combining (5.76)–(5.79), we obtain

$$\left[\widehat{\beta}^* - a_n^{-1}\rho(x)h^{p+1}\frac{\eta^{(p+1)}(x)}{(p+1)!}\mathbf{A}^{-1}\int u^{p+1}\mathbf{U}du\{1+o(1)\}\right]$$
$$\xrightarrow{D} N\left(0, \mathbf{A}^{-1}\mathbf{B}\mathbf{A}^{-1}\right).$$

Considering the marginal distribution of the above vector and using the definition of K_ν^*, we obtain Theorem 5.2. We omit the details of the calculations.

5.7 Bibliographic notes

A number of books on survival analysis have been published. Kalbfleisch and Prentice (1980), Miller (1981), Cox and Oakes (1984), Fleming and Harrington (1991) and Andersen, Borgan, Gill and Keiding (1993), among others, give a very detailed account of parametric and semiparametric techniques used in survival analysis. Kaplan and Meier (1958) introduced the product-limit estimator based on a maximum likelihood argument. The proportional hazards model in a general form with time-dependent explanatory variables was discussed by Cox (1972). The idea of partial likelihood was introduced by Cox (1975). The multiplicative intensity model was given in Aalen (1978). Early references on using counting process theory for survival analysis include Aalen (1976, 1978), Gill (1980) and Andersen and Gill (1982). Further references on developments in survival analysis can be found in, for example, the books mentioned above.

An early effort in linear regression with censored data is due to Miller (1976) who considered a modified residual sum of squares, based on the Kaplan-Meier estimator instead of on the usual empirical distribution function. But the proposed estimation procedure does not necessarily converge. Buckley and James (1979) then modified the normal equations rather than the residual sum of squares. An attempt to investigate the consistency of the Buckley-James estimator was made by James and Smith (1984). With a slight modification of the estimator Ritov (1990) and Lai and Ying (1991) established the asymptotic normality of this type of estimators. Efficiency and robustness were investigated by Ritov (1990). Another family of estimators, asymptotically equivalent to the former one, based on linear rank tests for the slope in a linear model, were introduced and studied by Tsiatis (1990). Zhou (1992) investigated the large sample properties of the censored data least squares estimators proposed by Leurgans (1987) and Zheng (1988). The asymptotic properties of the linear regression estimator based on a censoring unbiased transformation are further investigated by Lai, Ying and Zheng (1995). Papers dealing with nonparametric estimation of the regression function in the case of censored data are Doksum and Yandell (1982), Dabrowska (1987) and Zheng (1988) among others. Fan and Gijbels (1994) handle the censored regression problem by applying the local linear technique to the censoring unbiased transformed data.

Hazard rates have been studied extensively in the literature. Among recent papers are Müller and Wang (1990, 1994), Uzunoğullari and Wang (1992) and Hjort (1995). Padgett (1986) and Marron and Padgett (1987) give some ideas on how to select an optimal bandwidth when dealing with censored data. Estimation of functionals of the Kaplan-Meier estimator is discussed in, for example, Schick, Susarla and Koul (1988) and Gijbels and Veraverbeke (1991). A strong law under random censorship was established by Stute and Wang (1993). Li and Doss (1995) used local linear fits to estimate nonparametrically the conditional hazard function, the conditional cumulative hazard function and the conditional survival function.

Generalized linear models were introduced by Nelder and Wedderburn (1972) as a means to apply ordinary linear regression techniques to more general settings. Wedderburn (1974) suggested replacing the log-likelihood by a quasi-likelihood, requiring only the specification of the relationship between the conditional mean and variance of the response variable. Optimal properties of quasi-likelihood methods were obtained by, for example, Cox (1983) and Godambe and Heyde (1987). A comprehensive account of parametric generalized linear models can be found in the book by McCullagh and Nelder (1989). Extensions of the smoothing spline methodology to the setting of generalized linear models have been proposed by Green and Yandell (1985), O'Sullivan, Yandell and Raynor (1986), Cox and O'Sullivan (1990) and Gu (1990). Tibshirani and Hastie (1987) considered a nonparametric fit based on the running line smoother, whereas the methodology of Staniswalis (1989) relies on local constant fitting. Quasi-likelihood estimation using local constant fits was considered by Severini and Staniswalis (1994), and an extension of the estimators to semi-parametric settings can be found in Hunsberger (1994). The polynomial spline approach is studied in detail by Stone (1986, 1990a, 1994). Fan, Heckman and Wand (1995) investigate the asymptotic properties of the local quasi-likelihood method using local polynomial modelling.

Early papers on parametric quantile regression include Bickel (1973) and Koenker and Bassett (1978). Robust methods based on the idea of replacing the L_2-loss function by a nonquadratic loss function lead to the class of M-smoothers. Kernel estimators for the median using a local constant fit have been studied by, for example, Härdle (1984), Härdle and Gasser (1984), Tsybakov (1986), Truong (1989), Bhattacharya and Gangopadhyay (1990), Hall and Jones (1990) and Chaudhuri (1991). Robust kernel smoothers for estimation of derivatives have been investigated by, for example, Härdle and Gasser (1985) and Tsybakov (1986). Robust local polynomial fitting was studied by Tsybakov (1986), Fan, Hu and Truong (1994) and Welsh (1994), among others. Bootstrapping regression quantiles were discussed in Aerts, Janssen and Veraverbeke (1994). Robust smoothing spline estimators were investigated in detail by Koenker, Portnoy and Ng (1992) and Koenker, Ng and Portnoy (1994). Fan and Hall (1994), who studied nonparametric minimax efficiency of robust smoothers, were among the first to propose a theory on nonparametric minimax risk when the loss is not quadratic. Other references can be found in Section 6.3.

CHAPTER 6

Applications in nonlinear time series

6.1 Introduction

Previous chapters considered the applications of local polynomial fitting in the framework of independent observations. This independence assumption can be reasonable when data are collected from a human population or certain measurements. However, there are also many practical situations in which the independence assumption is not appropriate. In particular, when the data have been recorded over time such as the daily exchange rates discussed in Chapter 5, it is very likely that the present response values will be correlated with their past ones. This chapter focuses on how to apply the local polynomial modelling technique to explore important structural information contained in dependent data.

One important aspect of *nonlinear time series* is *nonlinear prediction*. It is related to regression problems for dependent data. These regression problems arise frequently from two kinds of applications. One is that the observations were collected dependently due to the nature of the experiments – random errors from two nearby design points can be highly correlated. The other occurs often in the prediction of a future event in which one uses the past values as covariate variables and the current value as the response variable. Sections 6.2 and 6.3 discuss respectively nonparametric regression and quantile regression for dependent observations. These regression problems play an important role in nonlinear prediction.

Another important issue in nonlinear time series analysis is to examine certain periodic patterns of a time series. This is usually studied via an estimation of the *spectral density* of a stationary time series. Various useful techniques for such estimation will be introduced in Section 6.4.

Associated with the nonlinear prediction is its *sensitivity to initial values* and *noise amplification*. This sensitivity measures how far apart the nonlinear prediction can be in short or medium term, given two nearby initial values. It indicates the instability of the nonlinear prediction near the observed value, since the value is frequently recorded with rounding errors. The noise amplification refers to the fact that the noise level of a stochastic system can be amplified or reduced through time evolution. Section 6.5 discusses how the sensitivity to initial values and noise amplification are related to the reliability of nonlinear prediction.

This chapter discusses only certain aspects of nonlinear time series analysis that have a nonparametric flavor. More comprehensive discussions can be found in Tong (1990). See also the dimensionality reduction principles discussed in Chapter 7.

6.2 Nonlinear prediction

The aim of this section is to show how nonparametric regression techniques can be used for nonlinear prediction. The key conditions that enable one to apply local polynomial fitting are *mixing conditions* which indicate the strength of dependence of two time events in medium or long terms. Local polynomial fitting can also be used to estimate *conditional densities* for nonlinear prediction. The conditional densities provide information not only about their centers – conditional mean regression functions, but also about the degree of spread around their centers and the shape of the density functions.

6.2.1 Mixing conditions

Let $\{(X_j, Y_j)\}$ be a stationary sequence of random vectors, and $\mathcal{F}_i^k$ be the σ-algebra of events generated by the random variables $\{(X_j, Y_j), i \leq j \leq k\}$. See Chow and Teicher (1988) for the definition of σ-algebra. Denote by $L_2(\mathcal{F}_i^k)$ the collection of all random variables which are $\mathcal{F}_i^k$-measurable and have finite second moment. The following three types of mixing conditions are frequently used in the literature. The stationary process $\{(X_j, Y_j)\}$ is called *strongly mixing* (Rosenblatt (1956b)) if

$$\sup_{A \in \mathcal{F}_{-\infty}^0, B \in \mathcal{F}_k^\infty} |P(AB) - P(A)P(B)| = \alpha(k) \to 0 \quad \text{as} \quad k \to \infty;$$

(6.1)

is said to be *uniformly mixing* (Ibragimov (1962)) if

$$\sup_{A \in \mathcal{F}^0_{-\infty}, B \in \mathcal{F}^\infty_k} |P(B|A) - P(B)| = \phi(k) \to 0 \quad \text{as} \quad k \to \infty; \quad (6.2)$$

and is called ρ-*mixing* (Kolmogorov and Rozanov (1960)) if

$$\sup_{U \in L_2(\mathcal{F}^0_{-\infty}), V \in L_2(\mathcal{F}^\infty_k)} |\text{corr}(U, V)| = \rho(k) \to 0 \quad \text{as} \quad k \to \infty, \quad (6.3)$$

where $\text{corr}(U, V)$ denotes the correlation coefficient between the random variables U and V. It is well known that these mixing coefficients $\alpha(\cdot)$, $\phi(\cdot)$ and $\rho(\cdot)$ satisfy

$$\alpha(k) \leq \frac{1}{4}\rho(k) \leq \frac{1}{2}\phi^{1/2}(k). \quad (6.4)$$

Other mixing coefficients can be found in Rosenblatt (1991).

The mixing conditions (6.1)–(6.3) indicate basically the maximum dependence, in a different mathematical sense, between two time events at least k steps apart. In particular, if the stationary sequence is ℓ-dependent, namely, (X_i, Y_i) depends only on previous ℓ observations, then all mixing coefficients (6.1)–(6.3) are zero for $k > \ell$. As shown in Theorem 6.1 below, local polynomial fitting techniques continue to apply under the weak dependence in medium or long term, namely, when k is large. The short term dependence does not have much effect on the local smoothing method. The reason is that for any two given random variables X_i and X_j and a point x, the random variables $K_h(X_i - x)$ and $K_h(X_j - x)$ are nearly uncorrelated as $h \to 0$. This property is, however, not shared by parametric estimators. For example, the variance of the sample mean $n^{-1} \sum_{i=1}^n X_i$ depends on all k-step correlation coefficients.

The key usage of mixing conditions is contained in the following lemma, due to Volkonskii and Rozanov (1959). The lemma shows that dependent random variables can be approximated by a sequence of independent random variables having the same marginal distributions. This can be seen by taking $V_j = \exp(it_j X_j)$. Then Lemma 6.1 becomes a statement about the characteristic functions of the random variables.

Lemma 6.1 *Let $V_1, \cdots, V_L$ be random variables measurable with respect to the σ-algebras $\mathcal{F}^{j_1}_{i_1}, \cdots, \mathcal{F}^{j_L}_{i_L}$ respectively with $i_{l+1} - j_l \geq$*

$w \geq 1$ and $|V_j| \leq 1$ for $j = 1, \cdots, L$. Then

$$\left| E \prod_{j=1}^{L} V_j - \prod_{j=1}^{L} E(V_j) \right| \leq 16(L-1)\alpha(w).$$

For other mixing coefficients, this lemma continues to apply due to (6.4).

6.2.2 Local polynomial fitting TIME SERIES

Consider observations $(X_1, Y_1), \cdots, (X_n, Y_n)$ that can be thought of as realizations from a stationary process. Of interest is to estimate

$$m(x) = E(Y_i | X_i = x) \text{ and its derivative } m^{(\nu)}(x).$$

One then approximates $m(x)$ as in (3.2) and fits locally a polynomial as in (3.3). Denote by $\widehat{\beta}(x)$ the solution to the weighted least squares problem (3.3). Then, an estimator for $m^{(\nu)}(x)$ is $\widehat{m}_\nu(x) = \nu! \widehat{\beta}_\nu(x)$.

The above setup is also applicable to nonlinear prediction. Suppose that a stationary time series $X_1, \cdots, X_n$ is given. Then, to perform a k-step forecasting, namely, to estimate $m(x) = E(X_{i+k} | X_i = x)$, we construct data as follows:

$$(X_i, Y_i), i = 1, \cdots, n-k \quad \text{with} \quad Y_i = X_{i+k}.$$

Thus, this nonlinear prediction problem is a special case of a nonparametric regression problem for dependent data.

We first state a theorem, which states that under certain mixing conditions, local polynomial estimators for dependent data have the same asymptotic behavior as for independent data. The joint asymptotic normality here is also applicable to independent observations and is a complement to Theorem 3.1. The proof of the theorem, due to Masry and Fan (1993), will be outlined in Section 6.6. Note that the bias arguments are unaffected, whereas the variance calculations are affected under dependence.

Let $f(x)$ be the density of X_1 and $\sigma^2(x) = \text{Var}(Y_1 | X_1 = x)$. Let S, S^* and c_p denote the same moment matrices and vector as those introduced before Theorem 3.1. Then we have the following result.

Theorem 6.1 *Under Condition 1 in Section 6.6, if* $h_n =$

$O(n^{1/(2p+3)})$, then as $n \to \infty$,

$$\sqrt{nh_n}\left[\text{diag}(1,\cdots,h_n^p)\{\widehat{\beta}(x)-\beta(x)\}-\frac{h_n^{p+1}m^{(p+1)}(x)}{(p+1)!}S^{-1}c_p\right]$$
$$\xrightarrow{D} N\{0,\sigma^2(x)S^{-1}S^*S^{-1}/f(x)\}$$

at x, a continuity point of $\sigma^2 f$, whenever $f(x) > 0$.

An immediate consequence of Theorem 6.1 is that the derivative estimator $\widehat{m}_\nu(x)$ based on the local polynomial fitting is asymptotically normal:

$$\sqrt{nh_n^{2\nu+1}}\left\{\widehat{m}_\nu(x)-m^{(\nu)}(x)-\int t^{p+1}K_\nu^*(t)dt\frac{\nu!m^{(p+1)}(x)}{(p+1)!}h_n^{p+1-\nu}\right\}$$
$$\xrightarrow{D} N\left\{0,\frac{(\nu!)^2\sigma^2(x)\int K_\nu^{*2}(t)dt}{f(x)}\right\}, \qquad (6.5)$$

where K_ν^* is the equivalent kernel given by (3.16). When $\nu = 0$, (6.5) gives the asymptotic normality of $\widehat{m}(x)$ itself.

As in all nonparametric regression problems, one has to select a smoothing parameter for the local polynomial regression. The methods introduced in Chapter 4 can be employed. In particular, we recommend using the refined bandwidth selector (4.21) when the underlying regression curve is reasonably smooth, and the variable bandwidth selector in Section 4.6.3 when the underlying curve is not very smooth.

To illustrate the effectiveness of the proposed methods, we generated a nonlinear time series from the model

$$X_t = 2X_{t-1}/(1+X_{t-1}^2) + \varepsilon_t, \quad t = 1,\cdots,100,$$

with X_t independent of ε_t, where $\{\varepsilon_t\}$ are independent and uniformly distributed over $[-1,1]$. This model was used by Härdle (1990, page 207) to illustrate the kernel regression method with a cross-validation bandwidth selector. The starting value X_0 was generated by running the process from X_{-99} to X_{-1} with the starting value $X_{-100} = 0$. Figure 6.1 (a) presents the simulated nonlinear time series. To obtain the one-step nonlinear prediction, we set $Y_i = X_{i+1}$, $i = 1,\cdots,99$. The scatter plot of (X_i, Y_i) is shown in Figure 6.1 (b). We then used the local linear regression estimate with the Epanechnikov kernel, along with the refined bandwidth (4.21), to smooth the scatter plot. The bandwidth $\widehat{h}_{1,0}^R = 0.600$ was selected. Figure 6.1 (b) presents the estimated one-step nonlinear prediction function, superimposed by the true nonlinear prediction function.

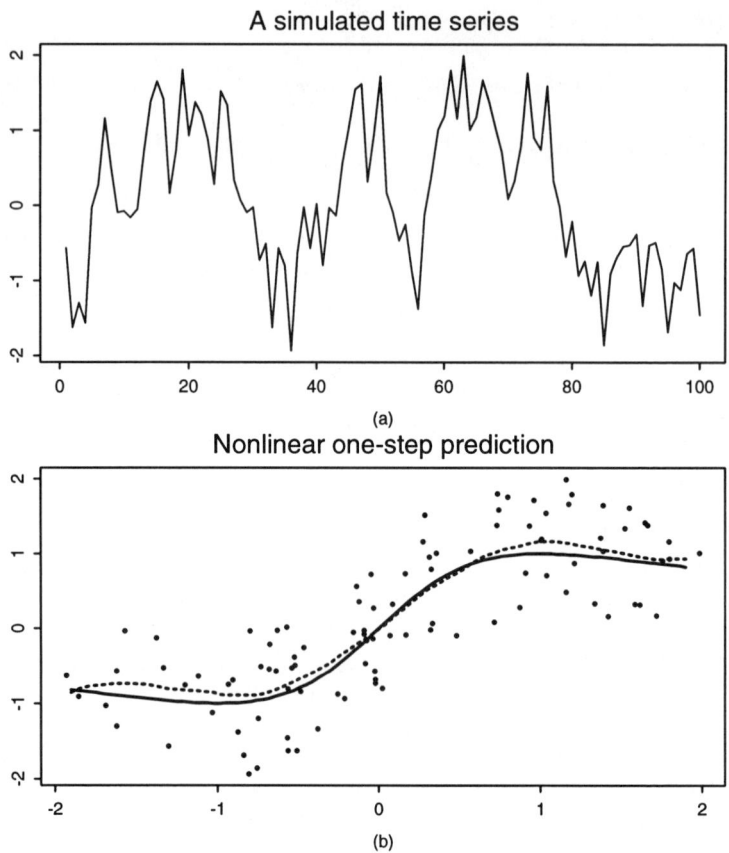

Figure 6.1. *(a) A simulated nonlinear time series. (b) Scatter plot of the lag one time series along with nonlinear prediction functions. Solid curve – true function; dashed curve – estimated function.*

We now illustrate the proposed nonlinear prediction procedure via the well known sunspots data set (see for example Waldmeir (1961)). Presented in Figure 6.2 (a) is the yearly average number of sunspots from 1700 to 1992. The *power spectrum* of this series is shown in Figure 6.7. Inspecting the power spectrum, there are two spikes near frequencies 0 and $2\pi/10$. This indicates that there is a smooth transition from X_t to X_{t-1} and that there is a 10-year cyclic pattern. Thus, we can predict X_t based on X_{t-1} and X_{t-10}. We first remove the '*seasonal effects*' by a linear regression of X_t on X_{t-10}, leading to the new series $Y_t = X_t - 0.903 X_{t-10}$. Figure

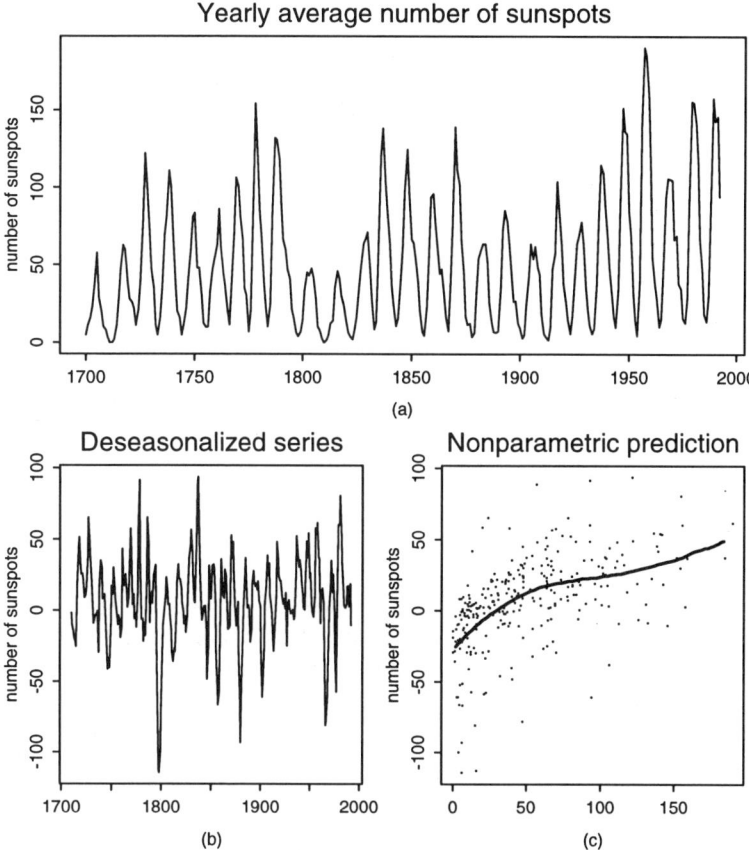

Figure 6.2. *(a) Yearly average number of sunspots from 1700 to 1992; (b) deseasonalized time series $Y_t = X_t - 0.903 X_{t-10}$; (c) scatter plot of Y_t against X_{t-1} along with estimated regression function $\widehat{f}$.*

6.2 (b) shows the 'deseasonalized series' $\{Y_t\}$. We now regress the variable $\{Y_t\}$ on X_{t-1} using the local linear regression estimator with the Epanechnikov kernel and the bandwidth 42.57 selected by the refined bandwidth selector (4.21). Figure 6.2 (c) depicts the resulting regression curve. Combining the two-step fitting, we obtain the following prediction equation

$$\widehat{X}_t = \widehat{f}(X_{t-1}) + \widehat{\alpha} X_{t-10}, \text{ with } \widehat{\alpha} = 0.903, \qquad (6.6)$$

where $\widehat{f}(\cdot)$ is the function presented in Figure 6.2 (c). This function $\widehat{f}$ appears to be nonlinear. Equation (6.6) exhibits a form of the partially linear model (7.12) discussed in Chapter 7. The parameters can also be estimated by using the methods discussed there.

6.2.3 Estimation of conditional densities

The mean regression function $m(x)$ summarizes only the 'center' of the conditional density of Y given $X = x$, denoted by $g(y|x)$. However, for a nonlinear time series, it is also very useful to examine the symmetry and multimodality of the estimated conditional density $\widehat{g}(\cdot|x)$. If the conditional density appears multimodal or highly skewed, it is not appropriate to summarize it via the conditional mean regression function.

Consider a stationary sequence $(X_1, Y_1), \cdots, (X_n, Y_n)$. To make a connection with the nonparametric regression problem, we note that

$$E\{K_{h_2}(Y - y)|X = x\} \approx g(y|x), \quad \text{as } h_2 \to 0, \qquad (6.7)$$

where K is a given density function and $K_h(\cdot) = K(\cdot/h)/h$. The left-hand side of (6.7) can be regarded as the regression of the variable $K_{h_2}(Y - y)$ on X. This leads us to consider the following locally weighted least squares regression problem: minimize

$$\sum_{i=1}^{n} \left\{ K_{h_2}(Y_i - y) - \sum_{j=0}^{p} \beta_j (X_i - x)^j \right\}^2 W_{h_1}(X_i - x), \qquad (6.8)$$

where $W(\cdot)$ is a kernel function. Let $\widehat{\beta}_j(x, y)$ be the solution to the least squares problem (6.8). Then, from Section 3.1, it is clear that $g^{(\nu)}(y|x) = \frac{\partial^\nu g(y|x)}{\partial x^\nu}$ can be estimated by $\nu! \widehat{\beta}_\nu(x, y)$, namely,

$$\widehat{g}_\nu(y|x) = \frac{\widehat{\partial^\nu g(y|x)}}{\partial x^\nu} = \nu! \widehat{\beta}_\nu(x, y). \qquad (6.9)$$

As usual, we write $\widehat{g}_0(\cdot|\cdot)$ as $\widehat{g}(\cdot|\cdot)$. This kind of idea is due to Fan, Yao and Tong (1996), who also implemented this idea in the multivariate setting.

By (3.11), it is clear that

$$\widehat{g}_\nu(y|x) = \nu! \sum_{i=1}^{n} W_\nu^n\{(X_i - x)/h_1\} K_{h_2}(Y_i - y),$$

where W_ν^n is given in (3.11) with K replaced by W. Thus, the estimator $\widehat{g}_\nu(y|x)$ admits a kernel form, but $W_\nu^n(\cdot)$ depends on

NONLINEAR PREDICTION 225

the design points $X_1, \cdots, X_n$ and the location x. We attribute, in Section 3.2.2, this property as the key to the design adaptation of local polynomial fitting.

When $K(\cdot)$ has mean zero, the local polynomial regression estimator $\widehat{m}(x)$ is just the 'gravity center' of the conditional density $\widehat{g}(\cdot|x)$:

$$\widehat{m}(x) = \int y\widehat{g}(y|x)dy.$$

Similarly, the local polynomial derivative estimator can be obtained via

$$\widehat{m}_\nu(x) = \int y\widehat{g}_\nu(y|x)dy.$$

We now briefly discuss the bandwidth selection problem. Since estimation of a bivariate conditional density function is computationally intensive, we focus on a simple bandwidth selector. The bandwidth h_2 is chosen by the normal reference rule (2.42), with σ being estimated by the standard deviation of the Y-variable. Now, for a given h_2 and y, problem (6.8) is the standard nonparametric regression problem discussed in Section 6.2.2. Thus, we can apply the bandwidth selectors proposed in Chapter 4. To save computation, we use the bandwidth selector (4.19) based on the residual squares criterion.

As an illustration, we consider the following simple quadratic model

$$X_t = 0.23X_{t-1}(16 - X_{t-1}) + 0.4\varepsilon_t \quad t \geq 1, \qquad (6.10)$$

where $\{\varepsilon_t\}$ are independent random variables having the same distribution as the sum of 48 independent random variables, each distributed uniformly on $[-0.25, 0.25]$. By the central limit theorem, ε_t can effectively be treated as a standard normal variable. However, it has bounded support, which is necessary for the stationarity of the time series (see Chan and Tong (1994)). A sample of size 1000 was drawn from model (6.10). The *skeleton* of model (6.10) appears chaotic. See the top panel of Figure 6.3 (a). The bottom panel of Figure 6.3 (a) shows a typical simulated time series.

We now consider estimating the conditional density for k-step ahead forecasting, namely the conditional density of X_{t+k} given X_t. We employed the local linear fit with both K and W Gaussian kernels. The bandwidths were selected by data according to the rules mentioned above. Figure 6.3 (b)–(d) depicts the estimated conditional densities for $k = 1, 2$ and 3. From model (6.10), it is

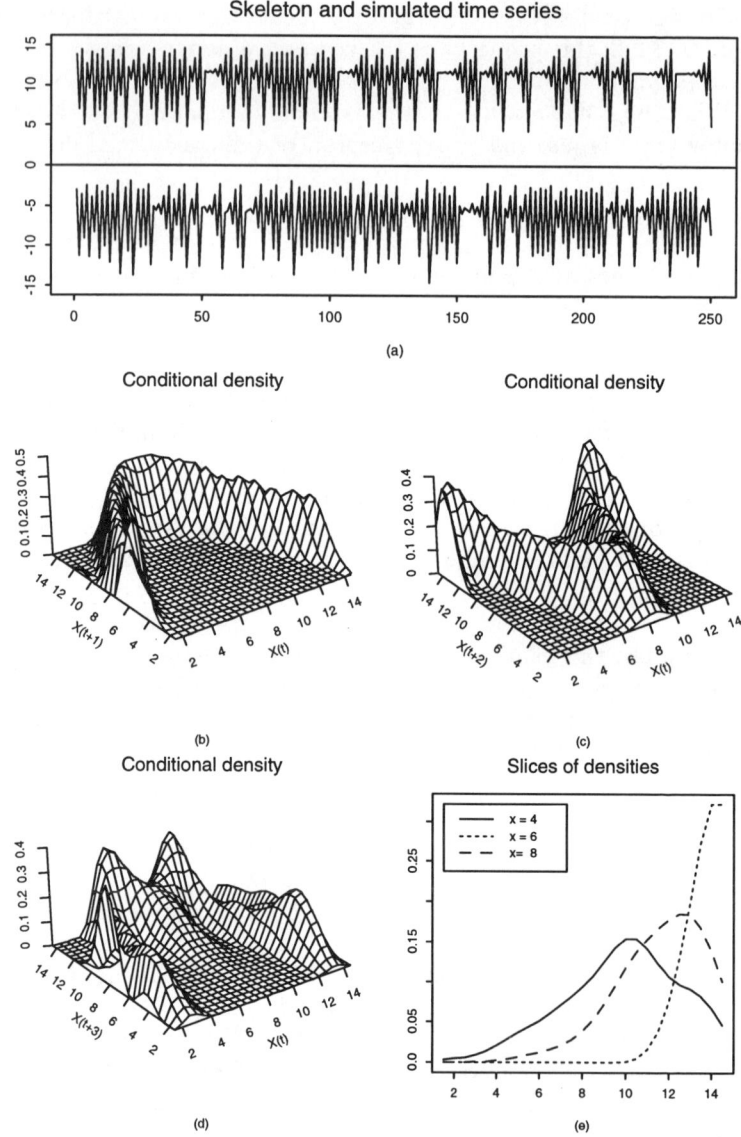

Figure 6.3. *(a) Plot of the skeleton $x_t = 0.23x_{t-1}(16 - x_{t-1})$ and a simulated series from model (6.10). Top panel – the skeleton; bottom panel – a simulated time series. (b)–(d) are the estimated conditional densities of X_{t+k} given X_t for $k = 1, 2$ and 3, respectively. (e) A few slices of conditional densities of X_{t+3} given $X_t = 4, 6$ and 8. From Fan, Yao and Tong (1996).*

clear that the conditional density of X_{t+1} given X_t is approximately normal with constant variance. This is consistent with the quadratic shape of the ridge shown in Figure 6.3 (b). For the two-step prediction, the shape of the conditional density cannot easily be seen from model (6.10). Nevertheless, with the proposed technique, we are able to show in Figure 6.3 (c) that the conditional density of X_{t+2} given X_t is unimodal. The ridge of the conditional density appears to be a fourth order polynomial function of x. Finally, we present the estimated conditional density of X_{t+3} given X_t in Figure 6.3 (d). Even with the known model (6.10), it is very difficult to compute this conditional density for the three-step prediction. The shape of this estimated conditional density is hard to examine. To have a closer look of the shape, we took three slices of the conditional density at $x = 4, 6$ and 8 and plotted them in Figure 6.3 (e). From this figure, it is clear that these conditional densities are unimodal.

The advantage of the conditional density method is that it not only gives a predicted value, but also displays other information such as the spread (uncertainty) of the prediction. Consider for example three-step ahead forecasting. Given the current value $x = 4$, the distribution of the future values is indicated by the solid curve in Figure 6.3 (e). Inspecting this curve, the predicted value is 10, give or take 2 or so. Similarly one can also zoom Figures 6.3 (b) and (c) via taking slices of the conditional densities.

The rest of this section presents the asymptotic properties of $\widehat{g}_\nu(y|x)$, which can be derived analogously to Theorem 6.1. For the sake of simplicity, we only discuss the two most useful cases: $p = 1, \nu = 0$ and $p = 2, \nu = 1$.

Let $\mu_K = \int t^2 K(t)dt$, $\nu_K = \int \{K(t)\}^2 dt$, $\mu_j = \int t^j W(t)dt$, and $\nu_j = \int t^j \{W(t)\}^2 dt$. Define

$$\vartheta_{n,0}(x,y) = \frac{h_1^2 \mu_2}{2} \frac{\partial^2 g(y|x)}{\partial x^2} + \frac{h_2^2 \mu_K}{2} \frac{\partial^2 g(y|x)}{\partial y^2} + o(h_1^2 + h_2^2),$$

$$\sigma_0^2(x,y) = \nu_K \nu_0 \frac{g(y|x)}{f(x)},$$

where $f(x)$ is the marginal density of X_1.

The following two theorems have been established by Fan, Yao and Tong (1996).

Theorem 6.2 *Under Condition 2 in Section 6.6, when the local linear fit $p = 1$ is used to estimate $g(y|x)$, we have*

$$\sqrt{nh_1 h_2}\{\widehat{g}(y|x) - g(y|x) - \vartheta_{n,0}\} \xrightarrow{D} N\left(0, \sigma_0^2\right),$$

provided that the bandwidths h_1 and h_2 converge to zero in such a way that $nh_1h_2 \to \infty$.

When the local linear fit is used to estimate $g(y|x)$, the asymptotic bias and variance are what is intuitively expected. The bias comes from the approximations in both x and y directions. The variance comes from the local conditional variance in the density estimation setting, which is $g(y|x)$.

To state the results for the local quadratic fit, we denote

$$\vartheta_{n,1}(x,y) = \frac{1}{2}\mu_K \frac{\partial^2 g(y|x)}{\partial y^2} h_2^2 + o(h_1^3 + h_2^2),$$

$$\sigma_1^2(x,y) = \frac{g(y|x)\nu_0\nu_K}{f(x)} \frac{\mu_4^2\nu_0 - 2\mu_2\mu_4\nu_2 + \frac{1}{2}\mu_2^2\nu_4}{(\mu_4 - \mu_2^2)^2},$$

$$\vartheta_{n,2}(x,y) = \frac{\mu_4}{6\mu_2} \frac{\partial^3 g(y|x)}{\partial x^3} h_1^2 + \frac{1}{2}\mu_K \frac{\partial^3 g(y|x)}{\partial x \partial y^2} h_2^2 + o(h_1^2 + h_2^2),$$

$$\sigma_2^2(x,y) = \frac{g(y|x)\nu_K}{f(x)} \frac{\nu_0\nu_2}{\mu_2^2}.$$

The constant factors that are related to the kernel W can be expressed in terms of the equivalent kernel W_ν^*, induced by the local quadratic fit.

Theorem 6.3 *Suppose that the bandwidths h_1 and h_2 converge to zero and that $nh_1^3 h_2 \to \infty$. Under Condition 2 in Section 6.6, when the local quadratic fit $p = 2$ is employed, the random variables*

$$\sqrt{nh_1 h_2}\{\widehat{g}(y|x) - g(y|x) - \vartheta_{n,1}\} \xrightarrow{D} N(0, \sigma_1^2)$$

and

$$\sqrt{nh_1^3 h_2}\{\widehat{g}_1(y|x) - \frac{\partial}{\partial x}g(y|x) - \vartheta_{n,2}\} \xrightarrow{D} N(0, \sigma_2^2).$$

Moreover, they are asymptotically jointly normal with covariance 0.

6.3 Percentile and expectile regression

In forecasting the economic climate such as the exchange rates in Section 5.5, one is not only interested in the expected value or median of a conditional distribution, but also in the low and high percentiles of the conditional distribution. These percentiles can be estimated via quantile regression introduced in Section 5.5. They can also be obtained by using conditional expectiles.

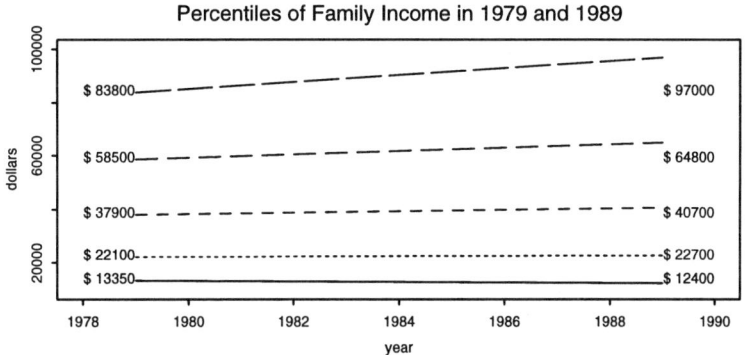

Figure 6.4. *Family income at selected points of the income ladder in 1979 and 1989 (in 1992 dollars). Solid curve – 10th percentile. From shortest to longest dash – 25th, 50th, 75th and 90th percentiles.*

Regression percentiles also provide informative summary statistics. From the yearly Current Population Survey conducted by the US government on a sample of 60,000 households, the sample quantiles can be computed. While information for years 1964 to 1989 should be available, we have at this time only data for 1979 and 1989. See Table 3 of Rose (1992). Figure 6.4 depicts the sample quantiles. The 10th and 90th percentiles at $x = 1979$ and $x = 1989$ indicate opposite trends. Figure 6.4 quantifies the extent to which the poor got poorer and the rich got richer during the Reagan administration (1981–1988).

6.3.1 Regression percentile

To estimate the conditional quantile $\xi_\alpha(x)$ based on a stationary sequence of data $(X_1, Y_1), \cdots, (X_n, Y_n)$, we apply, as in Section 5.5, local polynomial fitting to the following asymmetric least absolute deviation problem:

$$\min_\beta \sum_{i=1}^n \ell_\alpha \left(Y_i - \sum_{j=0}^p \beta_j (X_i - x)^j \right) K_h(X_i - x), \quad (6.11)$$

where $\ell_\alpha(t) = |t| + (2\alpha - 1)t$. The estimated conditional percentiles and their derivatives are

$$\widehat{\xi_\alpha^{(\nu)}}(x) = \nu! \widehat{\beta}_\nu(x). \quad (6.12)$$

For a univariate time series $X_1, \cdots, X_n$, we can estimate the conditional percentiles of X_{t+k} given X_t by using $Y_i = X_{i+k}$, from which a $(1-\alpha)100\%$ predictive interval $[\widehat{\xi}_{\alpha/2}(x), \widehat{\xi}_{1-\alpha/2}(x)]$ can be formed. In other words, given the current value $X_n = x$, with probability $(1-\alpha)$ this predictive interval contains the future value X_{n+k}.

The sampling property of $\widehat{\xi^{(\nu)}_\alpha}(x)$ was established in (5.60) for independent data. See also Section 5.5.2. For dependent data, we have not yet seen a formally published result. However, a closely related result is established in Yao and Tong (1995), which verifies the notion that short and medium term dependence does not have an adverse effect on the nonparametric technique. Indeed, as we explained in Section 6.2.1, the short and medium range dependence has little influence on a local smoothing method, since the random variables $K_h(X_i - x)$ and $K_h(X_j - x)$ are weakly dependent for any given $i \neq j$ and x. Thus, the percentile regression technique continues to apply to the dependent data under certain mixing conditions.

An illustration of this approach was given in Figure 5.11. To provide further insight, a sample of size 1000 was drawn from (6.10). Figure 6.5 presents its estimated conditional percentiles. These regression percentiles indicate not only the range of prediction, but also the homoscedasticity of the model. See also Section 5.5.2. Parallel quantile curves would suggest the homoscedasticity of $X(t+1)$ given $X(t)$.

6.3.2 Expectile regression

A closely related technique is the conditional expectile. Recall that

$$\xi_\alpha(x) = \operatorname{argmin}_\theta E\{\ell_\alpha(Y - \theta) | X = x\}, \qquad (6.13)$$

namely, the percentile solves the asymmetric least absolute deviation problem with

$$\begin{aligned}\ell_\alpha(t) &= |t| + (2\alpha - 1)t \\ &= 2\{\alpha |t| I(t > 0) + (1 - \alpha)|t| I(t \leq 0)\}, \qquad 0 \leq \alpha \leq 1.\end{aligned}$$

The discontinuity of the function $\ell'_\alpha(\cdot)$ at the origin makes the computation of the percentile not so appealing. A natural modification is to use the asymmetric square loss

$$Q_w(t) = wt^2 I(t > 0) + (1-w)t^2 I(t \leq 0), \qquad 0 \leq w \leq 1 \quad (6.14)$$

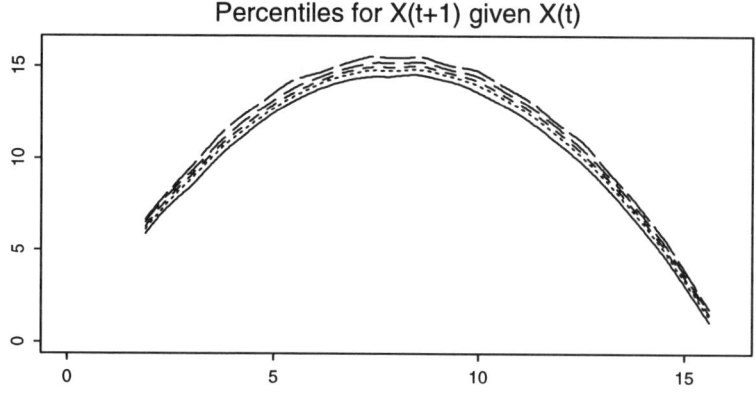

Figure 6.5. *Estimated conditional regression percentiles based on a simulated data set from (6.10) for* $\alpha = 0.05, 0.25, 0.5, 0.75$ *and* 0.95. *Results taken from Yao and Tong (1995).*

leading to

$$\tau_w(x) = \operatorname{argmin}_\theta E\{Q_w(Y-\theta)|X=x\}. \tag{6.15}$$

These quantities are called *expectiles* by Newey and Powell (1987), following earlier work by Aigner, Amemiya and Poirier (1976). Note that when $w = 0.5$, the expectile reduces to the mean regression function.

By taking the derivative with respect to θ in (6.15) and setting it to zero, we obtain the equation

$$\frac{E[|Y - \tau_w(x)|I\{Y \le \tau_w(x)\}]}{E\{|Y - \tau_w(x)|\}} = w. \tag{6.16}$$

The interpretation of $\tau_w(x)$ is as follows. For each strip near x, the average distance from the data Y_i below $\tau_w(x)$ to $\tau_w(x)$ is 100 w%. Thus, the expectile shares an interpretation similar to the percentile, replacing the distance by the number of observations.

The key advantage of the expectile is its computing expedience. Efron (1991) describes a very appealing algorithm for computing expectiles, starting from the least squares estimate with $w = 0.5$ and iterating only once or twice by using the Newton-Raphson method as w slowly deviates away from 0.5. The family of curves $\{\tau_w(\cdot)\}$ can be computed very rapidly. The disadvantage of this expectile method is that it does not have an interpretation as direct as the percentiles. Further, it is not as robust against outliers as the percentiles.

Local polynomial estimation of the expectiles is straightforward. One needs only to replace (6.11) by

$$\min_{\beta} \sum_{i=1}^{n} Q_w \left(Y_i - \sum_{j=0}^{p} \beta_j (X_i - x)^j \right) K_h(X_i - x), \quad (6.17)$$

leading to the estimators

$$\widehat{\tau_w^{(\nu)}}(x) = \nu! \widehat{\beta}_\nu(x). \quad (6.18)$$

Note that this is a special case of (5.57). The sampling property of this approach for dependent observations was established by Yao and Tong (1995). It basically says that the asymptotic normality given in (5.60) holds also for a stationary sequence of observations satisfying certain mixing conditions. The technical arguments are a hybrid of the proofs of Theorems 5.2 and 6.1, given respectively in Sections 5.6 and 6.6. Since the asymptotic normality result continues to hold for dependent data, the rule of thumb bandwidth selector introduced in Chapter 5 applies here too.

In the parametric linear model, Newey and Powell (1987) and Efron (1991) discuss how to calibrate $w = w(\alpha)$ so that $\tau_w(x)$ is effectively the same as the percentile $\xi_\alpha(x)$. In a nonparametric model, such a calibration is also possible, but w would usually depend on α as well as on x, unless the model is a location-scale model

$$Y_i = m(X_i) + \sigma(X_i)\varepsilon_i, \quad X_i \text{ and } \varepsilon_i \text{ are independent.}$$

In the latter case,

$$\tau_{w(\alpha)}(x) = \xi_\alpha(x), \quad w(\alpha) = \frac{\alpha \xi_\alpha^0 - E\{\varepsilon I(\varepsilon \leq \xi_\alpha^0)\}}{2E\{\varepsilon I(\varepsilon > \xi_\alpha^0)\} - (1 - 2\alpha)\xi_\alpha^0}, \quad (6.19)$$

where ξ_α^0 is the αth quantile of ε. See Newey and Powell (1987) and Yao and Tong (1995). As intuitively expected, $w(\alpha)$ is an increasing function of α. However, the location-scale models are usually not appropriate for nonlinear time series. For example, $Y_t = X_{t+2}$ and X_t in the two-step prediction do not satisfy a location-scale model even though model (6.10) is. For non-location-scale models, the function $w(\alpha, x)$ depends on unknown functions and has to be estimated. The computational expedience of using expectiles is weakened.

As an illustration, we use the simulated data from model (6.10) to estimate expectiles for the one-step, two-step and three-step prediction. The values of $w = 0.01, 0.2, 0.54, 0.88$ and 0.99 were chosen so that the resulting expectiles correspond to $\alpha = 0.05$, $0.25, 0.5, 0.75$ and 0.95 percentiles for X_{t+1} given X_t. The relation between w and α is given in (6.19). The expectiles for one-step prediction are the same as the percentiles in Figure 6.5. Figure 6.6 depicts the expectiles for two-step and three-step forecasting. It is clear from Figure 6.6 that the conditional distributions for two-step and three-step forecasting are heteroscedastic. This conclusion is consistent with our analyses in Figure 6.3. It reveals the fact that expectiles contain similar information as percentiles.

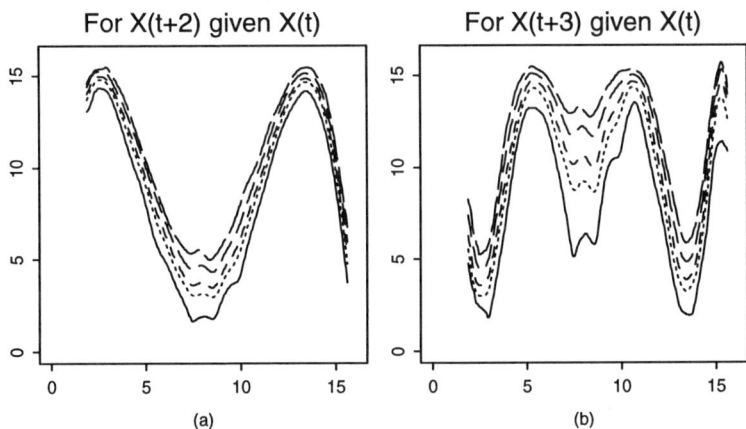

Figure 6.6. *Conditional expectiles for (a) two-step forecasting and (b) three-step prediction, with $w = 0.01, 0.2, 0.54, 0.88$ and 0.99. Results taken from Yao and Tong (1995).*

6.4 Spectral density estimation

Spectral density estimation is a very useful technique for examining certain cyclic patterns of a stationary time series. The raw material used to construct a spectral density estimate is the *periodogram*:

$$I^{(n)}(\lambda) = \frac{1}{2\pi n} |\sum_{t=1}^{n} e^{-i\lambda t} X_t|^2, \quad \lambda \in [-\pi, \pi], \qquad (6.20)$$

for an observed stationary time series $X_1, \cdots, X_n$ with mean zero. The periodogram is an asymptotically unbiased estimate of the *spectral density*

$$f(\lambda) = \frac{1}{2\pi} \sum_{u=-\infty}^{\infty} \gamma(u)\exp(-iu\lambda), \qquad \lambda \in [-\pi, \pi], \qquad (6.21)$$

where $\gamma(\cdot)$ is the *autocovariance function* defined as

$$\gamma(u) = E(X_s X_{s+u}), \quad u = 0, \pm 1, \pm 2, \ldots.$$

Note that $f(\lambda)$ is symmetric around 0 so that we need only focus on $f(\lambda)$ for $\lambda \in [0, \pi]$. The periodogram is, however, not a consistent estimate of $f(\lambda)$. See for example Brillinger (1981), Priestley (1981) and Brockwell and Davis (1989).

Consistent estimators of $f(\lambda)$ can be obtained via smoothing of the periodogram. Let $\lambda_k = 2\pi k/n$ be a *Fourier frequency*. Then, it is known that

$$I^{(n)}(\lambda_k) = f(\lambda_k) \cdot V_k + R_k, \qquad (6.22)$$

where R_k denotes a term that is asymptotically negligible, and the V_k's are independent random variables having the standard exponential distribution for $k = 1, \cdots, [(n-1)/2]$. Further, V_0 and $V_{[n/2]}$ (if n is even) have a χ_1^2-distribution. See, for example, Theorem 10.3.2 of Brockwell and Davis (1989). For convenience, we will drop $I^{(n)}(0)$ and $I^{(n)}(\pi)$ (when n is even) and only consider the periodogram

$$I^{(n)}(\lambda_k) \text{ for } k = 1, \cdots, N \quad \text{with } N = [(n-1)/2].$$

The effect of this is practically negligible because we drop at most two data points.

There are three possible ways to estimate f from the regression-type model (6.22). The first one is to smooth the logarithm of the periodogram using a least squares method. The log-periodogram admits the following regression model:

$$\begin{aligned} Y_k &= \log I^{(n)}(\lambda_k) = m(\lambda_k) + \varepsilon_k + r_k, \quad k = 1, \ldots, N, \\ \varepsilon_k &= \log(V_k) \text{ with density } \exp\{-\exp(x) + x\}, \end{aligned} \qquad (6.23)$$

where further $m(\lambda_k) = \log f(\lambda_k)$ and $r_k = \log[1 + R_k/\{f(\lambda_k)V_k\}]$ denotes an asymptotically negligible term. The resulting estimator is called a (least squares) *smoothed log-periodogram*. The second approach is to smooth the log-periodogram using a maximum local likelihood estimator. We refer to this estimator as a *local likelihood smoothed log-periodogram* or simply a local likelihood estimator. The likelihood function is constructed from the location model (6.23) by putting $r_k = 0$. This likelihood function is called the

Whittle likelihood (Whittle (1962)). Note that the Whittle likelihood based on model (6.22) is the same as that based on (6.23). Since the error distribution ε_k is known to be non-normal, the local likelihood smoothed log-periodogram is more efficient than the smoothed log-periodogram. The third approach is to smooth directly on the periodogram based on the heteroscedastic model (6.22). We will refer to the resulting estimator as the (least squares) *smoothed periodogram*. Most of the traditional approaches are based on this idea. See Brillinger (1981) and references therein. We will show that the local likelihood estimator has a smaller bias than the smoothed periodogram at peak regions, while maintaining the same asymptotic variance as the smoothed periodogram.

Based on the above discussions, we recommend using the local likelihood smoothed log-periodogram as a spectral density estimator.

6.4.1 Smoothed log-periodogram

We first note that

$$E(\varepsilon_k) = C_0 = -.57721, \quad \text{and} \quad \text{Var}(\varepsilon_k) = \pi^2/6, \qquad (6.24)$$

where C_0 is Euler's constant. See Davis and Jones (1968). To motivate an estimating scheme, we ignore the asymptotically negligible term r_k, leading to the approximated model

$$Y_k - C_0 = \log I^{(n)}(\lambda_k) - C_0 = m(\lambda_k) + (\varepsilon_k - C_0). \qquad (6.25)$$

Model (6.25) is a canonical nonparametric regression model with a uniform design density on $[0, \pi]$ and a homogeneous variance $\pi^2/6$. Thus, the local polynomial fitting technique and its bandwidth selection procedures in Chapters 3 and 4 can be employed.

We consider in particular the local linear fit. For a given λ, approximate $m(x)$ locally by

$$m(x) = a + b(x - \lambda) + O\{(x - \lambda)^2\}, \qquad (6.26)$$

where $a = m(\lambda)$ and $b = m'(\lambda)$. Then, the local linear regression estimator can be expressed as

$$\widehat{m}_{\text{LS}}(\lambda) = \widehat{a} = \sum_{j=1}^{N} K_n\left(\frac{\lambda - \lambda_j}{h}\right)(Y_j - C_0),$$

where with $S_{n,j}$ defined as in (3.10),

$$K_n(t) = \frac{1}{h} \cdot \frac{S_{n,2} - ht \cdot S_{n,1}}{S_{n,0} \cdot S_{n,2} - S_{n,1}^2} \cdot K(t).$$

The logarithms of the periodogram $\{I^{(n)}(\lambda_j)\}$ are neither exactly independent nor log-exponentially distributed. Nevertheless, the following theorem, proved by Fan and Kreutzberger (1995) and reproduced in Section 6.6, shows that $\widehat{m}_{\text{LS}}$ asymptotically behaves as if data were sampled from model (6.25). In particular, $\widehat{m}_{\text{LS}}$ possesses the property of boundary adaptation.

Theorem 6.4 *Under Condition 3 in Section 6.6, we have for each $0 < \lambda < \pi$,*

$$\sqrt{nh}\{\widehat{m}_{\text{LS}}(\lambda) - m(\lambda) - h^2 m''(\lambda)\mu_2(K)/2 + o(h^2)\}$$
$$\xrightarrow{D} N\{0, (\pi^2/6)\nu_0(K)\pi\}$$

and for a boundary point $\lambda_n^* = ch$, we have

$$\sqrt{nh}\{\widehat{m}_{\text{LS}}(\lambda_n^*) - m(\lambda_n^*) - h^2 m''(0+)\mu_2(K,c)/2 + o(h^2)\}$$
$$\xrightarrow{D} N\{0, (\pi^2/6)\nu_0(K,c)\pi\}$$

where with $\mu_{j,c} = \int_{-\infty}^{c} t^j K(t) dt$,

$$\mu_2(K,c) = \frac{\mu_{2,c}^2 - \mu_{1,c}\mu_{3,c}}{\mu_{0,c}\mu_{2,c} - \mu_{1,c}^2}, \quad \nu_0(K,c) = \frac{\int_{-\infty}^{c}(\mu_{2,c} - \mu_{c,1}t)^2 K^2(t)\,dt}{(\mu_{0,c}\mu_{2,c} - \mu_{1,c}^2)^2}.$$

Thus, the asymptotically optimal bandwidth, which minimizes the integrated asymptotic squared bias and variance, is given by

$$h_{\text{LS, OPT}} = \left[\frac{\nu_0(K)(\pi^2/6)\pi}{\mu_2^2(K)\int_0^\pi\{m''(\lambda)\}^2 d\lambda}\right]^{1/5} n^{-1/5}. \qquad (6.27)$$

This bandwidth can be estimated by using the refined bandwidth selector in Section 4.6.2, yielding an automatic procedure for estimating the spectral density.

As an illustration, we consider again the sunspots data presented in Section 6.2.2. Figure 6.7 summarizes the estimated log-spectral and spectral density using the least squares method $\widehat{m}_{\text{LS}}$ and the local likelihood method $\widehat{m}_{\text{LK}}$ to be introduced in Section 6.4.2. As anticipated, the periodogram is highly heteroscedastic, while the log-periodogram is homoscedastic. Note that there are two peaks around $\lambda = 0$ and $\lambda = 2\pi/10.8 \approx 0.58$, which inspired us to consider model (6.6).

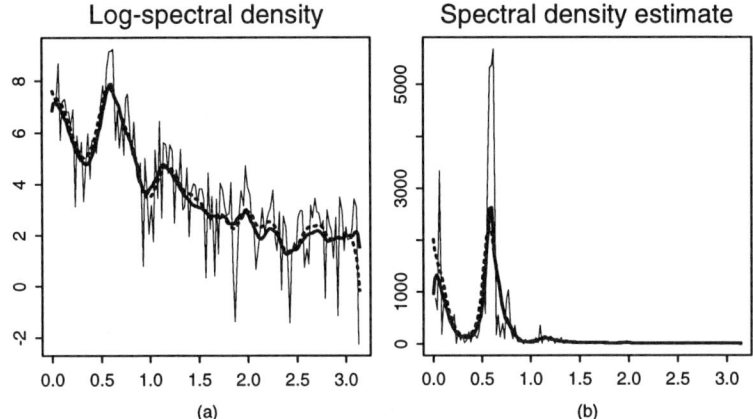

Figure 6.7. *Estimated (a) log-spectral density and (b) spectral density for sunspots data. Solid curve – local likelihood method; dashed curve – local least squares method; thin solid curve – log-periodogram $-C_0$ in (a) and periodogram in (b).*

6.4.2 Maximum local likelihood method

The smoothed periodogram estimator $\widehat{m}_{\text{LS}}$ is not efficient. In fact, the Fisher information for the location model (6.25) is 1, while the variance is $\pi^2/6 = 1.645$ (see (6.24)). Thus, the efficiency of the least squares method can be improved by using the likelihood method.

For each given λ, assuming model (6.25) and using the local linear model (6.26), we form the weighted log-likelihood as follows:

$$\mathcal{L}(a,b) = \sum_{k=1}^{N}[-\exp\{Y_k - a - b(\lambda_k - \lambda)\} + Y_k - a - b(\lambda_k - \lambda)]K_h(\lambda_k - \lambda),$$
(6.28)

where $K_h(\cdot) = K(\cdot/h)/h$. Let $\widehat{a}$ and $\widehat{b}$ be the maximizers of (6.28). The proposed local likelihood estimator for $m(x)$ is $\widehat{m}_{\text{LK}}(\lambda) = \widehat{a}$.

The weighted log-likelihood (6.28) is similar to the Whittle (1962) likelihood based on the exponential model

$$\exp(Y_k) \sim \text{Exponential}\{f(\lambda_k)\},$$

except that the kernel weight is introduced to localize the approximation (6.26). It is a strictly concave function so that there is a unique maximizer. The maximizer can be found by using the *Newton-Raphson* method or the *Fisher scoring* method using $\widehat{m}_{\text{LS}}(\lambda)$ as initial value.

The following result, due to Fan and Kreutzberger (1995), gives the sampling properties of $\widehat{m}_{\text{LK}}(\lambda)$. The proof is outlined in Section 6.6.

Theorem 6.5 *Under Condition 3 in Section 6.6, we have for each $0 < \lambda < \pi$,*

$$\sqrt{nh}\{\widehat{m}_{\text{LK}}(\lambda) - m(\lambda) - h^2 m''(\lambda)\mu_2(K)/2 + o(h^2)\}$$
$$\xrightarrow{D} N\{0, \nu_0(K)\pi\}$$

and for a boundary point $\lambda_n^* = ch$, we have

$$\sqrt{nh}\{\widehat{m}_{\text{LK}}(\lambda_n^*) - m(\lambda_n^*) - h^2 m''(0+)\mu_2(K,c)/2 + o(h^2)\}$$
$$\xrightarrow{D} N\{0, \nu_0(K,c)\pi\},$$

with $\mu_2(K,c)$ and $\nu_0(K,c)$ as in Theorem 6.4.

From Theorem 6.5, it is easily seen that the asymptotic variance of $\widehat{m}_{\text{LK}}$ is a factor of $\pi^2/6$ smaller than that of $\widehat{m}_{\text{LS}}$, while both estimators have the same asymptotic bias. In other words, $\widehat{m}_{\text{LS}}$ is asymptotically inadmissible.

An estimator for the spectral density f can easily be formed from the estimator for the log-spectral density. The maximum likelihood estimator for the spectral density is given by

$$\widehat{f}_{\text{LK}}(\lambda) = \exp\{\widehat{m}_{\text{LK}}(\lambda)\}.$$

It has the following sampling property:

$$\sqrt{nh}\{\widehat{f}_{\text{LK}}(\lambda) - f(\lambda) - h^2 m''(\lambda)f(\lambda)\mu_2(K)/2\}$$
$$\xrightarrow{D} N\{0, \nu_0(K)f^2(\lambda)\pi\}. \qquad (6.29)$$

We now deal with the bandwidth selection problem. From Theorem 6.5 and (6.27), the asymptotically optimal bandwidth for $\widehat{m}_{\text{LK}}$ is given by

$$h_{\text{LK, OPT}} = (6/\pi^2)^{1/5} \, h_{\text{LS, OPT}} = 0.9053 \, h_{\text{LS, OPT}}. \qquad (6.30)$$

Thus, an obvious estimator for $h_{\text{LK, OPT}}$ is

$$\widehat{h}_{\text{LK, OPT}} = 0.9053 \, \widehat{h}_{\text{LS, OPT}},$$

where $\widehat{h}_{\text{LS, OPT}}$ is the refined bandwidth selector given in Section 4.6.2.

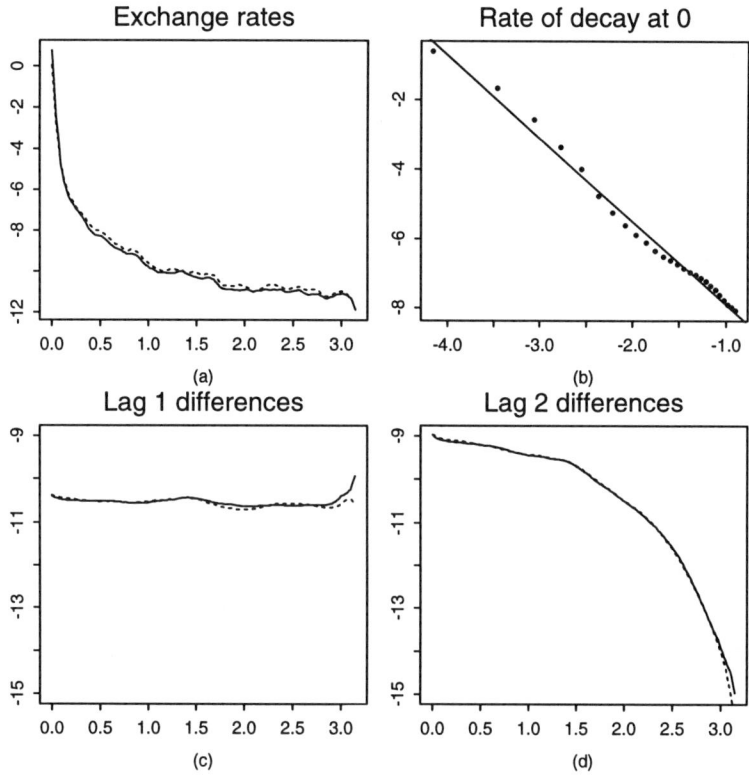

Figure 6.8. *Spectral density estimates for daily exchange rates between British pounds and US dollars. Estimated are log-spectral densities for (a) the daily exchange rates; (c) the lag 1 differences $\{X_t - X_{t-1}\}$ and (d) the lag 2 differences $\{X_t - X_{t-2}\}$. (b) Scatter plot of $\widehat{m}_{\mathrm{LK}}(\lambda_k)$ against $\log(\lambda_k)$ for $\lambda_k < 0.05$ and its least squares line. Solid curves – local likelihood method; dashed curve – the least squares method. From Fan and Kreutzberger (1995).*

We now apply the local likelihood method to the exchange rates example given in Section 5.5.2. Figure 6.8 gives the estimated log-spectral densities. For the original time series $\{X_t\}$, most of the energy is concentrated near the origin, indicating that the differences between the exchange rates for any two consecutive days are small. To understand the rate of decay of the power spectrum, we plot in Figure 6.8 (b) $\log(\lambda_j)$ against $\log\{\widehat{f}_{\mathrm{LK}}(\lambda_j)\}$ for $\lambda_j < 0.05$. We then fitted a least squares line through the data, and obtained the estimated intercept -10.29 and slope -2.39 with the estimated standard errors 0.10 and 0.05, respectively. This suggests that $f(\lambda) \approx \exp(-10.29)\lambda^{-2.39}$, as $\lambda \to 0$. Note that this function is not integrable at the origin, reflecting the volatility of the daily exchanges.

We next consider the lag 1 and 2 differences. Figure 6.8 (c) shows that the lag 1 series is simply white noise (constant spectrum) and hence is hard to predict. Figure 6.8 (d) indicates that the lag 2 series can reasonably be modelled by an AR(1) process as demonstrated in Section 5.5.2.

To better understand the performance of the proposed procedures, we use the $ARMA(p,q)$ model

$$X_t + a_1 X_{t-1} + \cdots + a_p X_{t-p} = \varepsilon_t + b_1 \varepsilon_{t-1} + \cdots + b_q \varepsilon_{t-q}$$

as testing examples, where $\varepsilon_t \sim N(0,1)$. This ARMA model has spectral density

$$\frac{1}{2\pi} \frac{|1 + b_1 \exp(-i\lambda) + \cdots + b_q \exp(-iq\lambda)|^2}{|1 + a_1 \exp(-i\lambda) + \cdots + a_p \exp(-ip\lambda)|^2}.$$

The four testing examples are

Example 1 The AR(3) model with $a_1 = -1.5$, $a_2 = 0.7$ and $a_3 = -0.1$.

Example 2 The MA(4) model with $b_1 = -0.3$, $b_2 = -0.6$, $b_3 = -0.3$ and $b_4 = 0.6$.

Example 3 The mixture model: $X_t = X_{t,1} + 4X_{t,2}$, where $X_{t,1}$ and $X_{t,2}$ are two independent time series, generated respectively from Examples 1 and 2.

Example 4 The AR(12) model with $a_4 = -0.9, a_8 = -0.7, a_{12} = 0.63$ and the rest of the coefficients zero.

Examples 1, 2 and 4 were used by Wahba (1980) to illustrate a least squares spline method. We simulated 400 times with $N = 250$. Figure 6.9 summarizes typical estimates with median performance in terms of mean absolute deviation errors. Evidently, both the local likelihood and least squares method perform very well. Note that the periodograms are highly heteroscedastic.

SPECTRAL DENSITY ESTIMATION

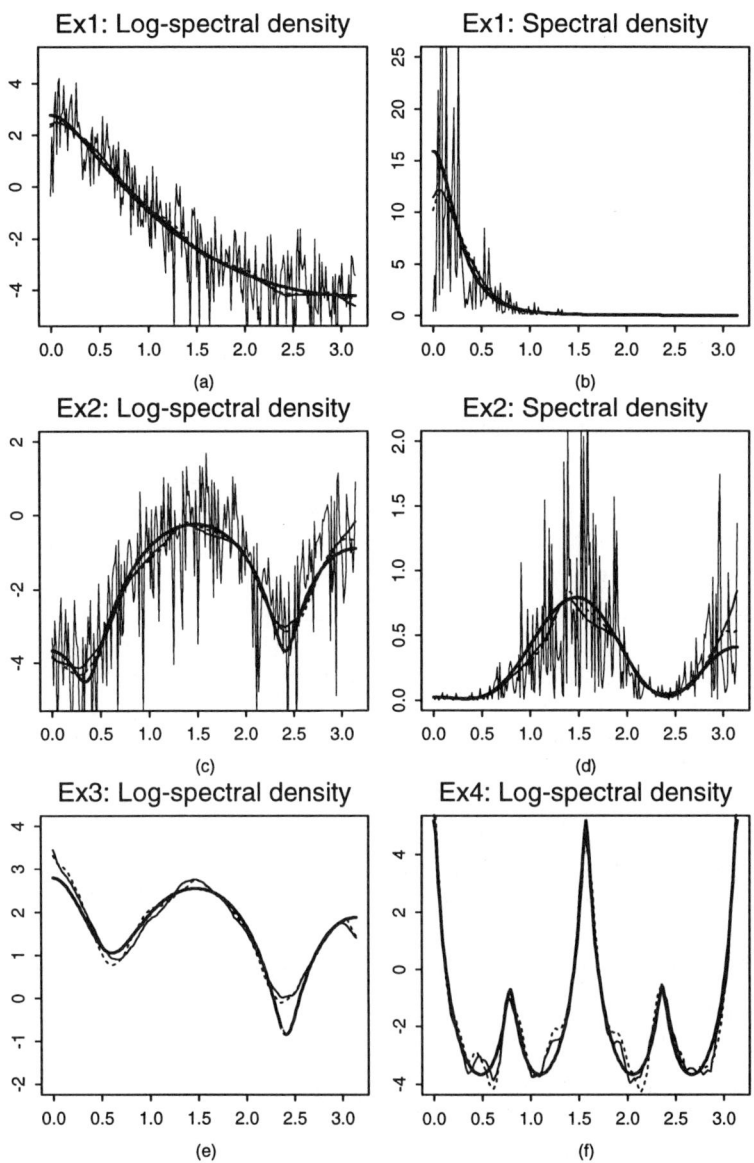

Figure 6.9. *Simulated Examples 1–4. Solid thick curves are the true functions. Solid and short-dashed curves are estimated functions based on respectively the local likelihood and least squares method. Solid thin curves are simulated periodograms or log-periodograms $- C_0$.*

6.4.3 Smoothed periodogram

Another possible way to estimate the spectral density is to smooth directly on the periodogram. Applying the local linear smoother directly to the data $\{\lambda_j, I^{(n)}(\lambda_j)\}$, we obtain

$$\widehat{f}_{\text{DLS}}(\lambda) = \sum_{j=1}^{N} K_n\left(\frac{\lambda - \lambda_j}{h}\right) I^{(n)}(\lambda_j), \qquad (6.31)$$

where the kernel function is given in Section 6.4.1. As mentioned before, most of the traditional methods use this idea of smoothing the periodogram directly and employ a kernel regression smoother.

We now offer a method of selecting the bandwidth. As shown in Figure 6.9, the spectral densities are usually very unsmooth and the periodograms are highly heteroscedastic. Therefore, we recommend using the variable bandwidth selector introduced in Section 4.6.3, resulting in an automatic procedure for estimating the spectral density. The performance of this approach is demonstrated in Figure 6.10, using the four examples in Section 6.4.2. The curves presented have median performance, in terms of mean absolute deviation loss, among 400 simulations with sample size $N = 250$.

Smoothed periodograms have a few drawbacks. Firstly, periodograms can often be highly heteroscedastic, making the task of local bandwidth selection hard. Secondly, the estimate can be unstable in the peak regions. In fact, the tail of the exponential distribution is not very light, and the variability of periodograms at peak regions is very large. Thus, outliers can often be observed around peaks, which seriously affect the least squares estimate. Thirdly, an asymptotic study shows that $\widehat{f}_{\text{DLS}}$ has a larger asymptotic bias than $\widehat{f}_{\text{LK}}$ at peak regions. Indeed, it can easily be shown that

$$\sqrt{nh}\{\widehat{f}_{\text{DLS}}(\lambda) - f(\lambda) - h^2 f''(\lambda)\mu_2(K)/2\} \xrightarrow{D} N\left\{0, \nu_0(K)f^2(\lambda)\pi\right\}. \qquad (6.32)$$

Comparing this with (6.29), and using

$$f''(\lambda) = f(\lambda)m''(\lambda) + f(\lambda)\{m'(\lambda)\}^2,$$

the bias of $\widehat{f}_{\text{DLS}}$ is larger than that of $\widehat{f}_{\text{LK}}$ at convex regions of m, namely $m''(\lambda) > 0$.

In conclusion, we recommend using the maximum local likelihood method for estimating spectral densities.

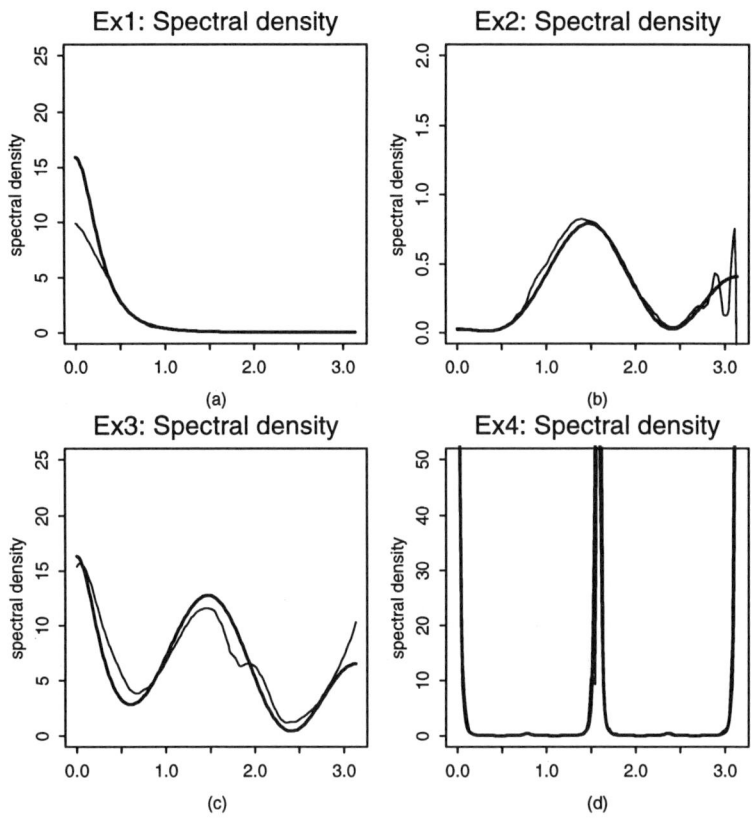

Figure 6.10. *Typical estimated spectral densities using smoothed periodogram $\widehat{f}_{\text{DLS}}$ with the variable bandwidth proposed in Fan and Gijbels (1995a). Solid thick curve – true spectral density; solid curve – estimated spectral density.*

6.5 Sensitivity measures and nonlinear prediction

Salient features of a nonlinear stochastic system include its sensitivity to initial values and noise amplification through time evolution. The *sensitivity* indicates how different two trajectories of the system can be given two nearby initial values. It monitors the instability of the prediction for the given initial values and hence the quality of prediction, since observations are always subject to rounding or measurement errors. *Noise amplification* refers to the fact that the nonlinear stochastic system will not be homoscedastic through time evolution: the noise level at each given point will be amplified or reduced. The noise level also affects the quality of prediction.

Take the simple nonlinear system (6.10) as an example. While the original model has a homogeneous error, as time evolves with two or three more steps, the conditional variance is no longer homogeneous. This can also be seen from Figure 6.6. The noise level for the two-step and three-step prediction given $X_t = 8$ is amplified through time evolution. Figure 6.6 also reveals the fact that the predictions at a region around $X_t = 12$ are unstable, since a small change in the initial value would lead to a significant change in prediction.

This section is based mainly on recent developments by Yao and Tong (1994a, b).

6.5.1 Sensitivity measures

A discrete-time dynamical *stochastic system* can be described by

$$\mathbf{X}_t = F(\mathbf{X}_{t-1}, \varepsilon_t), \tag{6.33}$$

where $\mathbf{X}_t$ denotes a state vector in $\mathbb{R}^d$, F is a real vector-valued function and ε_t is a noise process satisfying

$$E[\varepsilon_t | \mathbf{X}_0, \cdots, \mathbf{X}_{t-1}] = 0.$$

We assume that the series $\{\mathbf{X}_t\}$ is stationary. A nonlinear time series

$$X_t = f(X_{t-1}, \ldots, X_{t-p}) + \varepsilon_t, \tag{6.34}$$

with *embedding dimension d*, is included in model (6.33) by considering

$$\mathbf{X}_t = (X_t, X_{t-1}, \ldots, X_{t-p+1})^T.$$

The memory of the stochastic system (6.33) of its initial value, after a k-step time evolution, can be described via the conditional density $g_k(\mathbf{y}|\mathbf{x})$ of $\mathbf{X}_k$ given $\mathbf{X}_0 = \mathbf{x}$. For two nearby initial values $\mathbf{x}$ and $\mathbf{x} + \boldsymbol{\delta}$, how much can the difference between $g_k(\cdot|\mathbf{x})$ and $g_k(\cdot|\mathbf{x} + \boldsymbol{\delta})$ be? One way to manifest this sensitivity is to use the *Kullback-Leibler discriminant information*:

$$\int \{g_k(\mathbf{y}|\mathbf{x}+\boldsymbol{\delta}) - g_k(\mathbf{y}|\mathbf{x})\} \log\{g_k(\mathbf{y}|\mathbf{x}+\boldsymbol{\delta})/g_k(\mathbf{y}|\mathbf{x})\} d\mathbf{y}$$
$$= \boldsymbol{\delta}^T I_{1,k}(\mathbf{x})\boldsymbol{\delta} + o(\|\boldsymbol{\delta}\|^2),$$

as $\boldsymbol{\delta} \to 0$, where $\|\cdot\|$ is the Euclidean norm, and where, with

$\dot{g}_k(\mathbf{y}|\mathbf{x})$ denoting the partial derivatives with respect to $\mathbf{x}$,

$$I_{1,k}(\mathbf{x}) = \int \dot{g}_k(\mathbf{y}|\mathbf{x})\dot{g}_k^T(\mathbf{y}|\mathbf{x})/g_k(\mathbf{y}|\mathbf{x})dy. \tag{6.35}$$

Thus, the instability is monitored through the *Fisher information* matrix $I_{1,k}(\mathbf{x})$, when we treat the initial value as a parameter. Another possible measure of divergence is the L_2-distance between the conditional densities, leading to the sensitivity measure

$$I_{2,k}(\mathbf{x}) = \int \dot{g}_k(\mathbf{y}|\mathbf{x})\dot{g}_k^T(\mathbf{y}|\mathbf{x})dy. \tag{6.36}$$

One can also directly measure the sensitivity of the prediction function $m_k(\mathbf{x}) = E(\mathbf{X}_k|\mathbf{X}_0 = \mathbf{x})$. Under the additive error model

$$\mathbf{X}_t = m(\mathbf{X}_{t-1}) + \varepsilon_t, \tag{6.37}$$

by using the chain rule of differentiation, we find

$$m_k(\mathbf{x} + \boldsymbol{\delta}) - m_k(\boldsymbol{\delta}) = I_{3,k}(\mathbf{x})^T \boldsymbol{\delta} + o(\|\boldsymbol{\delta}\|),$$

where

$$I_{3,k}(\mathbf{x}) = E\Big\{\prod_{i=1}^{k} m'(\mathbf{X}_{i-1})|\mathbf{X}_0 = \mathbf{x}\Big\}. \tag{6.38}$$

The notion of sensitivity in a stochastic system is related to the k-step *Lyapounov exponent* for a deterministic system. See Yao and Tong (1994a).

When the system is multidimensional (i.e. $d > 1$), the task of estimating the sensitivity measures is not trivial. Therefore, we only consider the divergence in the marginal, rather than the joint, conditional distributions of $\mathbf{X}_k$ given $\mathbf{X}_0 = \mathbf{x}$. It is also of practical interest to concentrate only on the divergence in one particular component of the system, for example the first component as in the time series model (6.34).

The sensitivity measures can be estimated using a plug-in technique. Consider model (6.34). The sensitivity measures $I_{1,k}(\mathbf{x})$ and $I_{2,k}(\mathbf{x})$ can be estimated by the conditional density estimation techniques introduced in Section 6.2.3. Since the conditional density is integrated with respect to y, Fan, Yao and Tong (1996) recommend using the bandwidth h_2 half as large as the normal reference rule (2.42) and the bandwidth h_1 as in estimating conditional densities. They also show that the resulting estimators have good sampling properties. The sensitivity measure $I_{3,k}(\mathbf{x})$ can be estimated in a similar vein, using the local polynomial fitting technique in Section 6.2.2.

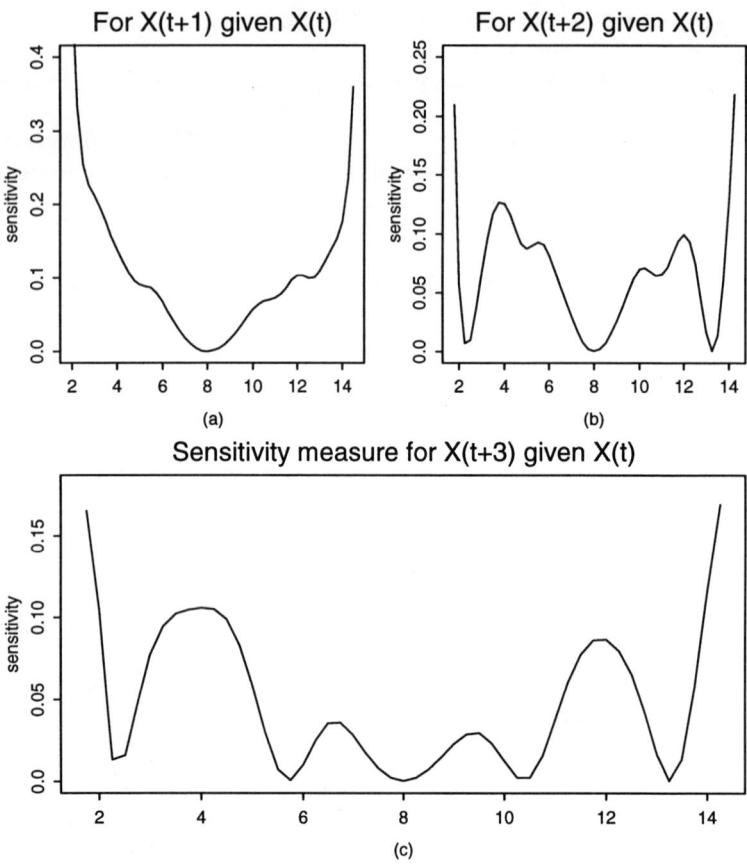

Figure 6.11. *Estimated sensitivity measures $I_{2,k}(x)$ based on a sample of $n = 1000$ from model (6.10) for (a) $k = 1$, (b) $k = 2$ and (c) $k = 3$. The results are from Fan, Yao and Tong (1996).*

As an illustration, we consider the quadratic map (6.10). Based on a sample of size 1000, we obtained the estimates for $I_{2,k}(\cdot)$. As shown in Figure 6.11, for $k = 1$, the least sensitivity location is at $x = 8$, while as x deviates away from 8, the sensitivity increases. This is consistent with Figure 6.5. For $k = 2$ and 3, the shape of the sensitivity curves is more complicated. It matches however the shape of the derivatives of the predicted curves given in Figure 6.6.

6.5.2 Noise amplification

To highlight how nonlinear dynamics amplify noise, we consider model (6.34) with $p = 1$. Assume that $\mathrm{Var}(\varepsilon_t) = \sigma_0$. Let f_k be the k-fold composition of f, i.e., $f_k = f \circ \cdots \circ f$ (k times). As $\sigma_0 \to 0$, it can be shown (see Yao and Tong (1994b)) that the conditional variance $\sigma_k(x) = \mathrm{Var}(X_k | X_0 = x)$ admits the following approximation

$$\sigma_k(x) = \lambda_k(x)\sigma_0\{1 + o(1)\}, \tag{6.39}$$

where

$$\lambda_k(x) = 1 + \sum_{t=1}^{k-1}\Big[\prod_{i=t}^{k-1} f'\{f_i(x)\}\Big]^2.$$

Thus, the noise-level after a k-step time evolution can be considerably large when the derivative map is large for a large range of values of x. On the other hand, it is possible to have $\sigma_k(x) > \sigma_{k+1}(x)$ – reduction of noise-level is also possible. Figure 6.6 illustrates, to some extent, this phenomenon.

6.5.3 Nonlinear prediction error

We now consider the error of nonparametric prediction under model (6.34). Assume that the sequence generated from model (6.34) is stationary. Let $\mathbf{X}_t = (X_t, \cdots, X_{t-p+1})^T$. A possible estimator for the k-step prediction function $m_k(\mathbf{x}) = E(X_{t+k}|\mathbf{X}_t = \mathbf{x})$ is the local linear regression smoother $\widehat{m}_k(\mathbf{x})$. It is also possible to use other smoothers.

To examine how much the round-off errors or measurement errors affect the nonlinear prediction, we assume that the true initial measurement is $\mathbf{x}$, but is recorded as $\mathbf{x} + \boldsymbol{\delta}$. Under certain mixing conditions, it can be shown that the prediction error is

$$\lim_{n\to\infty} E[\{X_{n+k} - \widehat{m}_k(\mathbf{x} + \boldsymbol{\delta})\}^2 | \mathbf{X}_n = \mathbf{x}]$$
$$= \sigma_k^2(\mathbf{x}) + \{\boldsymbol{\delta}^T I_{3,k}(\mathbf{x})\}^2 + R_k, \tag{6.40}$$

almost surely, where $R_k = o(\|\boldsymbol{\delta}\|^2)$. Thus, the prediction error comes from two sources: the conditional variance and the error due to rounding. The first component is related to noise amplification and the second component is connected to a sensitivity measure. These two features affect seriously nonlinear prediction in short or medium term. For long term prediction, due to noise amplification, the stochastic system is unlikely to have a good memory of its

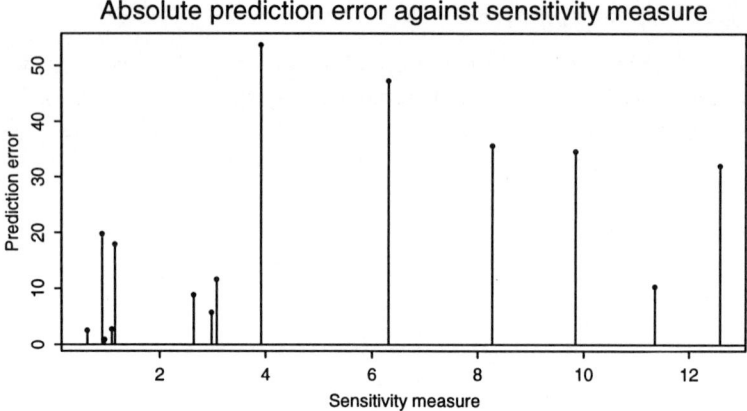

Figure 6.12. *Absolute values of prediction errors against the sensitivity measure for the sunspots data based on a nonparametric AR(4) model.*

Table 6.1. *Prediction of the annual number of sunspots.*

Year	True value	Prediction error	$\| \widehat{I}_{3,1} \|$
1979	155.4	−8.88	2.64
1980	154.7	−47.26	6.32
1981	140.5	−5.83	2.98
1982	115.9	−32.0	12.59
1983	66.6	2.80	1.10
1984	45.9	1.01	0.96
1985	17.9	17.94	1.16
1986	13.4	−2.57	0.64
1987	29.2	−19.73	0.92
1988	100.2	−53.67	3.92
1989	157.6	35.56	8.27
1990	142.6	34.51	9.84
1991	145.7	−11.63	3.08
1992	94.3	−10.40	11.35

Reproduced from Yao and Tong (1994b).

initial values. Therefore, the prediction is not feasible no matter which initial value is given.

We illustrate the above remarks using again the sunspots data. The prediction here is only one-step so that there is no serious noise amplification when model (6.34) holds. The reliability of the prediction is manifested mainly by the sensitivity measure $I_{3,k}(\mathbf{x})$. The data for years 1700–1978 were used to estimate the predictor function, and the data for years 1979–1992 were used for validating the quality of the prediction. In the local polynomial fitting, $d = 4$ and $h = 64.3$ were adopted. The results are summarized in Table 6.1. To help visualize the results, Figure 6.12 gives the scatter plot of the absolute prediction errors against the sensitivity measure. With a few exceptions, the reliability of prediction is fairly closely monitored by reference to $\|\widehat{I}_{3,1}\|$. Indeed, the correlation is about 0.48.

6.6 Complements

The key ideas of proving the theorems in Sections 6.2, 6.4 and 6.5 are the same. One tries to reduce the problem for dependent data to that for independent data, by using some mixing conditions. We first outline the proof of Theorem 6.1. We then give conditions for Theorems 6.2 and 6.3. The proof of these two theorems uses the same technique as that of Theorem 6.1. Finally, we prove Theorems 6.4 and 6.5. These proofs rely on a different technique.

Conditions and proof of theorem 6.1

Condition 1:
 (i) The kernel K is bounded with bounded support.
 (ii) $f_{X_0, X_\ell | Y_0, Y_\ell}(x_0, x_\ell | y_0, y_\ell) \leq A_1 < \infty, \forall \ell \geq 1$.
 (iii) For ρ-mixing processes we assume that

$$\sum_\ell \rho(\ell) < \infty, \quad EY_0^2 < \infty;$$

for strongly mixing processes we assume that for some $\delta > 2$ and $a > 1 - 2/\delta$,

$$\sum_\ell \ell^a [\alpha(\ell)]^{1-2/\delta} < \infty, \ E|Y_0|^\delta < \infty, \ f_{X_0|Y_0}(x|y) \leq A_2 < \infty.$$

 (iv) For ρ-mixing and strongly mixing processes we assume respectively that there exists a sequence of positive integers satisfying $s_n \to \infty$ and $s_n = o\{(nh_n)^{1/2}\}$ such that

$$(n/h_n)^{1/2} \rho(s_n) \to 0 \quad \text{and} \quad (n/h_n)^{1/2} \alpha(s_n) \to 0, \text{ as } n \to \infty.$$

Proof Let **X**, **y** and **W** be respectively the design matrix, the response vector and the weight matrix, as in Section 3.1. Denote by $\mathbf{m} = \{m(X_1), \cdots, m(X_n)\}^T$ and $\beta(x) = \{m(x), \cdots, m^{(p)}(x)/p!\}^T$. Then, we can write

$$\begin{aligned}\widehat{\beta}(x) - \beta(x) &= (\mathbf{X}^T\mathbf{W}\mathbf{X})^{-1}\mathbf{X}^T\mathbf{W}\{\mathbf{m} - \mathbf{X}\beta(x)\} \\ &\quad + (\mathbf{X}^T\mathbf{W}\mathbf{X})^{-1}\mathbf{X}^T\mathbf{W}(\mathbf{y} - \mathbf{m}) \\ &\equiv \mathbf{b} + \mathbf{t}.\end{aligned} \qquad (6.41)$$

The main idea is to show that the bias vector **b** converges in probability to a vector and that the centralized vector **t** is asymptotically normal.

We first establish the asymptotic behavior of the bias vector **b**. By Taylor's expansion of $m(X_i)$ around the point x, we have (see (4.4) with $a = 1$)

$$\mathbf{b} = S_n^{-1}\{\beta_{p+1}(S_{n,p+1}, \cdots, S_{n,2p+1})^T + o_P(h^{p+1})\}, \qquad (6.42)$$

where $h = h_n$, $S_n = \mathbf{X}^T\mathbf{W}\mathbf{X}$ and $S_{n,j}$ is defined in (3.10). Using the same argument as in computing $\text{Var}(Q_n)$ below, we can show that

$$S_{n,j} = nh^j f(x)\mu_j\{1 + o_P(1)\}. \qquad (6.43)$$

As in (3.60), we have by (6.42) that

$$\mathbf{b} = \beta_{p+1}(HSH)^{-1}Hc_p h^{p+1}\{1 + o_P(1)\}, \qquad (6.44)$$

with $H = \text{diag}(1, h, \cdots, h^p)$.

We next consider the joint asymptotic normality of **t**. By (6.43),

$$\mathbf{t} = f^{-1}(x)H^{-1}S^{-1}\mathbf{u}\{1 + o_P(1)\}, \qquad (6.45)$$

where $\mathbf{u} = n^{-1}H^{-1}\mathbf{X}^T\mathbf{W}(\mathbf{y} - \mathbf{m})$. Thus, we need to establish the asymptotic normality of **u**. Consider an arbitrary linear combination $\mathbf{c}^T\mathbf{u}$. Simple algebra shows that

$$Q_n \equiv \mathbf{c}^T\mathbf{u} = \frac{1}{n}\sum_{i=1}^n Z_i \qquad (6.46)$$

where with $C(u) = \sum_{j=0}^p c_j u^j K(u)$ and $C_h(u) = C(u/h)/h$,

$$Z_i = \{Y_i - m(X_i)\}C_h(X_i - x).$$

The problem reduces to proving the asymptotic normality of Q_n.

We will show that

$$\sqrt{nh}Q_n \xrightarrow{D} N\{0, \theta^2(x)\}, \qquad (6.47)$$

where
$$\theta^2(x) = \sigma^2(x)f(x)\int_{-\infty}^{\infty} C^2(x)dx = \sigma^2(x)f(x)\mathbf{c}^T S^*\mathbf{c}.$$

From this, it follows that
$$\sqrt{nh}\mathbf{u} \xrightarrow{D} N\{0, \sigma^2(x)f(x)S^*\}.$$

Hence,
$$\sqrt{nh}H\mathbf{t} \xrightarrow{D} N\{0, \sigma^2(x)f^{-1}(x)S^{-1}S^*S^{-1}\}.$$

Using this and (6.44), we obtain Theorem 6.1.

So far we have outlined a proof which is similar to that for independent observations. However, the proof of (6.47) requires some extra work. We divide the proof into two steps: computation of the variance of Q_n and showing the asymptotic normality of Q_n. Both steps rely strongly on the mixing conditions.

Computation of the variance of Q_n

First note that,
$$\text{Var}(Z_1) = \frac{1}{h}\left\{\theta^2(x) + o(1)\right\}. \tag{6.48}$$

By stationarity, we have
$$\text{Var}(Q_n) = \frac{1}{n}\text{Var}(Z_1) + 2\frac{1}{n}\sum_{\ell=1}^{n-1}(1 - \ell/n)\text{cov}(Z_1, Z_{\ell+1}).$$

Let $d_n \to \infty$ be a sequence of integers such that $d_n h_n \to 0$. Define
$$J_1 = \sum_{\ell=1}^{d_n-1}|\text{cov}(Z_1, Z_{\ell+1})|, \quad J_2 = \sum_{\ell=d_n}^{n-1}|\text{cov}(Z_1, Z_{\ell+1})|.$$

Let $B = \max_{X \in x \pm h} m(X)$. By conditioning on (Y_1, Y_ℓ) and using Condition 1 (ii), we obtain
$$\begin{aligned}
&|\text{cov}(Z_1, Z_\ell)| \\
&= |E[\{Y_1 - m(X_1)\}\{Y_\ell - m(X_\ell)\}C_h(X_1 - x)C_h(X_\ell - x)]| \\
&\leq A_1 E\{(|Y_1| + B)(|Y_\ell| + B)\}\left(\int_{-\infty}^{\infty}|C_h(u-x)|du\right)^2 \\
&\leq D,
\end{aligned}$$

for some $D > 0$. It follows that $J_1 \leq d_n D = o(1/h_n)$. We now consider the contribution of J_2. For ρ-mixing processes, we have from (6.48) that

$$J_2 \leq \text{Var}(Z_1) \sum_{\ell=d_n}^{\infty} \rho(\ell) = o(1/h_n).$$

For strongly mixing processes, we use Davydov's lemma (see Hall and Heyde (1980), Corollary A2) and obtain

$$|\text{cov}(Z_1, Z_{\ell+1})| \leq 8[\alpha(\ell)]^{1-2/\delta}[E|Z_1|^\delta]^{2/\delta}.$$

By conditioning on Y_1 and using Condition 1 (iii), we have

$$E|Z_1|^\delta \leq A_2 E(|Y_1| + B)^\delta \int_{-\infty}^{\infty} |C_h(x-u)|^\delta \leq D h_n^{-\delta+1},$$

for some $D > 0$. Combination of the last two inequalities leads to

$$\begin{aligned} J_2 &\leq \delta D^{2/\delta} h_n^{2/\delta - 2} \sum_{\ell=d_n}^{\infty} [\alpha(\ell)]^{1-2/\delta} \\ &\leq \delta D^{2/\delta} h_n^{2/\delta - 2} d_n^{-a} \sum_{\ell=d_n}^{\infty} \ell^a [\alpha(\ell)]^{1-2/\delta} \\ &= o(1/h_n) \end{aligned}$$

by taking $h^{1-2/\delta} d_n^a = 1$. This choice of d_n satisfies the requirement that $d_n h_n \to 0$. Using the properties of J_1 and J_2, we conclude that

$$\sum_{\ell=1}^{n-1} |\text{cov}(Z_1, Z_{\ell+1})| = o(1/h_n) \tag{6.49}$$

and that $n h_n \text{Var}(Q_n) \to \theta^2(x)$.

Asymptotic normality of Q_n

The proof of this step is the same for ρ-mixing and strongly mixing processes. We only concentrate on ρ-mixing processes.

We employ so-called small-block and large-block arguments. Partition the set $\{1, \cdots, n\}$ into subsets with large blocks of size $r = r_n$ and small blocks of size $s = s_n$. A large block is tagged by a smaller block. Let $k = k_n = [\frac{n}{r_n + s_n}]$ be the number of blocks. Let $Z_{n,i} = \sqrt{h} Z_{i+1}$. Then, $\sqrt{nh} Q_n = n^{-1/2} \sum_{i=0}^{n-1} Z_{n,i}$. By (6.48) and (6.49),

$$\text{Var}(Z_{n,0}) = \theta^2(x)\{1 + o(1)\}, \quad \sum_{\ell=1}^{n-1} |\text{cov}(Z_{n,0}, Z_{n,\ell})| = o(1). \tag{6.50}$$

Note that (6.50) implies that the $Z_{n,j}$'s are nearly independent.

Let the random variables η_j and ξ_j be the sum over the jth large-block and the jth small-block, respectively, namely

$$\eta_j = \sum_{i=j(r+s)}^{j(r+s)+r-1} Z_{n,i}, \quad \xi_j = \sum_{i=j(r+s)+r}^{(j+1)(r+s)-1} Z_{n,i},$$

and $\zeta_k = \sum_{i=k(r+s)}^{n-1} Z_{n,i}$ be the sum over the residual block. Then,

$$\sqrt{nh}Q_n = \frac{1}{\sqrt{n}} \left\{ \sum_{j=0}^{k-1} \eta_j + \sum_{j=0}^{k-1} \xi_j + \zeta_k \right\}$$

$$\equiv \frac{1}{\sqrt{n}} \{Q_n' + Q_n'' + Q_n'''\}.$$

We will show that as $n \to \infty$,

$$\frac{1}{n} E(Q_n'')^2 \to 0, \quad \frac{1}{n} E(Q_n''')^2 \to 0, \tag{6.51}$$

$$\left| E\left[\exp(itQ_n')\right] - \prod_{j=0}^{k-1} E\left[\exp(it\eta_j)\right] \right| \to 0, \tag{6.52}$$

$$\frac{1}{n} \sum_{j=0}^{k-1} E\left(\eta_j^2\right) \to \theta^2(x), \tag{6.53}$$

$$\frac{1}{n} \sum_{j=0}^{k-1} E\left[\eta_j^2 I\{|\eta_j| \geq \varepsilon\theta(x)\sqrt{n}\}\right] \to 0, \tag{6.54}$$

for every $\varepsilon > 0$. Statement (6.51) implies that the sums over small and residual blocks Q_n'' and Q_n''' are asymptotically negligible. Result (6.52) reveals that the summands in the large blocks $\{\eta_j\}$ in Q_n' are asymptotically independent, and (6.53) and (6.54) are the standard Lindeberg-Feller conditions for asymptotic normality of Q_n' under the independence assumption. Expressions (6.51)–(6.54) entail the asymptotic normality (6.47).

We now establish (6.51)–(6.54). We first choose the block sizes. Condition 1 (iv) implies that there exist constants $q_n \to \infty$ such that

$$q_n s_n = o(\sqrt{nh_n}); \quad q_n(n/h_n)^{1/2}\alpha(s_n) \to 0.$$

Define the large block size $r_n = [(nh_n)^{1/2}/q_n]$. Then, it can easily

be shown that
$$s_n/r_n \to 0, \quad r_n/n \to 0, \quad r_n/(nh_n)^{1/2} \to 0, \quad \frac{n}{r_n}\alpha(s_n) \to 0. \tag{6.55}$$

We now establish (6.51) and (6.53). First of all, by stationarity and (6.50), we find
$$\text{Var}(\xi_j) = s\theta^2(x)\{1 + o(1)\}.$$

By (6.50) and (6.55), we have
$$E(Q_n'')^2 = k\text{Var}(\xi_j) + O(n \sum_{j=0}^{n-1} |\text{cov}(Z_{n,0}, Z_{n,j})|) = o(n).$$

The same argument leads to the second part of (6.51) and (6.53).

Note that the indices in η_i and η_{i+1} are at least s_n apart. Hence, applying Lemma 6.1 with $V_j = \exp(it\eta_j)$, we find
$$\left| E\exp(itQ_n') - \prod_{j=0}^{k-1} E[\exp(it\eta_j)] \right| \leq 16k\alpha(s_n) \sim 16\frac{n}{r_n}\alpha(s_n),$$

which tends to zero by (6.55). This proves (6.52).

It remains to establish (6.54). We employ a truncation argument as follows. Let $Y_{L,i} = Y_i I\{|Y_i| \leq L\}$, where L is a fixed truncation point. Correspondingly, let us add the superscript L to indicate the quantities that involve $\{Y_{L,i}\}$ instead of $\{Y_i\}$. Then $Q_n = Q_n^L + \tilde{Q}_n^L$, where
$$\tilde{Q}_n^L = n^{-1} \sum_{j=1}^{n}(Z_j - Z_j^L).$$

Using the fact that $C(\cdot)$ is bounded (since K is bounded with compact support), we have $|Z_{n,j}^L| \leq D/h_n^{1/2}$, for some constant D. Then using (6.55) it follows that
$$\max_{0 \leq j \leq k-1} |\eta_j^L|/\sqrt{n} \leq Dr_n/\sqrt{nh_n} \to 0.$$

Hence, when n is large the set $\{|\eta_j^L| \geq \theta_L(x)\varepsilon\sqrt{n}\}$ becomes an empty set, hence (6.54) holds. Consequently, we have the following asymptotic normality:
$$\sqrt{nh_n}Q_n^L \xrightarrow{D} N\{0, \theta_L^2(x)\}. \tag{6.56}$$

In order to complete the proof, i.e. to establish (6.47), it suffices to show that as first $n \to \infty$ and then $L \to \infty$ we have
$$nh_n \text{Var}\left(\tilde{Q}_n^L\right) \longrightarrow 0. \tag{6.57}$$

Indeed, from this we proceed as follows:

$$\left| E\exp(it\sqrt{nh_n}Q_n) - \exp\{-t^2\theta^2(x)/2\} \right|$$
$$\leq E|\exp(it\sqrt{nh_n}Q_n^L)\{\exp(it\sqrt{nh_n}\tilde{Q}_n^L) - 1\}|$$
$$+ \left| E\exp(it\sqrt{nh_n}Q_n^L) - \exp\{-t^2\theta_L^2(x)/2\} \right|$$
$$+ \left|\exp\{-t^2\theta_L^2(x)/2\} - \exp\{-t^2\theta^2(x)/2\}\right|.$$

The first term is bounded by

$$E|\exp(it\sqrt{nh_n}\tilde{Q}_n^L) - 1| = O\{\text{Var}(\sqrt{nh_n}\tilde{Q}_n^L)\}.$$

Letting $n \to \infty$, the first term converges to zero by (6.57) as first $n \to \infty$ and then $L \to \infty$; the second term goes to zero by (6.56) for every $L > 0$; the third term goes to zero as $L \to \infty$ by the dominated convergence theorem. Therefore, it remains to prove (6.57). Note that $\tilde{Q}_n^L$ has the same structure as Q_n. Hence, by (6.50), we obtain

$$\lim_{n\to\infty} nh_n \text{Var}\left(\tilde{Q}_n^L\right) = \text{Var}(YI[|Y|>L]|X=x)f(x)\int_{-\infty}^{\infty} C^2(u)du.$$

By dominated convergence, the right-hand side converges to 0 as $L \to \infty$. This establishes (6.57) and completes the proof of Theorem 6.1.

Conditions for Theorems 6.2 and 6.3

Condition 2:

(i) The kernel functions W and K are symmetric and bounded with bounded supports.

(ii) The process $\{X_j, Y_j\}$ is ρ-mixing with $\sum_\ell \rho(\ell) < \infty$. Further, assume that there exists a sequence of positive integers $s_n \to \infty$ such that $s_n = o\{(nh_1h_2)^{1/2}\}$ and $\{n/(h_1h_2)\}^{1/2}\rho(s_n) \to 0$.

(iii) The function $g(y|x)$ has bounded and continuous third order derivatives at point (x, y), and $f(\cdot)$ is continuous at the point x.

(iv) The joint density of the distinct elements of $(X_0, Y_0, X_\ell, Y_\ell)$ ($\ell > 0$) is bounded by a constant independent of ℓ.

Conditions and proof for Theorems 6.4 and 6.5

Condition 3:
(i) The process $\{X_t\}$ is a linear Gaussian process given by
$$X_t = \sum_{j=-\infty}^{\infty} \psi_j Z_{t-j},$$
with $\sum_j |\psi_j|j^2 < \infty$, where $Z_j \sim$ iid $N(0,\sigma^2)$.
(ii) The spectral density function $f(\cdot)$ is positive on $[0,\pi]$.
(iii) The kernel function K is a symmetric probability density function and has compact support.
(iv) $(\log n)^4 h_n \to 0$ in such a way that $nh_n \to \infty$.

It follows from Condition 3(i) that the spectral density function
$$f_X(\lambda) = |\psi(\lambda)|^2 f_Z(\lambda) = \frac{\sigma^2}{2\pi}|\psi(\lambda)|^2$$
has a bounded second derivative, where
$$\psi(\lambda) = \sum_{j=-\infty}^{\infty} \psi_j \exp(-ij\lambda).$$

We first give two lemmas showing that R_k in (6.22) and r_k in (6.23) are negligible.

Lemma 6.2 *Under Conditions 3(i) and (ii), we have*
$$\max_{1\le k\le N} |R_k| = O\left(\frac{\log n}{\sqrt{n}}\right),$$
almost surely, where $N = [(n-1)/2]$.

Proof Let $J_X(\lambda)$ and $J_Z(\lambda)$ denote the discrete Fourier transforms of $\{X_t\}$ and $\{Z_t\}$:
$$J_X(\lambda) = n^{-1/2} \sum_{t=1}^n X_t \exp(-i\lambda t).$$

Define
$$U_{nj}(\lambda) = \sum_{t=-j+1}^{n-j} Z_t \exp(-i\lambda t) - \sum_{t=1}^n Z_t \exp(-i\lambda t),$$
$$Y_n(\lambda) = n^{-1/2} \sum_{j=-\infty}^{\infty} \psi_j \exp(-i\lambda j) U_{nj}(\lambda).$$

COMPLEMENTS   257

Then, it can be computed (see page 347 of Brockwell and Davis (1989)) that with $\lambda = \lambda_k$,

$$R_k = \psi(\lambda)J_Z(\lambda)Y_n(-\lambda) + \psi(-\lambda)J_Z(-\lambda)Y_n(\lambda) + |Y_n(\lambda)|^2, \quad (6.58)$$

and that $\{J_Z(\lambda_k)\}$ and $\{Y_n(\lambda_k)\}$ are independently Gaussian distributed with mean 0 and variance respectively $O(1)$ and $O(n^{-1})$. Recall that the maximum of n i.i.d. Gaussian white noise is asymptotically equal to $\sqrt{2\log n}$ almost surely. It follows that

$$\max_k |J_Z(\lambda_k)| = O(\sqrt{\log n}) \quad \text{and} \quad \max_k |Y_n(\lambda_k)| = O(\sqrt{\frac{\log n}{n}}),$$

almost surely. Substituting these into (6.58) and using the fact that

$$\max_\lambda |\psi(\lambda)| < \infty,$$

we obtain Lemma 6.2.

Lemma 6.3 *Under Conditions 3(i) and (ii), we have for any sequence a_n,*

$$r_k \leq O_P(\frac{\log n}{\sqrt{n}})\frac{I(V_k > a_n)}{V_k} + O_P(\log n)I(V_k \leq a_n),$$

uniformly for $1 \leq k \leq [(n-1)/2]$, where V_k are i.i.d. random variables, having the standard exponential distribution.

Proof Recall that

$$r_k = \log\Big\{1 + \frac{R_k}{f(\lambda_k)V_k}\Big\}.$$

Using the inequality $\log(1 + x) \leq x$ for $x > 0$ and dividing the state space into $V_k > a_n$ and $V_k \leq a_n$ for a given sequence a_n, we get that

$$r_k \leq \frac{|R_k|}{f(\lambda_k)V_k}I(V_k > a_n) + \log\Big\{1 + \frac{\max_k |R_k|}{\min_\lambda f(\lambda)\min_k V_k}\Big\}I(V_k \leq a_n). \quad (6.59)$$

Obviously,

$$P(\min_{1 \leq k \leq N} V_k > n^{-2}) = \exp(-N/n^2) \to 1.$$

Thus, $(\min_k V_k)^{-1} = O_P(n^2)$. Substituting this term into (6.59) and using Lemma 6.2, we obtain

$$r_k \leq O_P(\frac{\log n}{\sqrt{n}})I(V_k > a_n)V_k^{-1} + O_P(\log n)I(V_k \leq a_n).$$

This completes the proof of Lemma 6.3.

Proof of Theorem 6.4 We will show that

$$\widehat{m}_{LS}(\lambda) = \sum_{j=1}^{N} K_n\left(\frac{\lambda - \lambda_j}{h}\right) Y_j' + O_P(\frac{\log^2 n}{\sqrt{n}}), \qquad (6.60)$$

where $Y_j' = m(\lambda_j) + \varepsilon_j'$ with $\varepsilon_j' = \varepsilon_j - C_0$. Applying standard nonparametric regression theory to the first term (see Theorem 5.2) we obtain the result.

We now establish (6.60). Using Lemma 6.3, the remainder term in (6.60) is bounded by

$$\sum_{j=1}^{N} \left|K_n\left(\frac{\lambda - \lambda_j}{h}\right) r_j\right| = O_P(\frac{\log n}{\sqrt{n}}) T_{n,1} + O_P(\log n) T_{n,2}, \qquad (6.61)$$

where

$$T_{n,1} = \sum_{j=1}^{N} \left|K_n\left(\frac{\lambda - \lambda_j}{h}\right)\right| I(V_j > a_n) V_j^{-1}$$

$$T_{n,2} = \sum_{j=1}^{N} \left|K_n\left(\frac{\lambda - \lambda_j}{h}\right)\right| I(V_j \le a_n).$$

Note that when $a_n \to 0$,

$$E\{I(V_j > a_n) V_j^{-1}\} \le \int_{a_n}^{1} t^{-1} dt + \int_{1}^{\infty} \exp(-t) dt = O(\log a_n^{-1}),$$

and

$$EI(V_j \le a_n) = 1 - \exp(-a_n) = O(a_n).$$

Translating (5.70) into the current setting,

$$\sum_{j=1}^{N} \left|K_n\left(\frac{\lambda - \lambda_j}{h}\right)\right| = 2 + o_P(1).$$

Using the last three expressions, we find

$$E|T_{n,1}| = O(\log a_n^{-1}), \quad \text{and} \quad E|T_{n,2}| = O(a_n).$$

Substituting these into (6.61), we conclude that (6.61) is of order

$$O_P(\frac{\log a_n^{-1} \log n}{\sqrt{n}}) + O_P(a_n \log n) = O_P(\frac{\log^2 n}{\sqrt{n}}), \qquad (6.62)$$

by taking $a_n = n^{-1}$.

Proof of Theorem 6.5 The idea of the proof is to reduce the problem for dependent data to that for i.i.d. exponential distributions. The latter was proved in Section 5.6.

Let $\widehat{\boldsymbol{\beta}} = a_n^{-1}[\widehat{a} - m(\lambda), h\{\widehat{b} - m'(\lambda)\}]^T$, where $a_n = (nh)^{-1/2}$. Define

$$L_k(Y_k, \boldsymbol{\beta}) = -\exp\{Y_k - \bar{m}(\lambda, \lambda_k) - a_n \boldsymbol{\beta}^T \boldsymbol{\lambda}_k\}$$
$$+ Y_k - \bar{m}(\lambda, \lambda_k) - a_n \boldsymbol{\beta}^T \boldsymbol{\lambda}_k,$$

where $\bar{m}(\lambda, \lambda_k) = m(\lambda) + m'(\lambda)(\lambda_k - \lambda)$ and $\boldsymbol{\lambda}_k = \{1, (\lambda_k - \lambda)/h\}^T$. Then, it can easily be seen via a linear transform that $\widehat{\boldsymbol{\beta}}$ maximizes

$$\sum_{k=1}^{N} L_k(Y_k, \boldsymbol{\beta}) K_h(\lambda_k - \lambda),$$

or equivalently $\widehat{\boldsymbol{\beta}}$ maximizes

$$\ell_n(\boldsymbol{\beta}) = h \sum_{k=1}^{N} \{L_k(Y_k, \boldsymbol{\beta}) - L_k(Y_k, 0)\} K_h(\lambda_k - \lambda).$$

Let $Y'_k = m(\lambda_k) + \varepsilon_k$, the main term of (6.23). Then, we can write

$$\ell_n(\boldsymbol{\beta}) = \ell_{1,n}(\boldsymbol{\beta}) + U_n$$

where $\ell_{1,n}(\boldsymbol{\beta})$ is defined in the same way as $\ell_n(\boldsymbol{\beta})$ with Y_k replaced by Y'_k, and

$$U_n = -h \sum_{k=1}^{N} R_k \Big[\exp\{-\bar{m}(\lambda, \lambda_k) - a_n \boldsymbol{\beta}^T \boldsymbol{\lambda}_k\}$$
$$- \exp\{-\bar{m}(\lambda, \lambda_k)\} \Big] K_h(\lambda_k - \lambda).$$

By using Taylor's expansion and Lemma 6.2, for each fixed $\boldsymbol{\beta}$,

$$U_n = O_P(h \cdot a_n \cdot n \cdot \log n / \sqrt{n}) = o_P(1).$$

Thus, we have

$$\ell_n(\boldsymbol{\beta}) = \ell_{1,n}(\boldsymbol{\beta}) + o_P(1). \tag{6.63}$$

Now, observe that $\ell_{1,n}(\boldsymbol{\beta})$ is also the local likelihood based on $\exp(Y'_k)$, which are i.i.d. exponentially distributed. Thus, it is a special case of Theorem 5.2. As shown in the equation preceding

(5.76), we have for each β,

$$\ell_{1,n}(\beta) = \mathbf{W}_n^T \beta + \frac{1}{2}\beta^T A \beta + o_P(1), \qquad (6.64)$$

where $A = -\pi^{-1}\mathrm{diag}\{1, \mu_2(K)\}$ and

$$\mathbf{W}_n - a_n^{-1}\left\{\frac{m''(x)}{2}h^2\pi^{-1}\{\mu_2(K), 0\}^T + o_P(h^3)\right\} \xrightarrow{D} N(0, B) \qquad (6.65)$$

with $B = \pi^{-1}\mathrm{diag}\{\nu_0(K), \int t^2 K^2(t)dt\}$. Combining (6.63) and (6.64), we obtain

$$\ell_n(\beta) = \mathbf{W}_n^T \beta + \frac{1}{2}\beta^T A \beta + o_P(1).$$

As noted before, $\mathcal{L}(a,b)$ is a concave function. By the quadratic approximation lemma given in Section 5.6, we conclude that

$$\widehat{\beta} = -A^{-1}\mathbf{W}_n + o_P(1).$$

The result follows from the last equality by taking its first component.

6.7 Bibliographic notes

Kernel regression smoothers for dependent data have been extensively studied in the literature. A comprehensive summary of the area can be found in Rosenblatt (1991). The asymptotic mean squared errors and strong rates of convergence have been derived under various mixing conditions. See for example Rosenblatt (1969), Robinson (1983, 1986), Roussas (1990), Truong (1991) and Truong and Stone (1992). Azzalini (1984) and Hart and Wehrly (1986) studied nonparametric regression problems with repeated measurements having a certain dependence structure. Collomb (1985) established the uniform almost sure convergence of window and k-nearest neighbor autoregression estimators. Robustified kernel regression estimators have been studied by Collomb and Härdle (1986) and Truong (1992). Roussas and Tran (1992) derived the asymptotic normality of kernel regression smoothers with variable bandwidths. Cheng and Robinson (1991) investigated the behavior of a kernel density estimator for long memory processes.

The suggestion that an improved spectral estimate might be obtained by smoothing the periodogram was made by Daniels (1946). See also Bartlett (1948, 1950). Recent advances in smoothing techniques enrich the techniques of spectral density estimation. Wahba

(1980) used smoothing splines to smooth a log-periodogram. Extensive efforts have been made in selecting appropriate smoothing parameters for spectral density estimators. See for example Swanepoel and van Wyk (1986), Beltrão and Bloomfield (1987), Hurvich and Beltrão (1990) and Franke and Härdle (1992). Based on the penalized Whittle likelihood, Pawitan and O'Sullivan (1994) use smoothing splines to estimate the spectral density. Kooperberg, Stone and Truong (1995a, b) develop log-spline spectral density estimates allowing for a line spectrum.

The notion of sensitivity measures for stochastic systems is inspired by the theory of chaos. See Eckmann and Ruelle (1985) and Drazin and King (1992). The key notions in chaos theory have been increasingly gaining attention in the statistical literature. See, for example, Nychka, Ellner, Gallant and McCaffrey (1992), Smith (1992), Wolff (1992), Cutler (1993) and Hall and Wolff (1995).

The stationarity and ergodicity of nonlinear autoregressive models have been studied by Tjøstheim (1990), Chen and Tsay (1993), Meyn and Tweedie (1993, 1994), Chan and Tong (1994) and An and Huang (1996), among others.

CHAPTER 7

Local polynomial regression for multivariate data

7.1 Introduction

In the previous chapters of this book we mainly focused on the univariate situation, having a response variable Y and a univariate covariate X. Often however the behavior of Y cannot be explained by one single covariate, and more covariates have to be taken into consideration. In *multivariate regression problems* one of the tasks is to study the structural relationship between the response variable Y and the vector of covariates $\mathbf{X} = (X_1, \cdots, X_d)^T$ via

$$m(\mathbf{x}) = E(Y|\mathbf{X} = \mathbf{x}), \qquad (7.1)$$

with $\mathbf{x} = (x_1, \cdots, x_d)^T$ and $m(\mathbf{x}) = m(x_1, \cdots, x_d)$.

In the *multiple linear regression model* it is assumed that the conditional mean relationship between the response and each of the predictors is linear, i.e. that $m(\mathbf{x})$ is of a linear form. In *parametric generalized linear models* the unknown regression function $m(\mathbf{x})$ is modelled linearly via a known link function g. An example is the logistic regression model, popularly used for binary response data, in which the logit of the regression function is modelled linearly. But there is not always evidence for such a linear relationship and hence there is a need for more flexible models. Further, there are a growing number of enormous (in millions) data sets in finance, business, economics, public health, etc. that allow one to implement nonparametric techniques for exploring fine structural architecture, even when there are several covariates.

The most flexible models do not make any assumption about the form of the d-variate function $m(\mathbf{x})$. The problem then is to fit a d-dimensional surface to the observed data $\{(\mathbf{X}_i^T, Y_i) : i = 1, \cdots, n\}$, where $\mathbf{X}_i = (X_{i1}, \cdots, X_{id})^T$, $i = 1, \cdots, n$. An obvious approach is to try to generalize the univariate smoothing techniques, such as those described in Sections 2.2–2.6, to this multivariate situation. For kernel estimators and locally weighted polynomial fitting this generalization is quite straightforward as can be seen from Section 7.8. In that section we discuss multivariate locally weighted linear fitting. A technical difficulty popping up here is to define neighborhoods in this d-dimensional setting. Here a d-dimensional kernel function K and a smoothing matrix B – the bandwidth matrix – come into the scene. The choice of these two quantities raises some important issues such as for example 'Should the amount of smoothing be the same in each direction or not?'. This question is related to the question of whether the variation in the surface is similar with respect to all covariates. See for example Silverman (1986), Hastie and Tibshirani (1990), Scott (1992) and Wand and Jones (1993), among others for a discussion on these issues. The generalization of the univariate smoothing spline technique to two or more dimensions is less straightforward, but two possible generalizations have been proposed in the literature: the so-called *thin plate splines*, discussed in detail in Green and Silverman (1994), and the *multivariate tensor-product splines* studied by for example Friedman (1991).

Although generalizations of most of the univariate smoothing techniques to multivariate *surface smoothing* appear to be feasible, there is a serious problem arising: the so-called *curse of dimensionality* as it was termed by Bellman (1961). This problem refers to the fact that a local neighborhood in higher dimensions is no longer local: a neighborhood with a fixed percentage of data points can be very big and far from what is understood by the term 'local'. Or to put it another way: if a local neighborhood contains 10 data points along each axis, then there are 10^d data points in the corresponding d-dimensional neighborhood. As a consequence much larger data sets are needed even when d is moderate, and such large data sets are often not available in practical situations. The curse of dimensionality problem has been illustrated clearly in many books, among which are Silverman (1986), Härdle (1990), Hastie and Tibshirani (1990) and Scott (1992).

Due to the curse of dimensionality, surface smoothing techniques are in practice not very useful when there are more than two or three predictor variables. Several approaches have been proposed to deal with the curse of dimensionality problem. They all involve some dimensionality reduction process. In this chapter we discuss some of these approaches such as *additive modelling, partially linear modelling, modelling with interactions* and *sliced inverse regression*.

See Sections 7.2–7.5 respectively. In each of these dimensionality reduction principles univariate smoothers are used at certain stages. Hence univariate local polynomial regression estimators can serve as a building block. This will be spelled out in Sections 7.6 and 7.7.

There are several other approaches aimed at overcoming the curse of dimensionality problem. It is beyond the scope of this book to discuss them all. Among other approaches are the *Classification and Regression Trees* (CART) described in detail in Breiman, Friedman, Olshen and Stone (1993), and *projection pursuit regression* introduced by Friedman and Stuetzle (1981).

FAN and Gijbels
7.2 Generalized additive models

Consider the general multiple regression model

$$Y = m(\mathbf{X}) + \varepsilon, \qquad (7.2)$$

where $E(\varepsilon) = 0$, $\text{Var}(\varepsilon) = \sigma^2$ and ε is independent of the vector of covariates $\mathbf{X}$. In the linear multiple regression model the regression function $m(\cdot)$ is assumed to be linear and hence additive in the predictors. In *additive models* the linearity assumption is dropped and the additivity feature is retained, leading to the following model

$$Y = \alpha + \sum_{j=1}^{d} g_j(X_j) + \varepsilon, \qquad (7.3)$$

where $g_1, \cdots, g_d$ are unknown univariate functions. To avoid free constants in the functions and hence to ensure identifiability it is required that they satisfy

$$E\{g_j(X_j)\} = 0, \qquad j = 1, \cdots, d. \qquad (7.4)$$

This implies that $E(Y) = \alpha$.

The additive model (7.3) is a special case of the so-called *projection pursuit* model which is of the form

$$Y = \sum_{\ell=1}^{L} g_\ell(\boldsymbol{\alpha}_\ell^T \mathbf{X}) + \varepsilon,$$

where $\boldsymbol{\alpha}_\ell$, $\ell = 1, \cdots, L$, denote the direction vectors onto which the observations are projected. The contributions of the one-dimensional projections $\boldsymbol{\alpha}_\ell^T \mathbf{X}$, $\ell = 1, \cdots, L$, are then modelled additively via univariate functions g_ℓ, $\ell = 1, \cdots, L$.

How do we fit the additive model (7.3)? Note first of all that when the additive model (7.3) is correct then

$$E\{Y - \alpha - \sum_{j \neq k} g_j(X_j)|X_k\} = g_k(X_k), \qquad k = 1, \cdots, d.$$

This immediately suggests an iterative algorithm for computing all univariate functions $g_1, \cdots, g_d$. Indeed, for given α and given functions $g_j, j \neq k$, the function g_k can be obtained via a simple univariate regression fit based on the observations $\{(X_{ik}, Y_i) : i = 1 \cdots, n\}$. Denote the univariate smoother of g_k by $\mathcal{S}_k$. Note that here any univariate regression smoothing technique, such as for example a Gasser-Müller estimator, a cubic smoothing spline, a wavelet shrinkage estimator or <u>local polynomial regression estimators</u>, can be used. In order to meet condition (7.4) the resulting estimate $\widehat{g}_k(\cdot)$ of $g_k(\cdot)$, obtained by using the smoother $\mathcal{S}_k$, is replaced by its centralized version

$$\widehat{g}_k^*(\cdot) = \widehat{g}_k(\cdot) - \frac{1}{n} \sum_{j=1}^n \widehat{g}_k(X_{jk}). \qquad (7.5)$$

An initial choice of the univariate functions, say g_k^0, is needed as well as an iteration scheme. This leads to the so-called *backfitting algorithm* which reads as follows

Step 1. Initialization: $\widehat{\alpha} = n^{-1} \sum_{i=1}^n Y_i, \widehat{g}_k = g_k^0, k = 1, \cdots, d$.

Step 2. For each $k = 1, \cdots, d$, obtain

$$\widehat{g}_k = \mathcal{S}_k \{Y - \widehat{\alpha} - \sum_{j \neq k} \widehat{g}_j(X_j)|X_k\}$$

and obtain $\widehat{g}_k^*(\cdot)$ as in (7.5).

Step 3. Keep cycling Step 2 until convergence.

The idea of the above algorithm is to carry through a fit, calculate *partial residuals* from that fit and refit again. The term *backfitting* refers to this action, and was used by Friedman and Stuetzle (1981). The estimator $n^{-1} \sum_{i=1}^n Y_i$ of α in the above algorithm follows naturally from the fact that $E(Y) = \alpha$. As initial functions $g_1^0, \cdots, g_d^0$ one could use for example the fits resulting from a linear regression fit of Y on the predictors X_k.

When model (7.3) does not hold, the above algorithm is expected to give the estimates that are the best additive approximation to the regression surface. See Breiman and Friedman (1985) for further discussions.

Note that additive models overcome the curse of dimensionality problem since the fitting procedure is built up on univariate smoothers. The price paid for this is a loss in flexibility of the model. What is the effective number of parameters used in fitting an additive model? Denote by ℓ_n the average effective number of parameters used for each univariate nonparametric regression fit. Then the effective number of parameters for the additive fit is $\ell_n d + 1 - d = (\ell_n - 1)d + 1$, where $-d$ comes from the fact that the unknown functions should satisfy constraints (7.4).

The above backfitting algorithm was introduced by Breiman and Friedman (1985) as the *Alternating Conditional Expectation* (ACE) algorithm in a more general setting which allows for an unknown transformation $\theta(Y)$ of the reponse variable Y, i.e.

$$\theta(Y) = \sum_{j=1}^{d} g_j(X_j) + \varepsilon.$$

Breiman and Friedman (1985) study the ACE-procedure, an iterative algorithm for finding the functions $\theta(\cdot)$ and $g_1(\cdot), \cdots, g_d(\cdot)$.

There is of course a question of convergence of the backfitting algorithm, or of the ACE-algorithm in general. Further, if the algorithm converges, does it converge to a unique solution? See Chapter 5 of Hastie and Tibshirani (1990) for a discussion on these issues.

The additive model (7.3) generalizes the linear multiple regression model with the linear function replaced by an additive sum of univariate functions. Generalized linear models generalize the linear multiple regression model by introducing a link function that links the mean regression function to a linear space of the predictors and by allowing error distributions to be in the exponential family. In the same spirit one can generalize the additive model to a *Generalized Additive Model* (GAM) which links the mean regression function to an additive sum of univariate functions:

$$g\{m(\mathbf{X})\} = \alpha + \sum_{j=1}^{d} g_j(X_j), \qquad (7.6)$$

where $g(\cdot)$ is a known link function and $g_1, \cdots, g_d$ are unknown univariate functions. As before the univariate functions satisfy condition (7.4). Generalized additive models were introduced by Hastie and Tibshirani (1986, 1987) and studied in detail by Stone (1985, 1986) and Hastie and Tibshirani (1990).

Since generalized additive models are a generalization of generalized linear models on the one hand and additive models on the other, the algorithm for fitting a generalized additive model consists of a combination of the basic ideas underlying the algorithm for fitting a parametric generalized linear model and the backfitting algorithm for additive models. We start with describing the algorithm for fitting a parametric generalized linear model with canonical link function g:

$$g\{m(\mathbf{X})\} = \mathbf{X}^T\beta = \theta(\mathbf{X}),$$

(see also Section 5.4). The estimates for β in this model are defined by score equations and a typical way of solving these equations is the *Fisher scoring method*. This method consists of a Newton-Raphson algorithm where the observed information matrix (the Hessian matrix) is replaced by the expected information matrix, i.e. Fisher's information matrix. A procedure equivalent to the Fisher scoring method is the following *adjusted dependent variable procedure* which is a kind of iteratively reweighted least squares procedure. The procedure consists basically of two steps: an initialization step and an updating step using weighted least squares regression. The updating step is repeated until convergence. In the initialization step one starts with some initial vector β^0 and from this the initial predictors

$$\eta^0 = (\eta_1^0, \cdots, \eta_n^0)^T \equiv (\mathbf{X}_1^T\beta^0, \cdots, \mathbf{X}_n^T\beta^0)^T \qquad (7.7)$$

and the initial fitted values

$$(\mu_1^0, \cdots, \mu_n^0) \equiv \{g^{-1}(\mathbf{X}_1^T\beta^0), \cdots, g^{-1}(\mathbf{X}_n^T\beta^0)\} \qquad (7.8)$$

are computed. This ends the initialization step. The updating step then starts with constructing an *adjusted dependent variable*

$$Z_i = \eta_i^0 + (Y_i - \mu_i^0)\left(\frac{\partial \eta_i}{\partial \mu_i}\right)_0 \qquad (7.9)$$

and associated weights

$$w_i^{-1} = \left(\frac{\partial \eta_i}{\partial \mu_i}\right)_0^2 V_i^0, \qquad (7.10)$$

where $V_i^0 = V(\mu_i^0)$, with $V(\cdot)$ as in Section 5.4.2. Note that the idea behind (7.9) is a simple one-term Taylor expansion of the function $g(Y_i)$ around the given value μ_i^0 assumed to be close to Y_i. Further the weight w_i^{-1} is simply the variance of the variable Z_i given the current values η_i^0 and μ_i^0. After constructing all adjusted dependent variables $Z_1, \cdots, Z_n$, one regresses Z_i on $\mathbf{X}_i = (X_{i1}, \cdots, X_{id})^T$

with weight w_i to obtain a new revised estimate of β. This updating step is repeated until the change in the deviance function, as defined in (5.50) in Section 5.4, is sufficiently small. See Hastie and Tibshirani (1990) for more details of this algorithm and its equivalence with other algorithms.

We now discuss the modification of the above algorithm for fitting a generalized additive model. In the initialization step we need initial choices of α and the functions $g_1, \cdots, g_d$. Since the model implies that $E[g\{m(\mathbf{X})\}] = \alpha$ an initial choice α^0 for α is simply given by its estimator $g(n^{-1}\sum_{i=1}^{n} Y_i)$. Further one can put $g_1^0 = \cdots = g_d^0 = 0$ initially. With these initial quantities we compute as before the initial predictors

$$\eta_i^0 = \alpha^0 + \sum_{j=1}^{d} g_j^0(X_{ij}), \qquad i = 1, \cdots, n,$$

and the initial fitted values

$$\mu_i^0 = g^{-1}(\eta_i^0), \qquad i = 1, \cdots, n,$$

by anology with (7.7) and (7.8) respectively, but now adapted to the generalized additive model (7.6). From these initial predictors and fitted values the (modified) adjusted dependent variable Z_i is computed via (7.9). Then an additive regression model for the adjusted dependent variable Z_i and the covariate $\mathbf{X}_i$, with weight w_i computed as in (7.10), is fitted using for example the backfitting algorithm. This leads to the updated functions $g_1^1, \cdots, g_d^1$ and the updated quantities $\eta^1 = (\eta_1^1, \cdots, \eta_n^1)^T$ and $\mu^1 = (\mu_1^1, \cdots, \mu_n^1)^T$. The iteration is carried through until convergence is reached, using for example the convergence criterion

$$\Delta(\eta^1, \eta^0) = \frac{\sum_{j=1}^{d} \|g_j^1 - g_j^0\|}{\sum_{j=1}^{d} \|g_j^0\|},$$

where $\|\cdot\|$ denotes an appropriate norm on the function space containing the g_j's.

In summary the algorithm for generalized additive models differs from that for generalized linear models in defining the adjusted dependent variable Z_i and in fitting an additive model instead of a linear regression model.

The number of effective parameters used in fitting a generalized additive model is similar to that for fitting an additive model.

We now illustrate the use of a generalized additive model for data from the Coronary Risk-Factor Study (CORIS) baseline sur-

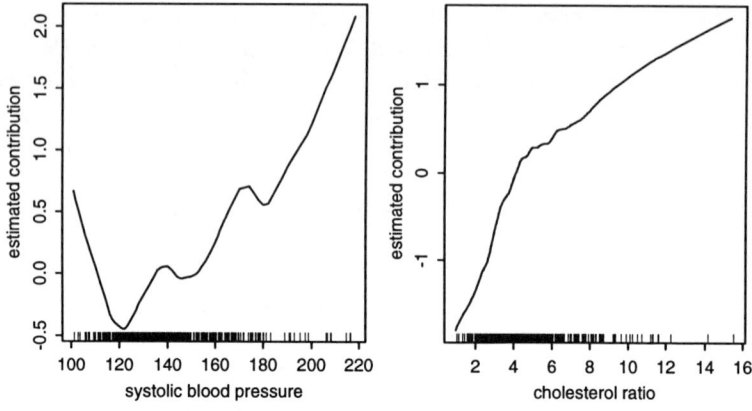

Figure 7.1. *The estimated additive contributions for systolic blood pressure (left panel); and cholesterol ratio (right panel) after fitting model (7.11).*

vey. This survey took place in rural South Africa (see Rousseauw et al. (1983)), and the data have been analyzed by Hastie and Tibshirani (1987). The aim of the study is to identify risk factors for the incidence of myocardial infarction (MI). Denote by Y the binary response variable representing the absence ($Y = 0$) or presence ($Y = 1$) of myocardial infarction. Several covariates (possible risk factors) were recorded. In this illustration we only focus on two covariates: the systolic blood pressure (covariate X_1) and the cholesterol ratio (covariate X_2). This cholesterol ratio is the ratio of total cholesterol to HDL cholesterol. The latter cholesterol level is known as the 'good' cholesterol and hence small values of the cholesterol ratio are healthier. The aim is to describe the mean regression function $m(X_1, X_2) = E(Y|X_1, X_2) = P(Y = 1|X_1, X_2)$ in terms of the predictors X_1 and X_2. Since the response variable Y is binary a sensible choice for a model would be a logistic regression model, which would model the transformed conditional probability $P(Y = 1|X_1, X_2)$ linearly in the two covariates, using a logit link function. In a generalized additive logistic regression model this sum of linear terms is replaced by an additive sum of terms, i.e.

$$\text{logit}\{m(X_1, X_2)\} \equiv \log\left\{\frac{m(X_1, X_2)}{1 - m(X_1, X_2)}\right\} = \alpha + g_1(X_1) + g_2(X_2), \tag{7.11}$$

with $g_1(\cdot)$ and $g_2(\cdot)$ unknown univariate functions satisfying the conditions $E\{g_1(X_1)\} = E\{g_2(X_2)\} = 0$. For the logit link function $g(t) = \log\{t/(1-t)\}$, it is easy to show that the adjusted dependent variable, as defined in (7.9), is given by

$$Z_i = \eta_i^0 + (Y_i - \mu_i^0)/\{\mu_i^0(1 - \mu_i^0)\}.$$

For the Bernoulli model the variance function equals $V(x) = x(1-x)$ (see Section 5.4) and therefore the weights as defined in (7.10) are given by

$$w_i^{-1} = 1/\{\mu_i^0(1 - \mu_i^0)\}.$$

The analysis was carried through by Hastie and Tibshirani (1990). We reproduce their results by using the GAM function provided in the S-Plus software (see Hastie (1992)) with LOESS as the basic smoother (see Sections 2.4 and 7.8.1). For the smoothing parameter in the LOESS procedure we took $d = 1/3$. This parameter determines the size of the neighborhood as explained in Section 2.4. Figure 7.1 shows the estimated additive contribution $\hat{g}_1(X_1)$ of the systolic blood pressure covariate as well as the estimated additive contribution $\hat{g}_2(X_2)$ of the cholesterol ratio.

To verify the additivity of the model, we divided the data into four groups according to the quartiles of the cholesterol ratio observations. An additive model (7.11) was fitted to each of the four groups. In Figure 7.2 we depict the resulting estimated additive contributions for systolic blood pressure obtained for each of the four groups. In the same figure we also plot the estimated additive contribution for systolic blood pressure from fitting model (7.11) to the whole group. From Figure 7.2 we can see that the shape of the estimated additive contribution for systolic blood pressure in the group with high cholesterol ratio differs from the shape of the estimated curves for the other groups. This suggests that there might be an interaction between X_1, the systolic blood pressure, and X_2, the cholesterol ratio, for large values of X_2. See Section 7.4 for modelling of interactions.

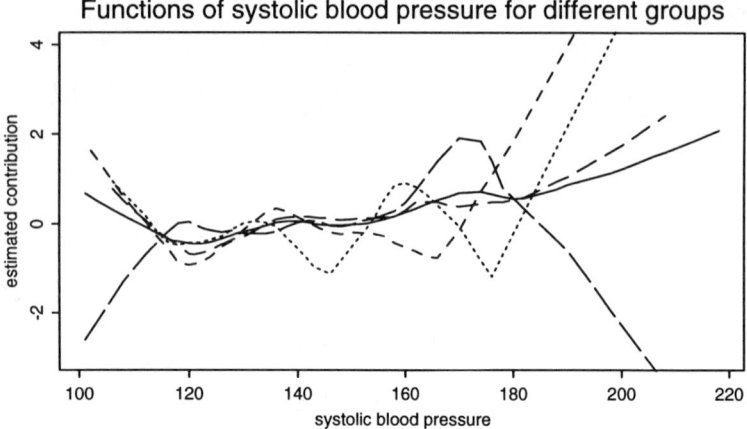

Figure 7.2. *Verifying the additivity of the model by subdividing into four groups according to the cholesterol ratio. Solid curve: estimated additive contribution of* X_1 *when fitting model (7.11) to the whole group; dashed curves: estimated additive contribution of* X_1 *for each of the four groups. From the shortest to longest dash: the first, second, third and fourth quartile group.*

7.3 Generalized partially linear single-index models

The additive and generalized additive models described in Section 7.2 generalize respectively the multiple linear regression model and the generalized linear model by replacing the linear function by an additive sum of nonparametric univariate functions. So each covariate X_i, $i = 1, \cdots, d$, is modelled in a similar way via a nonparametric univariate function. But in many practical situations the response variable Y can depend linearly on certain covariates while depending nonlinearly on other covariates. This would require a certain flexibility to model some covariates linearly and others nonlinearly, say nonparametrically in general. This is the basic idea behind the class of *Generalized Partially Linear Single-Index Models* (**GPLSIM**) described in this section. The effective number of parameters for these models is rather small, and hence they provide an appealing method for dimension reduction, particularly when the data set is small or when the dimension is high. This type of model also incorporates neatly discrete and continuous covariates. These issues will become clear in Section 7.3.3. The class of generalized partially linear single-index models includes the *partially linear models* or *partial spline models* and the so-called *single-index models*. We first explain briefly the last two classes of models, and then discuss in detail the new more general class of models.

7.3.1 Partially linear models

Suppose that the response variable Y depends linearly on the vector of covariates $\mathbf{Z} = (Z_1, \cdots, Z_q)^T$ and nonlinearly on a scalar covariate X. Then an appropriate model would be

$$Y = \eta_0(X) + \mathbf{Z}^T \beta_0 + \varepsilon, \qquad (7.12)$$

where $\eta_0(\cdot)$ is an unknown univariate function and β_0 is a q-dimensional parameter vector. As an example one could think of the following situation in agricultural field trials. Of interest is to investigate the relationship between the yield on the one hand and the treatments and the fertility on the other. Measures are conducted on several lots; let X denote the spatial location of the lot. The environmental factors such as soil nutrients, moisture, sunlight, etc. can be quite different from lot to lot. A model such as (7.12) then assumes linear treatment effects and a nonlinear fertility effect (determined by the environmental factors).

We refer to model (7.12) as a *partially linear model*, or a *partial spline model* as it is known in the spline literature. The additive model (7.3) is in fact a further generalization of the partially linear model (7.12).

Fitting a partially linear model is quite straightforward and can make use of any univariate smoothing technique. Suppose that an i.i.d. sample $(X_1, \mathbf{Z}_1^T, Y_1), \cdots, (X_n, \mathbf{Z}_n^T, Y_n)$ from the population $(X, \mathbf{Z}^T, Y)$ is available. Then, use of a penalized likelihood approach would lead to choosing β_0 and $\eta_0(\cdot)$ that minimize

$$\sum_{i=1}^n \{Y_i - \eta_0(X_i) - \mathbf{Z}_i^T \beta_0\}^2 w_i + \lambda \int \{\eta_0''(x)\}^2 dx, \qquad (7.13)$$

which is in the same spirit as the penalized least squares regression described in Section 2.6, but adapted to the current situation, and allowing for weights w_i. Other fitting procedures using smoothing spline techniques have been proposed but we do not elaborate on this. A nice and thorough discussion of partial spline models is provided in Green and Silverman (1994).

Using local polynomial regression techniques to fit a partially linear model would lead to minimization of

$$\sum_{i=1}^{n}\{Y_i - \sum_{j=0}^{p}\beta_j^*(X_i - x)^j - \mathbf{Z}_i^T\boldsymbol{\beta}_0\}^2 K_h(X_i - x), \qquad (7.14)$$

simultaneously in $(\beta_0^*, \cdots, \beta_p^*)^T$ and $\boldsymbol{\beta}_0$, leading to the estimates $(\widehat{\beta}_0^*, \cdots, \widehat{\beta}_p^*)^T$ and $\widehat{\boldsymbol{\beta}}_0$. The minimization problem (7.14) is in the same spirit as the locally weighted polynomial regression problem described in Section 2.3. An estimator for $\eta_0(x)$ is then $\widehat{\beta}_0^*$. Note that in (7.14) only local data points are used. A more efficient estimator for $\boldsymbol{\beta}_0$ requires the use of all data points and a proposal here is to regress $\{Y_i - \widehat{\eta}_0(X_i)\}$ on $\{\mathbf{Z}_i\}$ to obtain a more efficient estimator for $\boldsymbol{\beta}_0$. More details of this kind of fitting procedure are provided in Section 7.3.3 for the more general class of generalized partially linear single-index models.

Partially linear models generalize the multiple linear regression model. A similar kind of generalization of a generalized linear model leads to the model

$$g\{m(X, \mathbf{Z})\} = \eta_0(X) + \mathbf{Z}^T\boldsymbol{\beta}_0, \qquad (7.15)$$

where $g(\cdot)$ is a known link function.

The partially linear models (7.12) and (7.15) overcome the curse of dimensionality problem since they only involve nonparametric smoothing in one dimension.

7.3.2 Single-index models

Another popular way to overcome the dimensionality problem is to first project all covariates $\mathbf{X} = (X_1, \cdots, X_d)^T$ onto a linear space spanned by the covariates and then to fit a nonparametric curve to their linear combinations. This dimension reduction principle leads to the single-index model

$$Y = \eta_0(\mathbf{X}^T\boldsymbol{\alpha}_0) + \varepsilon, \qquad (7.16)$$

where the task is now to estimate the unknown univariate function $\eta_0(\cdot)$ and the d-dimensional projection vector $\boldsymbol{\alpha}_0$. Clearly the scale of $\mathbf{X}^T\boldsymbol{\alpha}_0$ in $\eta_0(\mathbf{X}^T\boldsymbol{\alpha}_0)$ can be chosen arbitrarily, and for identifiability purposes we impose the constraint $\|\boldsymbol{\alpha}_0\| = 1$ and the first nonzero element is positive.

Suppose that an i.i.d. random sample $(\mathbf{X}_1^T, Y_1), \cdots, (\mathbf{X}_n^T, Y_n)$ from $(\mathbf{X}^T, Y)$ is available. For a given value $\boldsymbol{\alpha}$ the regression function

$$\eta(u|\boldsymbol{\alpha}) \equiv E(Y|\mathbf{X}^T\boldsymbol{\alpha} = u) \qquad (7.17)$$

can be estimated using any nonparametric univariate smoother, such as for example a local linear regression estimator. Denote the resulting estimator by $\hat{\eta}(u|\boldsymbol{\alpha})$. Since $\eta(u|\boldsymbol{\alpha}_0) = \eta_0(u)$, as is easily seen from model (7.16), an estimator for $\boldsymbol{\alpha}_0$ can be obtained by minimizing a measure of the distance $\eta(\cdot|\boldsymbol{\alpha}) - \eta_0(\cdot)$. Härdle, Hall and Ichimura (1993) suggest estimating the function $\eta(\cdot|\boldsymbol{\alpha})$ by a Nadaraya-Watson type estimator (see Section 2.2) and they use cross-validation techniques to define an appropriate empirical measure of distance, namely

$$S(\boldsymbol{\alpha}, h) = \sum_{i=1}^{n} \{Y_i - \hat{\eta}_{-i}(\mathbf{X}_i^T\boldsymbol{\alpha}|\boldsymbol{\alpha})\}^2 w_i,$$

where w_i are weights, $\hat{\eta}_{-i}(\cdot|\boldsymbol{\alpha})$ is the leave-one-out version of $\hat{\eta}(\cdot|\boldsymbol{\alpha})$, and h is the bandwidth used in $\hat{\eta}(\cdot|\boldsymbol{\alpha})$. The bandwidth h and the direction $\boldsymbol{\alpha}$ can be chosen simultaneously by minimizing $S(\boldsymbol{\alpha}, h)$. Once an estimator $\hat{\boldsymbol{\alpha}}_0$ for $\boldsymbol{\alpha}_0$ is obtained the model fitting is completed by calculating a univariate nonparametric smoother for $\eta_0(\cdot)$ based on $(\mathbf{X}_1^T\hat{\boldsymbol{\alpha}}_0, Y_1), \cdots, (\mathbf{X}_n^T\hat{\boldsymbol{\alpha}}_0, Y_n)$.

Another approach to fit model (7.16) is based on the so-called (weighted) *average derivative*

$$\delta_w = E\{\nabla m(\mathbf{X})w(\mathbf{X})\}, \qquad (7.18)$$

where $w(\cdot)$ is a weight function and $\nabla m(\mathbf{x})$ denotes the gradient of the d-variate function $m(\cdot)$. For model (7.16)

$$m(\mathbf{X}) = E(Y|\mathbf{X}) = \eta_0(\mathbf{X}^T\boldsymbol{\alpha}_0)$$

and consequently

$$\delta_w = E\{\eta_0'(\mathbf{X}^T\boldsymbol{\alpha}_0)w(\mathbf{X})\}\boldsymbol{\alpha}_0.$$

Hence an estimator for the weighted average derivative δ_w would provide also an estimator for the projection vector $\boldsymbol{\alpha}_0$ up to a scaling factor. Then, as before, an estimator for η_0 is obtained using any nonparametric univariate smoothing technique.

Estimation of δ_w can be obtained via a plug-in method. Denote by $\hat{m}_1(\cdot)$ any nonparametric estimator of the d-variate function $\nabla m(\cdot)$, and let $\hat{m}_{1,-i}(\cdot)$ denote the leave-one-out estimator based on the reduced sample $\{(\mathbf{X}_j^T, Y_j) : j = 1, \cdots, n, \ j \neq i\}$. The average derivative δ_w can then be estimated directly by

$$\widehat{\delta}_w = \frac{1}{n} \sum_{i=1}^n \widehat{m}_{1,-i}(\mathbf{X}_i) w(\mathbf{X}_i). \tag{7.19}$$

Other estimators for the average derivative δ_w have been proposed in the literature. For example Härdle and Stoker (1989) propose an estimator for the average derivative relying on an estimate for the marginal density of $\mathbf{X}$. See Härdle (1990) and Newey and Stoker (1993) for a nice discussion of average derivative estimation and its applications. The average derivative method is also applicable to multi-index models. See Samarov (1993) and Wong and Shen (1995).

7.3.3 Generalized partially linear single-index models

We now introduce a broader class of models, which includes generalized linear models, partially linear models and single-index models. Suppose that a response variable Y has to be predicted from the covariates $(\mathbf{X}, \mathbf{Z})$, where $\mathbf{X}$ and $\mathbf{Z}$ are possibly vector valued, say $\mathbf{X} = (X_1, \cdots, X_d)^T$ and $\mathbf{Z} = (Z_1, \cdots, Z_q)^T$. We first recall briefly the concept of generalized linear models (see also Section 5.4). Suppose that the conditional density of Y given $(\mathbf{X}, \mathbf{Z}) = (\mathbf{x}, \mathbf{z})$ belongs to a canonical exponential family

$$f_{Y|\mathbf{X}\mathbf{Z}}(y|\mathbf{x}, \mathbf{z}) = \exp\left[y\theta(\mathbf{x}, \mathbf{z}) - \mathcal{B}\{\theta(\mathbf{x}, \mathbf{z})\} + \mathcal{C}(y)\right], \tag{7.20}$$

for known functions $\mathcal{B}(\cdot)$ and $\mathcal{C}(\cdot)$. For ease of notation we drop the dispersion parameter ϕ and its associated function $a(\phi)$. As can be seen from the examples in Section 5.4.1 the factor $a(\phi)$ is usually a constant. In a parametric generalized linear model the unknown regression function

$$m(\mathbf{x}, \mathbf{z}) = E(Y|\mathbf{X} = \mathbf{x}, \mathbf{Z} = \mathbf{z}) = \mathcal{B}'\{\theta(\mathbf{x}, \mathbf{z})\}$$

is modelled linearly via a known link function g by

$$g\{m(\mathbf{X}, \mathbf{Z})\} = \gamma_0 + \mathbf{X}^T \boldsymbol{\alpha}_0 + \mathbf{Z}^T \boldsymbol{\beta}_0. \tag{7.21}$$

See also (5.48). The choice $g = (\mathcal{B}')^{-1}$ is termed the canonical link function.

In many practical situations some of the covariates have a linear effect on the response variable Y while other covariates will possibly have a highly nonlinear effect. This inspires us to consider a *generalized partially linear single-index model*

$$g\{m(\mathbf{X}, \mathbf{Z})\} = \eta_0(\mathbf{X}^T \boldsymbol{\alpha}_0) + \mathbf{Z}^T \boldsymbol{\beta}_0, \tag{7.22}$$

with $\|\alpha_0\| = 1$ and the first nonzero element of α_0 positive, for identifiability reasons.

When $\beta_0 = 0$, or equivalently when there are no predictors $\mathbf{Z}$, model (7.22) is simply a generalized linear model with an *unknown* link function. The problem of missing link functions in generalized linear models has been considered by Weisberg and Welsh (1994). Similarly, model (7.22) includes the single-index model (7.16) which only specifies the mean regression function and not the error distribution. Another special case occurs when $\mathbf{X}$ is scalar, so that $d = 1$ and $\alpha_0 = 1$. Then, model (7.22) reduces to the partially linear model (7.15).

Model (7.22) can easily incorporate interactions or partial interactions among $\mathbf{X}$ and $\mathbf{Z}$ by forming either a new longer $\mathbf{X}$ or $\mathbf{Z}$ vector. See also Section 7.4.2.

The dimensionality reduction principle behind model (7.22) is clear: due to a projection on a linear space spanned by the covariates $(X_1, \cdots, X_d)^T$ we are only left with nonparametric estimation of a univariate function. The above class of generalized partially linear single-index models is quite attractive because of its model parsimony. What can we say about the effective number of parameters needed to fit such a model? If estimation of the nonparametric component requires ℓ_n parameters, then fitting a generalized partially linear single-index model (7.22) would involve about $(\ell_n + d + q - 1)$ parameters, where the -1 comes from the constraint on α_0. Note that this total number of parameters grows very slowly when the dimensionality parameters d and q increase. This is an appealing feature for high-dimensional data analysis. In contrast, the generalized additive model (7.6), with $q = 0$, has effective number of parameters approximately equal to $(\ell_n - 1)d + 1$, which can increase rather rapidly when d increases. The fully nonparametric model would have effective number of parameters ℓ_n^d. Even though ℓ_n differs from model to model, the above comparisons provide some intuition on the complexity of each class of models. These effective numbers of parameters also give a feeling on when a model can be over-parametrized for a given amount of data.

In order to fit model (7.22) one needs to determine estimation procedures for the unknown parameter vectors α_0 and β_0 and for the unknown univariate function $\eta_0(\cdot)$. Such an estimation procedure then also provides an alternative estimation procedure for partially linear models, models with missing link function and single-index models.

As in Section 5.4 we will work in a more general setting where only the relationship between the conditional variance function $\text{Var}(Y|\mathbf{X} = \mathbf{x}, \mathbf{Z} = \mathbf{z})$ and the conditional mean function $m(\mathbf{x}, \mathbf{z})$ is specified. In this situation, estimation of the regression function can be done by replacing the conditional log-likelihood $\ln f_{Y|\mathbf{X}\mathbf{Z}}(y|\mathbf{x}, \mathbf{z})$ by a quasi-likelihood function $Q\{m(\mathbf{x}, \mathbf{z}), y\}$, as mentioned in Section 5.4. Suppose that the conditional variance is modelled as

$$\text{Var}(Y|\mathbf{X} = \mathbf{x}, \mathbf{Z} = \mathbf{z}) = \sigma^2 V\{m(\mathbf{x}, \mathbf{z})\}$$

for some known positive function $V(\cdot)$. The factor σ^2 simply reflects a factor similar to $a(\phi)$ involving the dispersion parameter (see also Section 5.4). Then the corresponding quasi-likelihood function $Q(w, y)$ satisfies

$$\frac{\partial}{\partial w} Q(w, y) = \frac{y - w}{V(w)}.$$

We now discuss an estimation procedure for fitting a generalized partially linear single-index model (7.22). Since the univariate function η_0 has to be estimated nonparametrically it is natural to consider a local quasi-likelihood such as in Section 5.4. On the other hand, efficient estimation of the parameter vectors $\boldsymbol{\alpha}_0$ and $\boldsymbol{\beta}_0$ would require using all data points. How do we meet both of these demands? In determining a local quasi-likelihood we approximate the unknown function $\eta_0(\cdot)$ locally by a linear function

$$\eta_0(v) \approx \eta_0(u) + \eta_0'(u)(v - u) \equiv a + b(v - u),$$

for v in a neighborhood of u, and where $a = \eta_0(u)$ and $b = \eta_0'(u)$. Let K be a symmetric probability density function, and define as before its rescaled version $K_h(t) = K(t/h)/h$.

Denote the observed sample by $\{(\mathbf{X}_i^T, \mathbf{Z}_i^T, Y_i) : i = 1, \cdots, n\}$. The proposed estimation procedure then reads as follows. With any given $\boldsymbol{\alpha}$, first find $\widehat{a}$, $\widehat{b}$ and $\widehat{\boldsymbol{\beta}}$ to maximize the *local quasi-likelihood*

$$\sum_{i=1}^{n} Q\left[g^{-1}\left\{a + b(\mathbf{X}_i^T \boldsymbol{\alpha} - u) + \mathbf{Z}_i^T \boldsymbol{\beta}\right\}, Y_i\right] K_h(\mathbf{X}_i^T \boldsymbol{\alpha} - u). \quad (7.23)$$

Note the analogy with the local quasi-likelihood in (5.53) in Chapter 5. Let $\widehat{\eta}(u) \equiv \widehat{\eta}(u; h, \boldsymbol{\alpha}) = \widehat{a}$ and $\widehat{\boldsymbol{\beta}} \equiv \widehat{\boldsymbol{\beta}}(u; h, \boldsymbol{\alpha})$. Having obtained the function $\widehat{\eta}(u; h, \boldsymbol{\alpha})$, estimate the global parameters $\boldsymbol{\alpha}$ and $\boldsymbol{\beta}$ by maximizing the *global quasi-likelihood*

$$\sum_{i=1}^{n} Q\left[g^{-1}\left\{\widehat{\eta}(\mathbf{X}_i^T \boldsymbol{\alpha}; h, \boldsymbol{\alpha}) + \mathbf{Z}_i^T \boldsymbol{\beta}\right\}, Y_i\right] \quad (7.24)$$

GENERALIZED PARTIALLY LINEAR SINGLE-INDEX MODELS 279

with respect to α and β.

In summary the basic ideas of the estimation procedure are as follows: estimate $\eta_0(\cdot)$ locally via (7.23), then use all the data and (7.24) to estimate (α_0, β_0), with $\widehat{\eta}(\cdot)$ replacing $\eta_0(\cdot)$.

The implementation of the above estimation procedure needs some further thought. Note first of all that if α_0 is known or if $\mathbf{X}$ is a scalar and hence $\alpha_0 = 1$ then expressions (7.23) and (7.24) determine a fully specified, noniterative algorithm. We now describe two approaches to implement the estimation procedure: the *one-step algorithm* and the *fully iterated algorithm*. The one-step algorithm is simple computationally, particularly when α_0 is known, while the fully iterated algorithm is preferable from the point of view of asymptotic efficiency as was found by Carroll, Fan, Gijbels and Wand (1995).

The one-step algorithm reads as follows:

Step 1. Fit a parametric generalized linear model to obtain the initial estimates $(\widehat{\alpha}_1, \widehat{\beta})$ and set $\widehat{\alpha} = \widehat{\alpha}_1 / \|\widehat{\alpha}_1\|$.

Step 2. With the estimated $\widehat{\alpha}$, find $\widehat{\eta}(u; h, \widehat{\alpha})$ by maximizing the local quasi-likelihood (7.23).

Step 3. With the estimated $\widehat{\eta}(u; h, \widehat{\alpha})$, update $\widehat{\alpha}$ and $\widehat{\beta}$ by maximizing the global quasi-likelihood (7.24).

Step 4. Continue Steps 2 and 3 until convergence.

Step 5. Fix (α, β) at its estimated value from Step 4. The final estimate of $\eta_0(\cdot)$ is $\widehat{\eta}(u; h, \widehat{\alpha}, \widehat{\beta}) = \widehat{a}$ where $(\widehat{a}, \widehat{b})$ maximizes

$$\sum_{i=1}^n Q\left[g^{-1}\left\{a + b(\mathbf{X}_i^T\widehat{\alpha} - u) + \mathbf{Z}_i^T\widehat{\beta}\right\}, Y_i\right] K_h(\mathbf{X}_i^T\widehat{\alpha} - u). \tag{7.25}$$

The name 'one-step' algorithm comes from the fact that the algorithm is noniterative when α_0 is known. When α_0 is unknown, the above algorithm is iterative. However, one can make it noniterative by maximizing (7.23) with respect to (a, b, α, β) and then maximizing (7.24) in (α, β). We would expect that this noniterative algorithm is less efficient than our defined one-step algorithm.

The fully iterated algorithm differs from the one-step algorithm in that (7.23) is maximized over a and b instead of over (a, b, β). Thus, this algorithm always requires iteration even when α_0 is known.

The details of the fully iterated algorithm are

Step 1. Fit a parametric generalized linear model to obtain the initial estimates $(\widehat{\alpha}_1, \widehat{\beta})$ and set $\widehat{\alpha} = \widehat{\alpha}_1/\|\widehat{\alpha}_1\|$.

Step 2. With the estimated $\widehat{\alpha}$ and $\widehat{\beta}$ find $\widehat{\eta}(u; h, \widehat{\alpha}, \widehat{\beta}) = \widehat{a}$ by solving (7.25).

Step 3. With the estimated $\widehat{\eta}(u; h, \widehat{\alpha}, \widehat{\beta})$, update $(\widehat{\alpha}, \widehat{\beta})$ by maximizing

$$\sum_{i=1}^{n} Q\left[g^{-1}\left\{\widehat{\eta}(\mathbf{X}_i^T \alpha; h, \alpha, \beta) + \mathbf{Z}_i^T \beta\right\}, Y_i\right], \qquad (7.26)$$

with respect to α and β.

Step 4. Continue Steps 2 and 3 until convergence.

Step 5. Fix (α, β) at its estimated value from Step 4. The final estimate of $\eta_0(\cdot)$ is $\widehat{\eta}(u; h, \widehat{\alpha}, \widehat{\beta}) = \widehat{a}$ where $(\widehat{a}, \widehat{b})$ is obtained by solving (7.25).

Note that both the one-step algorithm and the fully iterated algorithm involve choices of the bandwidth or smoothing parameter h. A first choice of this kind arises in Steps 2 and 3 of the algorithms. There, given an initial value of α_0 (and β_0 for the fully iterated algorithm) a nonparametric estimate of $\eta_0(\cdot)$ based on the local quasi-likelihood is calculated. The resulting estimate is then used to update the previous values of the parameter vectors. Finally, after cycling between Steps 2 and 3, one obtains reliable estimators for the parameter vectors α_0 and β_0. Since this is the final goal here the bandwidth parameter h in the local quasi-likelihood is taken to be a bandwidth that is optimal for estimating (α_0, β_0). In particular, the bandwidth should be small enough so that $\widehat{\alpha}_0$ and $\widehat{\beta}_0$ would have a negligible bias. The bias of $\widehat{\alpha}_0$ and $\widehat{\beta}_0$ is typically of order $O(h_n^2)$ and in order to achieve the parametric rate of convergence $n^{1/2}$ the formal requirement for a negligible bias is that $h_n^2 = o(n^{-1/2})$. With this choice of bandwidth, the variance of $\widehat{\eta}_0(\cdot)$ will be large, but this effect is averaged out in the global estimation of α_0 and β_0, and the $\sqrt{n}$-consistency of $\widehat{\alpha}_0$ and $\widehat{\beta}_0$ is retained. See Carroll, Fan, Gijbels and Wand (1995) for more details.

The situation is quite different however in Step 5 of the algorithms where the local quasi-likelihood is also used, but now the focus is on estimation of the nonparametric part $\eta_0(\cdot)$. Logically here h should be taken to be optimal for estimation of $\eta_0(\cdot)$ when α_0 and β_0 are known. The question remains of how to select such optimal bandwidths. This issue is discussed in the recent paper by Carroll, Fan, Gijbels and Wand (1995).

The proposed estimation procedure uses the idea of local linear approximation, where the local neighborhood is determined by a kernel function K and a constant bandwidth h. There is no need however to restrict to local linear approximation or to a constant bandwidth. Instead of using local linear approximations one can rely more generally on local polynomial approximations. The theoretical results obtained for the estimation procedure with local linear approximation (see Carroll, Fan, Gijbels and Wand (1995)) easily generalize to local polynomial fitting. Similarly it is not necessary to restrict to constant bandwidths. One can also opt for a data-driven local bandwidth. Here the methods of Chapter 4 can be applied.

The asymptotic distribution theory for the one-step and the fully iterated estimator, based on local linear approximation, has been established by Carroll, Fan, Gijbels and Wand (1995). That paper provides asymptotic expressions for the variances and covariances of all estimators and also describes an approach to deriving estimated standard errors. Some conclusions drawn from that theoretical study are:

(i) estimation of the nonparametric part $\eta_0(\cdot)$ of model (7.22) can be done just as well as if the parameters (α_0, β_0) were known;

(ii) the parametric parts can be estimated at the usual parametric rate of convergence;

(iii) the fully iterative estimator can have a smaller variance-covariance matrix than the one-step estimator, but for the important least squares case with weakly correlated **X** and **Z** both estimators share the same variance-covariance matrix;

(iv) estimation of the nonparametric part in the model has an effect on the distribution of the estimates for the parametric parts;

(v) best estimation of the parametric parts requires undersmoothing of the nonparametric part.

Since the fully iterated method is more efficient, we will use this estimator for the following illustration. Details of its implementation, which we have omitted for brevity, can be found in Carroll, Fan, Gijbels and Wand (1995).

We now illustrate the generalized partially linear single-index model using data from the Coronary Risk Factor Study (CORIS) survey, described briefly in Section 7.2. The response variable Y is the binary variable indicating the presence or absence of myocardial infarction. Here we consider four covariates: the systolic blood pressure (X_1), tobacco (X_2), the cholesterol ratio (X_3) and family history (Z). The variables X_1 and X_3 were described in Section 7.2. The tobacco variable X_2 attempts to measure the total tobacco (in kilograms) consumed in a subject's lifetime, and equals the average consumption per day multiplied by the period of use. The family history variable Z is a 0–1 variable, where 1 indicates that a family member of the subject has had heart disease, and 0 means that there has been no occurence of heart disease in the family so far.

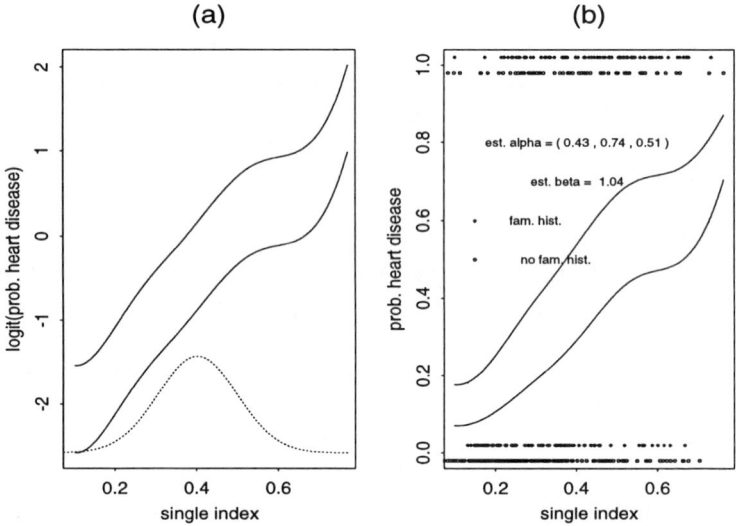

Figure 7.3. *Fit of a generalized partially linear single-index model for the CORIS data. (a) Solid curves correspond to estimates of* logit$\{P(heart\ disease)\}$ *for those with* $Z = 1$ *(upper curve) and those with* $Z = 0$ *(lower curve) against the estimated single-index* $0.43X_1 + 0.74X_2 + 0.51X_3$. *The dotted curve is the kernel weight used in the local linear fitting process. (b) Estimates of P(heart disease) for those with* $Z = 1$ *(upper curve) and those with* $Z = 0$ *(lower curve) against the single index. Courtesy of Dr. Matt P. Wand.*

We fit the generalized partially linear single-index model (7.22) using the logit link. The estimated coefficients and their estimated standard errors are presented in Table 7.1. Figure 7.3 summarizes the findings. The result does not seem to departure significantly from those obtained from ordinary logistic regression. See Table 7.1 for details. Thus, we have validated the logistic regression model by embedding it in a larger class of models. Note that all of the covariates are statistically significant. In particular, for patients with a family history of heart disease the odds of heart disease increase by a factor of approximately exp(1.04).

Table 7.1. *Summary of CORIS study.*

	systolic	tabacco	cholrat	family hist.
Ordinary logistic	0.29	0.74	0.59	1.10
std. err.	0.13	0.11	0.13	0.22
GPLSIM	0.43	0.74	0.51	1.04
std. err.	0.13	0.11	0.15	0.22

7.4 Modelling interactions

In all of the models discussed in Sections 7.2 and 7.3 only the influence of each covariate separately has been taken into account. But covariates may affect each other, and this calls for modelling *interaction terms* next to the *main effect terms* which have been included so far. There are several approaches to model interactions and in this section we restrict the discussion to some of these. In particular, we describe how to fit generalized additive models and generalized partially linear single-index models with interaction terms. We also indicate the use of adaptive regression spline techniques in modelling interactions. In the classification and regression trees technique, interactions are inherently included. See Breiman, Friedman, Olshen and Stone (1993).

7.4.1 Interactions in generalized additive models

The basic ideas for fitting a generalized additive model consist of those for fitting an additive model and of those for fitting a generalized linear model. See Section 7.2. Therefore, for ease of presentation, we restrict the discussion on interactions to least squares type procedures and indicate briefly later on extensions to likelihood-based procedures. Furthermore we only discuss the incorporation of *two-term interactions* describing the influence between two covariates. The extension to interactions involving more than two covariates is straightforward.

Consider a model with componentwise additive *main terms* and pairwise *interaction terms*

$$Y = \alpha + \sum_{j=1}^{d} g_j(X_j) + \sum_{1 \leq j < k \leq d} g_{jk}(X_j, X_k) + \varepsilon, \qquad (7.27)$$

where g_j, $j = 1, \cdots, d$, are unknown univariate functions, and g_{jk}, $1 \leq j < k \leq d$ are unknown bivariate functions. As in Section 7.2 constraints (7.4) are imposed on the univariate functions. To make model (7.27) identifiable however, conditions on the bivariate functions g_{jk} are also needed: indeed we can add an arbitrary function $h(X_j)$ to $g_j(X_j)$ and substract it from $g_{jk}(X_j, X_k)$. So, in addition to the constraints on the main effects

$$E\{g_j(X_j)\} = 0, \qquad j = 1, \cdots, d, \qquad (7.28)$$

the following requirements are imposed on the interactions

$$E\{g_{jk}(X_j, X_k)|X_j\} = E\{g_{jk}(X_j, X_k)|X_k\} = 0, \qquad 1 \leq j < k \leq d. \tag{7.29}$$

It is easily seen that with these constraints $E(Y) = \alpha$ in model (7.27).

Finding interactions is an integral part of statistical data analyses. However, the cost of the flexible model (7.27) can also be high: there are about $(\ell_n - 1)^2 d(d - 1)/2 + d(\ell_n - 1) + 1$ effective parameters in model (7.27), assuming that each coordinate effectively uses ℓ_n parameters. If $d = 4$ and $\ell_n = 11$, one uses effectively 641 parameters in model (7.27). Thus, the curse of dimensionality is still inherited in this interaction model. However, when ℓ_n is small, the model is feasible. Section 7.4.2 provides another scheme of modelling interactions which has fewer effective parameters.

How do we fit an additive model with interactions? Note first of all that if model (7.27) is correct then

$$E\{Y - \alpha - \sum_{j \neq k} g_j(X_j) - \sum_{1 \leq j < \ell \leq d} g_{j\ell}(X_j, X_\ell)|X_k\} = g_k(X_k)$$
$$\text{for all } k = 1, \cdots, d, \qquad (7.30)$$

and moreover

$$E\{Y - \alpha - \sum_{j=1}^{d} g_j(X_j) - \sum_{\substack{1 \le j < k \le d \\ (j,k) \ne (j_0, k_0)}} g_{jk}(X_j, X_k) | X_{j_0}, X_{k_0}\}$$
$$= g_{j_0 k_0}(X_{j_0}, X_{k_0}) \quad \text{for all } 1 \le j_0 < k_0 \le d. \quad (7.31)$$

As before this set of equations suggests an iterative algorithm for computing all univariate and bivariate functions. Given α, the univariate functions g_j, $j \ne k$, and the bivariate functions $g_{j\ell}$, $1 \le j < \ell \le d$, the unknown function g_k can be estimated using any univariate smoother $\mathcal{S}_k$ based on the observations $\{(X_{ik}, Y_i) : i = 1, \cdots, n\}$, as indicated by (7.30). To meet constraint (7.28) the resulting estimate $\widehat{g}_k(\cdot)$ is centralized as in (7.5). Further, the set of equations (7.31) reveals that given α, all univariate functions g_j, $j = 1, \cdots, d$, and the bivariate functions g_{jk}, $1 \le j < k \le d$, $(j, k) \ne (j_0, k_0)$, an estimate $\widehat{g}_{j_0 k_0}$ for the unknown bivariate function $g_{j_0 k_0}$ can be obtained by applying any bivariate smoother, say $\mathcal{S}_{j_0 k_0}$, based on the observations $\{(X_{ij_0}, X_{ik_0}, Y_i) : i = 1, \cdots, n\}$. It is important to note here that in principle any bivariate smoother such as a bivariate kernel estimate or a bivariate local linear regression estimate (see Section 7.8) can be used. How do we make sure that the estimate $\widehat{g}_{j_0 k_0}(\cdot)$ meets condition (7.29)? A simply way is to fit an additive model with componentwise terms $h_{j_0}(X_{j_0})$ and $h_{k_0}(X_{k_0})$ to $\widehat{g}_{j_0 k_0}$, resulting in the estimated curves $\widehat{h}_{j_0}(\cdot)$ and $\widehat{h}_{k_0}(\cdot)$, and then to put

$$\widehat{g}^*_{j_0 k_0}(x_1, x_2) = \widehat{g}_{j_0 k_0}(x_1, x_2) - \widehat{h}_{j_0}(x_1) - \widehat{h}_{k_0}(x_2), \quad (7.32)$$

and to replace $\widehat{g}_\ell(\cdot)$ by $\widehat{g}_\ell(\cdot) - \widehat{h}_\ell(\cdot)$, $\ell = j_0, k_0$.

We are now ready to formulate an algorithm for fitting model (7.27). The main difference with fitting a componentwise additive model is that the algorithm for fitting (7.27) will require both univariate and bivariate smoothing techniques. Details of the algorithm for fitting model (7.27) read as follows:

Step 1. Initialization: $\widehat{\alpha} = n^{-1} \sum_{i=1}^{n} Y_i$, $\widehat{g}_k = g^0_k$, $k = 1, \cdots, d$, $\widehat{g}_{j_0 k_0} = g^0_{j_0 k_0}$, $1 \le j_0 < k_0 \le d$.

Step 2. For each $k = 1, \cdots, d$

$$\widehat{g}_k = \mathcal{S}_k\{Y - \widehat{\alpha} - \sum_{j \ne k} \widehat{g}_j(X_j) | X_k\},$$

and obtain $\widehat{g}^*_k(\cdot)$ as indicated in (7.5).

Step 3. Keep cycling Step 2 until convergence.

Step 4. For each pair (j_0, k_0), $1 \leq j_0 < k_0 \leq d$,

$$\widehat{g}_{j_0 k_0} = \mathcal{S}_{j_0 k_0} \{Y - \widehat{\alpha} - \sum_{j=1}^{d} \widehat{g}_j(X_j)$$
$$- \sum_{\substack{1 \leq j < k \leq d \\ (j,k) \neq (j_0, k_0)}} \widehat{g}_{jk}(X_j, X_k) | X_{j_0}, X_{k_0}\},$$

and obtain $\widehat{g}^*_{j_0 k_0}(\cdot)$ as indicated in (7.32).

Step 5. Keep cycling Steps 2–4 until convergence.

We now turn to the discussion of fitting a generalized additive model with two-term interactions, i.e. a model of the form (see (7.6))

$$g\{m(\mathbf{X})\} = \alpha + \sum_{j=1}^{d} g_j(X_j) + \sum_{1 \leq j < k \leq d} g_{jk}(X_j, X_k), \qquad (7.33)$$

where g_j, $j = 1, \cdots, d$, are unknown univariate functions and g_{jk}, $1 \leq j < k \leq d$, are unknown bivariate functions, satisfying constraints (7.28) and (7.29). The algorithm for fitting model (7.33) relies on the adjusted dependent variable procedure (for fitting a generalized linear model) and an appropriate backfitting algorithm. See also Section 7.2. The main difference from fitting a generalized additive model is that now the backfitting part of the algorithm in the inner loop involves applications of univariate as well as bivariate smoothing techniques.

The result of fitting an additive model, with or without interactions, is a d-dimensional function and in the case $d = 2$ a three-dimensional plot can be used to display the fitted surface. Another possibility is to use conditional plots, which present a few slices of the surface, or contour plots.

Interactions between a continuous and a categorical covariate are particularly easy to handle: the function describing the influence of the continuous covariate changes according to the value of the categorical covariate. This then leads to separate fitted curves, again according to the value of the categorical covariate. For example in the case of two covariates X_1 and X_2, with X_1 a continuous covariate and X_2 a binary covariate, such a simple interaction is described in the mean regression function as follows

$$m(x_1, x_2) = \alpha + g_{x_2}(x_1) = h_{x_2}(x_1), \qquad x_2 = 0, 1. \qquad (7.34)$$

MODELLING INTERACTIONS 287

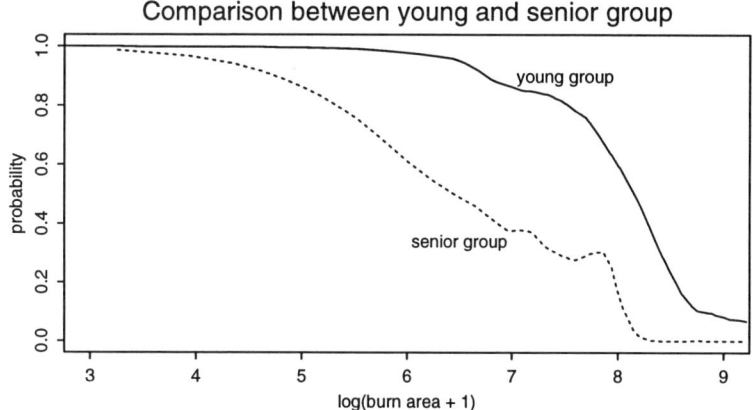

Figure 7.4. *Fitting model (7.34) to the burn data with the continuous covariate log(area of third degree burn +1) and the binary covariate indicating the age group. Solid curve: estimated curve for the age group 17–50; dotted curve: estimated curve for the age group 51–85.*

The functions $h_0(\cdot)$ and $h_1(\cdot)$ can be easily estimated by applying a univariate smoothing technique to the data with $X_{i2} = 0$ and the data with $X_{i2} = 1$ respectively. We illustrate this procedure for the burn data set described in Section 2.1. The response variable is the binary variable indicating survival or death after a burn injury. The continuous covariate X_1 is the logarithm of the area of third degree burn +1, and the categorical covariate X_2 describes the age group to which the patient belongs. As in the illustration in Section 4.8 the age groups 17–50 and 51–85 are considered. We fitted model (7.34) using the backfitting algorithm for the generalized additive model with the LOESS smoother with smoothing parameter $d = 0.4$ as the basic smoother. The result is shown in Figure 7.4.

7.4.2 Interactions in generalized partially linear single-index models

In this section we discuss briefly how to incorporate two-term interactions, such as the ones described in Section 7.4.1, in a generalized partially linear single-index model (7.22). An advantage of this approach, as indicated in Section 7.4.1, is that it copes well with the curse of dimensionality.

A simple type of interaction would be similar to the ones described in the last paragraph of Section 7.4.1 (see model (7.34)), leading to a model of the form

$$g\{m(\mathbf{x},\mathbf{z})\} = \eta_{z_1}(\mathbf{x}^T\boldsymbol{\alpha}_0) + \mathbf{z}_2^T\boldsymbol{\beta}_0, \qquad (7.35)$$

where $\mathbf{z} = (z_1, \mathbf{z}_2)$ with z_1 binary and $\mathbf{z}_2$ arbitrary. In Section 7.3.3 we discussed in detail the estimation procedure for fitting a generalized partially linear single-index model (7.22). The key quantities appearing in the fitting procedure are the local quasi-likelihood (7.23), for estimation of the nonparametric part, and the global quasi-likelihood (7.24) for estimation of the parametric part. The estimation method described in Section 7.3.3 can be easily modified to handle a model with interactions such as (7.35). By analogy with the ideas used in fitting model (7.34) the local quasi-likelihood (7.23) is replaced by

$$\sum_{i=1}^{n} Q\left[g^{-1}\left\{a_0 + b_0(\boldsymbol{\alpha}^T\mathbf{X}_i - u) + \boldsymbol{\beta}^T\mathbf{Z}_{2i}\right\}, Y_i\right]$$
$$\times K_{h_0}(\boldsymbol{\alpha}^T\mathbf{X}_i - u)I(Z_{i1} = 0)$$
$$+ \sum_{i=1}^{n} Q\left[g^{-1}\left\{a_1 + b_1(\boldsymbol{\alpha}^T\mathbf{X}_i - u) + \boldsymbol{\beta}^T\mathbf{Z}_{2i}\right\}, Y_i\right]$$
$$\times K_{h_1}(\boldsymbol{\alpha}^T\mathbf{X}_i - u)I(Z_{i1} = 1), \qquad (7.36)$$

where h_0 and h_1 are bandwidths for η_0 and η_1, respectively. The estimators for η_0 and η_1 are then respectively

$$\widehat{\eta}_0(u) = \widehat{a}_0 \quad \text{and} \quad \widehat{\eta}_1(u) = \widehat{a}_1.$$

The global quasi-likelihood (7.24) is modified along the same lines leading to

$$\sum_{i=1}^{n} Q\left[g^{-1}\left\{\widehat{\eta}_0(\mathbf{X}_i^T\boldsymbol{\alpha}; h, \boldsymbol{\alpha}) + \mathbf{Z}_{2i}^T\boldsymbol{\beta}\right\}, Y_i\right] I(Z_{i1} = 0)$$
$$+ \sum_{i=1}^{n} Q\left[g^{-1}\left\{\widehat{\eta}_1(\mathbf{X}_i^T\boldsymbol{\alpha}; h, \boldsymbol{\alpha}) + \mathbf{Z}_{2i}^T\boldsymbol{\beta}\right\}, Y_i\right] I(Z_{i1} = 1). \qquad (7.37)$$

The effective number of parameters for fitting a model such as (7.35) would be $2\ell_n + d + q - 1$, with ℓ_n the average number of parameters for fitting $\eta_0(\cdot)$ or $\eta_1(\cdot)$.

Model (7.22) also allows modelling interactions of the following form

$$g\{m(\mathbf{x},\mathbf{z})\} = \eta_0\{\mathbf{x}^T\boldsymbol{\alpha}_0 + (\mathbf{x}^T, \mathbf{z}^T)\Lambda(\mathbf{x}^T, \mathbf{z}^T)^T\} + \mathbf{z}^T\boldsymbol{\beta}_0, \qquad (7.38)$$

MODELLING INTERACTIONS 289

where Λ is a $(d+q) \times (d+q)$ parameter matrix describing the two-term interactions between the covariates $X_1, \cdots, X_d, Z_1, \cdots, Z_q$. Note that by forming a new and longer $\mathbf{X}$ vector, model (7.38) can be included in model (7.22). The effective number of parameters for fitting such a model would be about $\ell_n + d + (d+q) \times (d+q-1)/2 + q - 1 = \ell_n + (d+q-1)(d+q+2)/2$, where ℓ_n is the effective average number of parameters required for estimating $\eta_0(\cdot)$. Further, with (7.22) we can also model interactions of the form

$$g\{m(\mathbf{x},\mathbf{z})\} = \eta_0\{\mathbf{x}^T\boldsymbol{\alpha}_0\} + (\mathbf{x}^T,\mathbf{z}^T)\Lambda(\mathbf{x}^T,\mathbf{z}^T)^T + \mathbf{z}^T\boldsymbol{\beta}_0, \quad (7.39)$$

by forming a longer $\mathbf{Z}$ vector.

When appropriate one could consider a model with selected partial interaction terms, leading to a reduction of the number of effective parameters.

7.4.3 Multivariate adaptive regression splines

In this section we report very briefly on an approach to modelling interactions using spline smoothing techniques (as discussed in Section 2.6).

In an *additive regression spline model* the mean regression function $m(\mathbf{X}) = E(Y|\mathbf{X})$ is of the form

$$m(\mathbf{X}) = \alpha + \sum_{j=1}^{d} \mathbf{B}_j^T(X_j)\boldsymbol{\theta}_j, \quad (7.40)$$

where $\mathbf{B}_j(\cdot) = \{B_{j1}(\cdot), \cdots, B_{j\ell_j}(\cdot)\}^T$ denotes the vector of ℓ_j basis functions associated with the covariate X_j. This set of basis functions could consist of for example truncated power functions (leading to a so-called power basis) or B-splines. See Section 2.6. The parameter vector $\boldsymbol{\theta}_j = (\theta_{j1}, \cdots, \theta_{j\ell_j})^T$ describes the influence of the covariate X_j on the mean regression function. Model (7.40) can be written as

$$m(\mathbf{X}) = \alpha + \sum_{j=1}^{d} \sum_{k=1}^{\ell_j} B_{jk}(X_j)\theta_{jk},$$

which is a parametric regression model with number of parameters $L \equiv 1 + \sum_{j=1}^{d} \ell_j$ for given sets of basis functions $\mathbf{B}_1(\cdot), \cdots, \mathbf{B}_d(\cdot)$. Note that the constant basis function is always involved due to the constant α. So, L also refers to the number of basis functions involved.

For the set of basis functions one can use for example B-splines or power functions with an associated sequence of knots. This then brings up the question of how to choose the number of knots and the location of the knots for each set of basis functions $\mathbf{B}_j$. A method for this is described in Section 2.6.1. Using a linear spline basis, Friedman and Silverman (1989) provided an algorithm to optimize over the number and the locations of the knots in an adaptive way. With such an adaptive procedure for choosing the basis functions, fitting an additive regression spline model (7.40) results in an adaptive additive regression spline procedure.

Friedman (1991) enlarged the flexibility of the adaptive additive regression spline procedure by allowing the L basis functions to be a *multivariate tensor-product spline basis*. This results in the so-called *multivariate adaptive regression splines*. A tensor-product spline for two covariates would correspond to a product of two sets of one-dimensional basis functions, one for each covariate. With this generalization to tensor-product bases, interaction terms are incorporated automatically in the procedure. See Friedman (1991) for a thorough discussion of multivariate adaptive regression splines and in particular their flexibility features. See also Stone (1990b, 1991, 1994) and the references therein.

7.5 Sliced inverse regression

In Sections 7.2 and 7.3 we discussed some approaches to deal with the curse of dimensionality problem in multivariate regression problems. All these approaches focus on the modelling aspect. Sliced inverse regression is a data-analytic tool which aims at a reduction of the dimensionality without a complicated model-fitting process. The sliced inverse regression technique was studied in detail by Li (1991) and Duan and Li (1991). In this section we present the main ideas of this dimension-reduction procedure. For a more detailed introduction to the subject we refer to the above two papers.

When dealing with a high-dimensional vector of covariates $\mathbf{X} = (X_1, \cdots, X_d)^T$ an ideal situation arises when the interesting features of all covariates could be retrieved from a few low-dimensional projections. These low-dimensional projections would capture all structural information about the response variable Y. Such an ideal situation can be described as

$$Y = g(\mathbf{X}^T \boldsymbol{\beta}_1, \cdots, \mathbf{X}^T \boldsymbol{\beta}_L, \varepsilon), \qquad (7.41)$$

$1 \leq L \leq d$, where $\beta_1, \cdots, \beta_L$ denote d-dimensional column vectors, ε is independent of $\mathbf{X}$ and g is an arbitrary unknown function defined on $I\!R^{L+1}$. When L is considerably smaller than d the goal of *data reduction* is achieved since (7.41) reveals that looking at the low-dimensional projections $(\mathbf{X}^T\beta_1, \cdots, \mathbf{X}^T\beta_L)$ will be sufficient. Any linear combination of the column vectors $\beta_1, \cdots, \beta_L$ is called an *effective dimension-reduction (e.d.r.) direction*. The linear space generated by the β's is called the *effective dimension-reduction (e.d.r.) space*, and is denoted by $\mathcal{B}$. The main focus is now on estimation of the effective dimension-reduction directions and the linear space generated by them. After determining this space, the data can be projected into this smaller space, bringing the investigator to a better, lower-dimensional, position.

We first outline the main ideas of sliced inverse regression and give additional explanation on each in subsequent separate paragraphs. The main question is how to estimate the e.d.r. directions. The key to this is to rely on the *inverse regression* $E(\mathbf{X}|Y)$, i.e. to regress $\mathbf{X}$ against Y. In a later paragraph we explain how precisely the inverse regression helps in finding the e.d.r. directions. The immediate benefit from using the inverse regression $E(\mathbf{X}|Y)$ instead of the *forward regression* $E(Y|\mathbf{X})$ is clear: one can side-step the dimensionality problem since the inverse regression can be carried out by regressing each coordinate of $\mathbf{X}$ against Y. Hence, one is essentially dealing with a one-dimension to one-dimension regression problem. Li (1991) proposes a simple algorithm for carrying out the inverse regression fit. The basic idea is to use a local average regression smoother as follows. Slice the range of the one-dimensional variable Y into several intervals, and partition the whole data into several slices according to the division of the Y-range. In each slice the mean vector of the $\mathbf{X}$'s is calculated, resulting in the so-called slice means. Note that these slice means are Nadaraya-Watson estimates using a uniform kernel. Then, in order to locate the most important L-dimensional subspace for tracking the inverse regression curve $E(\mathbf{X}|Y)$ a principal component analysis is carried through to these slice means.

We now provide more explanation of these key ideas. First it should be noted that the individual vectors $\beta_1, \cdots, \beta_L$ in (7.41) are not identifiable (unless of course additional conditions are imposed on the function g). To see this suppose that we have L linearly independent e.d.r. directions, and denote by $\mathbf{A}$ the $L \times L$ matrix containing the coefficients of the linear combinations. Then it is easy to see that if $(\beta_1, \cdots, \beta_L)$ and g satisfy (7.41), then

$$(\tilde{\beta}_1, \cdots, \tilde{\beta}_L) = (\beta_1, \cdots, \beta_L)\mathbf{A}$$

(i.e. $\tilde{\beta}_j$ is the jth column of $(\beta_1, \cdots, \beta_L)\mathbf{A}$) and

$$\tilde{g}(\bullet, \varepsilon) = g(\bullet A^{-1}, \varepsilon)$$

also satisfy (7.41). The unidentifiability of the individual vectors $\beta_1, \cdots, \beta_L$ means that conditioning on $(\mathbf{X}^T\beta_1, \cdots, \mathbf{X}^T\beta_L)$ is equivalent to conditioning on any nondegenerate affine transformation of this vector. Further, it implies that only the effective dimension-reduction space $\mathcal{B}$ can be identified.

It will be convenient to look at the standardized version of $\mathbf{X}$. Denote the covariance matrix of $\mathbf{X}$ by $\Sigma_\mathbf{X}$ and let

$$\mathbf{Z} = \Sigma_\mathbf{X}^{-1/2}\{\mathbf{X} - E(\mathbf{X})\} \qquad (7.42)$$

denote the standardized version of $\mathbf{X}$. Then (7.41) translates into

$$Y = \tilde{g}(\mathbf{Z}^T\boldsymbol{\eta}_1, \cdots, \mathbf{Z}^T\boldsymbol{\eta}_L, \varepsilon), \qquad (7.43)$$

with $\boldsymbol{\eta}_j = \Sigma_\mathbf{X}^{1/2}\beta_j$, $j = 1, \cdots, L$. Any vector in the linear space generated by the $\boldsymbol{\eta}_j$'s is called a *standardized effective dimension-reduction direction*.

We now explain how the inverse regression problem comes into the scene. The inverse regression curve $E(\mathbf{X}|Y = y)$ is a function of y and the center of this curve is $E(\mathbf{X})$. The centered inverse regression curve

$$E(\mathbf{X}^T|Y = y) - E(\mathbf{X}^T)$$

is a curve in $\mathbb{R}^d$. Suppose now that the distribution of $\mathbf{X}$ is such that

$\forall \mathbf{b}^T \in \mathbb{R}^d$, the conditional expectation $E(\mathbf{X}^T\mathbf{b}|\mathbf{X}^T\beta_1, \cdots, \mathbf{X}^T\beta_L)$ is linear in $\mathbf{X}^T\beta_1, \cdots, \mathbf{X}^T\beta_L$.

This is for example the case when the distribution of $\mathbf{X}$ is elliptically symmetric. Details of the properties of this kind of distributions can be found in Fang, Kotz and Ng (1990). An example of such an elliptically symmetric distribution is the d-variate normal distribution. Li (1991) showed that under the ideal situation (7.41) and the above condition on the distribution of $\mathbf{X}$, the centered inverse regression curve

$$E(\mathbf{X}|Y = y) - E(\mathbf{X})$$

is contained in the linear subspace spanned by $\Sigma_\mathbf{X}\beta_1, \cdots, \Sigma_\mathbf{X}\beta_L$. This fact together with (7.43) then reveals that the standardized inverse regression curve $E(\mathbf{Z}|Y)$ is contained in the linear subspace generated by the standardized effective dimension-reduction directions $\boldsymbol{\eta}_1, \cdots, \boldsymbol{\eta}_L$. Hence, the covariance matrix $\operatorname{Cov}\{E(\mathbf{Z}|Y)\}$ is degenerate in any direction orthogonal to the $\boldsymbol{\eta}_\ell$'s, $\ell = 1, \cdots, L$. Therefore the eigenvectors $\boldsymbol{\eta}_1, \cdots, \boldsymbol{\eta}_L$ associated with the largest L eigenvalues of $\operatorname{Cov}\{E(\mathbf{Z}|Y)\}$ are the standardized e.d.r. directions. Transforming back to the original scale

$$(\Sigma_{\mathbf{X}}^{-1/2} \boldsymbol{\eta}_1, \cdots, \Sigma_{\mathbf{X}}^{-1/2} \boldsymbol{\eta}_L) = (\boldsymbol{\beta}_1, \cdots, \boldsymbol{\beta}_L),$$

completes the search for the effective dimension-reduction space $\mathcal{B}$.

The following algorithm, as formulated by Li (1991), carries through all the above steps. For an observed sample $\{(\mathbf{X}_i^T, Y_i) : i = 1, \cdots, n\}$, the algorithm reads as follows

Step 1. Standardize $\mathbf{X}$ by an affine transformation, as in (7.42), to get

$$\widetilde{\mathbf{X}}_i = \widehat{\Sigma}_{\mathbf{X}}^{-1/2}(\mathbf{X}_i - \overline{\mathbf{X}}), \qquad i = 1, \cdots, n,$$

where $\overline{\mathbf{X}}$ and $\widehat{\Sigma}_{\mathbf{X}}$ denote respectively the sample mean and the sample covariance matrix of $\mathbf{X}_1, \cdots, \mathbf{X}_n$.

Step 2. Divide the range of Y into S slices denoted by $I_1, \cdots, I_S$. Let

$$\widehat{P}_s = n^{-1} \sum_{i=1}^{n} I\{Y_i \in I_s\}$$

be the proportion of Y_i's that fall in the slice I_s, $s = 1, \cdots, S$.

Step 3. Compute within each slice I_s the sample mean vector of the $\widetilde{\mathbf{X}}_i$'s:

$$\widehat{m}_s = (n\widehat{P}_s)^{-1} \sum_{i: Y_i \in I_s} \widetilde{\mathbf{X}}_i.$$

Step 4. Conduct a (weighted) principal component analysis for the data $\widehat{m}_s$, $s = 1, \cdots, S$, in the following way: form the weighted covariance matrix $\widehat{V} = \sum_{s=1}^{S} \widehat{P}_s \widehat{m}_s \widehat{m}_s^T$, and find the eigenvalues and the eigenvectors.

Step 5. Denote the eigenvectors associated with the largest L eigenvalues by $\widehat{\boldsymbol{\eta}}_1, \cdots, \widehat{\boldsymbol{\eta}}_L$ and compute

$$\widehat{\boldsymbol{\beta}}_\ell = \widehat{\Sigma}_{\mathbf{X}}^{-1/2} \widehat{\boldsymbol{\eta}}_\ell, \qquad , \ell = 1, \cdots, L,$$

to transform back to the original scale.

The weighted version of principal component analysis is used to take care of the unequal sample sizes in the slices.

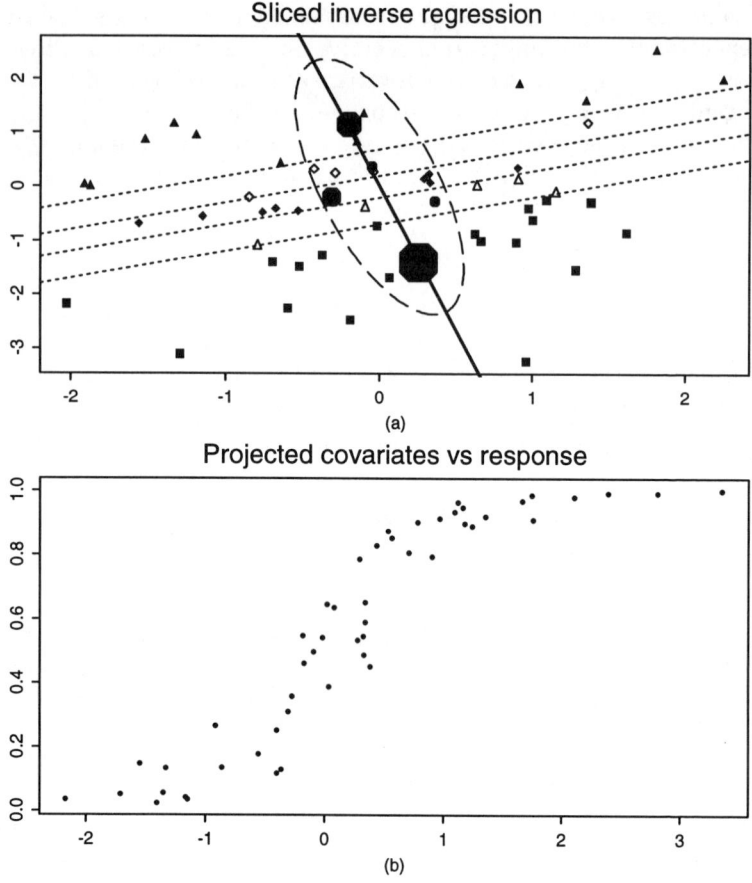

Figure 7.5. *Illustration of sliced inverse regression.*

We illustrate the sliced inverse regression procedure for dimension reduction with a very simple example. A sample of size 50 is drawn from the model

$$Y = \frac{\exp(X_1 - 2X_2)}{1 + \exp(X_1 - 2X_2)} + \varepsilon,$$

where X_1 and X_2 are independent standard normally distributed random variables. For clarity of visualization we take the error random variable ε to be equal to zero. In this example we divide the range of Y, which is $[0, 1]$, into five equally sized slices. Slicing the data into five pieces and then projecting them into the X-plane is equivalent to partitioning the X-plane into five pieces. Since the function $y = [\exp(x_1 - 2x_2)]/[1 + \exp(x_1 - 2x_2)]$ is monotone in the argument $x_1 - 2x_2$, the separation lines are very easy to compute, and are depicted in Figure 7.5 (a). The observations (X_1, X_2) in each of the slices are plotted using different characters in each slice. The sample mean vector (gravity center) in each slice is depicted by the hexagonal character, with the size of the hexagon being proportional to the number of observations in the corresponding slice. The weighted covariance matrix of these sample means was computed as well as the two eigenvectors. The solid line in Figure 7.5 (a) indicates the direction $\widehat{\beta}_1 = (0.186, -0.982)^T$ of the eigenvector with the largest eigenvalue. The ellipse used to circle the five gravity centers enhances the clarity of the principal component analysis. We projected the observations $\{(X_{i1}, X_{i2}) : i = 1, \cdots, n\}$ using the direction $\widehat{\beta}_1$, and plot the projected observations against the response Y in Figure 7.5 (b). The model obtained from this analysis is

$$Y = \widehat{f}(0.186 X_1 - 0.982 X_2) + \varepsilon,$$

where the regression curve $\widehat{f}$ can be estimated from Figure 7.5 (b).

In the above approach we obtained the standardized e.d.r. directions $\eta_1, \cdots, \eta_L$ via the covariance matrix $\text{Cov}\{E(\mathbf{Z}|Y)\}$. An alternative approach is to rely on the conditional covariance. Indeed, it is easy to see that

$$E\{\text{Cov}(\mathbf{Z}|Y)\} = \text{Cov}(\mathbf{Z}) - \text{Cov}\{E(\mathbf{Z}|Y)\} = \mathbf{I} - \text{Cov}\{E(\mathbf{Z}|Y)\},$$

where $\mathbf{I}$ denotes the $d \times d$ identity matrix, and now the task is to find the eigenvectors of $E\{\text{Cov}(\mathbf{Z}|Y)\}$ with the L smallest eigenvalues, resulting in the standardized e.d.r. directions.

Other possible approaches for estimating the β's can be found in Samarov (1993) and Wong and Shen (1995), among others. Those approaches use forward regression and are based on average derivatives.

7.6 Local polynomial regression as a building block

All estimation or fitting procedures discussed in Section 7.2 involve at a certain stage the use of a univariate smoothing technique, and a local polynomial regression estimator would serve well. The estimation procedure proposed for the generalized partially linear single-index models in Sections 7.3 and 7.4.2 also uses the idea of local polynomial approximation. For modelling interactions in generalized additive models a bivariate smoothing technique is needed, and here local polynomial fitting in two dimensions, as discussed in Section 7.8, can serve as a building block.

7.7 Robustness

The estimation procedures described in Sections 7.2 and 7.4 can also be robustified. An attempt to do this would be to focus on estimation of the function

$$m_\ell(x_1, \cdots, x_d) = \mathrm{argmin}_a E\{\ell(Y-a)|X_1 = x_1, \cdots, X_d = x_d\}, \tag{7.44}$$

where $\ell(\cdot)$ is a loss function, usually more resistant to outliers than the L_2-loss function $\ell(t) = t^2$. See also Section 5.5. The function $m_\ell(x_1, \cdots, x_d)$ is a multivariate extension of the function defined in (5.59). Note that if we take $\ell(t) = t^2$, then the function defined in (7.44) corresponds to the conditional mean function $E(Y|X_1 = x_1, \cdots, X_d = x_d)$. We now indicate how the estimation procedures for additive models described in Sections 7.2 and 7.4.1 can be modified for estimation of the function $m_\ell(x_1, \cdots, x_d)$.

Following an additive modelling approach we would estimate the parameter α and the functions $g_1(\cdot), \cdots, g_d(\cdot)$ which minimize the error not explained by the additive functions, i.e. which minimize

$$E\ell\{Y - \alpha - \sum_{j=1}^{d} g_j(X_j)\}, \tag{7.45}$$

with respect to α and $g_1, \cdots, g_d$, subject to constraints (7.4). Recall the algorithm for fitting an additive model as described in Section 7.2. The algorithm for estimating α and $g_1(\cdot), \cdots g_d(\cdot)$ in the current context follows the same basic lines, and we only indicate the differences. In the initialization step we estimate α by $\hat{\alpha}$ the minimizer of the quantity $\sum_{i=1}^{n} \ell(Y_i - \alpha)$, and put $g_1^0 = \cdots = g_d^0 = 0$. In the backfitting step we now use the estimator (5.58) instead of the usual local polynomial regression estimator based on the L_2-loss. When $\ell(t) = |t| + (2\alpha - 1)t$, the above algorithm leads to an estimate of the best additive approximation functions to the conditional α^{th} quantile function of Y given $X_1, \cdots, X_d$. These estimated quantities provide useful tools for constructing predictive intervals and verifying heteroscedasticity for high-dimensional data (see also Section 5.5).

The extension of the above method to two-term interaction nonparametric regression is quite straightforward. Recall the interaction model (7.27). In the current context this leads us to estimate the parameter α, the univariate functions $g_1(\cdot), \cdots, g_d(\cdot)$ and the bivariate functions $g_{jk}(\cdot,\cdot)$, $1 \leq j < k \leq d$, which minimize the error not explained by the main effects and the interactions

$$E\ell\{Y - \alpha - \sum_{j=1}^{d} g_j(X_j) - \sum_{1 \leq j < k \leq d} g_{jk}(X_j, X_k)\}, \qquad (7.46)$$

subject to the constraints (7.28) and (7.29). The algorithm carrying through this estimation task is a modification of the algorithm for fitting an additive model with interaction terms as described in Section 7.4.1. The major modification appears in the backfitting part of the algorithm, where the univariate and bivariate local polynomial regression estimators are replaced by their robust versions, i.e. using a loss function $\ell(\cdot)$ instead of the L_2-loss. The ideas mentioned here have not been investigated yet, and further research is needed.

7.8 Local linear regression in the multivariate setting

In this section we describe the generalization of local polynomial fitting to the case of multivariate covariates. For simplicity of presentation we restrict ourselves here to local linear fitting, i.e. $p = 1$, and focus only on estimation of the regression function $m(\mathbf{x})$ defined in (7.1), meaning that we take $\nu = 0$ in our standard notation of Chapter 3. The resulting estimator is referred to as the *multivariate local linear regression estimator*. We define the estimator in Section 7.8.1 and give expressions for bias and variance in Section 7.8.2. A brief discussion on the choice of an optimal weight function is provided in Section 7.8.3. Finally, some asymptotic minimax properties of the multivariate local linear regression estimator are presented in Section 7.8.4. All discussions in this section are kept rather short, and results are presented in a similar way to the univariate case studied in detail in Chapter 3.

In principle the methodology generalizes to using local polynomial fitting but we do not elaborate on this point.

7.8.1 Multivariate local linear regression estimator

Given multivariate covariates $\mathbf{X}$ and a univariate response Y, it is of interest to estimate the mean regression function $m(\mathbf{x}) = E(Y|\mathbf{X} = \mathbf{x})$. First we have to adjust the notation of Chapter

K has all the same marginals K₀

3 to this multivariate setting. Let K be a d-variate nonnegative kernel function. For simplicity we assume that K is a multivariate probability density function, such that $\int K(\mathbf{u})d\mathbf{u} = 1$ and $\int \mathbf{u} K(\mathbf{u})d\mathbf{u} = 0$. Further we assume that K has compact support and that

$$\int u_i u_j K(\mathbf{u}) d\mathbf{u} = \delta_{ij} \mu_2(K),$$

$\mu_2(K) = \int u^2 K(u) du$

with $\mu_2(K) \geq 0$. In other words, the mean of the density function $K(\cdot)$ is zero and the covariance matrix of K is $\mu_2(K)\mathbf{I}_d$, with $\mathbf{I}_d$ the $d \times d$ identity matrix. Define

$$K_B(\mathbf{u}) = \frac{1}{|B|} K(B^{-1}\mathbf{u}),$$

$B = \begin{pmatrix} b & 0 \\ 0 & b \end{pmatrix}$, $|B| = b^2$

where B is a nonsingular $d \times d$ matrix, the *bandwidth matrix*, and $|B|$ denotes its determinant. The observations are given by $\{(\mathbf{X}_i^T, Y_i) : i = 1, \cdots, n\}$, with $\mathbf{X}_i = (X_{i1}, \cdots, X_{id})^T$. Let $\mathbf{x}^T = (x_1, \cdots, x_d)$ be a point in $\mathbb{R}^d$. Using a local linear approximation the multivariate version of (3.3) reads as follows: minimize

$$\sum_{i=1}^{n} \{Y_i - \beta_0 - \sum_{j=1}^{d} \beta_j(X_{ij} - x_j)\}^2 K_B(\mathbf{X}_i - \mathbf{x}), \qquad (7.47)$$

with respect to $\beta = (\beta_0, \cdots, \beta_d)^T$, where now

$$\beta_0 = m(\mathbf{x}) \quad \text{and} \quad \beta_j = \frac{\partial m}{\partial x_j}(\mathbf{x}), \quad j = 1, \cdots, d. \qquad (7.48)$$

Denote by $\widehat{\beta} = (\widehat{\beta}_0, \cdots, \widehat{\beta}_d)^T$ the estimator of $(\beta_0, \cdots, \beta_d)^T$ resulting from (7.47).

As in Chapter 3 the solution to this weighted least squares regression problem is given by

$$\widehat{\beta} = (\mathbf{X}_D^T \mathbf{W} \mathbf{X}_D)^{-1} \mathbf{X}_D^T \mathbf{W} \mathbf{y}, \qquad (7.49)$$

where

$$\mathbf{X}_D = \begin{pmatrix} 1 & X_{11} - x_1 & \cdots & X_{1d} - x_d \\ 1 & X_{21} - x_1 & \cdots & X_{2d} - x_d \\ \vdots & \vdots & & \vdots \\ 1 & X_{n1} - x_1 & \cdots & X_{nd} - x_d \end{pmatrix}$$

denotes the design matrix of problem (7.47) and

$$\mathbf{W} = \text{diag}\{K_B(\mathbf{X}_i - \mathbf{x})\}$$

is the $n \times n$ matrix of weights. Further, we have adopted the notation $\mathbf{y} = (Y_1, \cdots, Y_n)^T$ from Chapter 3. The estimates of the

d-variate regression function m and its d first-order partial derivatives are given by

$$\widehat{m}(\mathbf{x}) = \widehat{\beta}_0, \quad \text{and} \quad (\widehat{\frac{\partial m}{\partial x_j}})(\mathbf{x}) = \widehat{\beta}_j, \ j = 1, \cdots, d. \qquad (7.50)$$

A robustified version of the above multivariate local linear regression estimation procedure, called *LOESS* was proposed by Cleveland and Devlin (1988). See also Cleveland, Grosse and Shyu (1992). This robust fitting procedure is an extension of the LOWESS procedure described in Section 2.4. We only highlight the main differences appearing in the multivariate setting. The bandwidth matrix B is taken to be of the form $B = h\mathbf{I}_d$. The diagonality of the smoothing matrix means that smoothing is done along the coordinate axes of the covariates. Further, the use of equal diagonal elements implies that the amount of smoothing is the same in each direction. If the covariates are measured on a different scale it is suggested that we work with their standardized versions. The multivariate kernel function is taken to be of the form

$$K(\mathbf{u}) = K(u_1, \cdots, u_d) = W\{(\sum_{j=1}^{d} u_j^2)^{1/2}\}, \qquad (7.51)$$

where $W(\cdot)$ is a univariate kernel function. Cleveland and Devlin (1988) take $W(\cdot)$ to be the tricube kernel. The basic setup for the LOESS procedure is very similar to that for the univariate version LOWESS described in Section 2.4. Recall that in the univariate procedure the distance function

$$\rho(u, v) = |u - v|$$

was used to determine the distance between points in the one-dimensional space $I\!R$. In the multivariate setting a distance between points in the d-dimensional space $I\!R^d$ is needed, and here the *Euclidean distance*

$$\rho(\mathbf{u}, \mathbf{v}) = \{\sum_{j=1}^{d} (u_j - v_j)^2\}^{1/2} \qquad (7.52)$$

is a candidate and is used by Cleveland and Devlin (1988). Further, LOESS, as LOWESS, uses a nearest neighbor type of bandwidth. More precisely the neighborhood of a particular observation $\mathbf{X}_k$ is determined by its associated bandwidth h_k which is the rth smallest number among $\rho(\mathbf{X}_k, \mathbf{X}_j)$, for $j = 1, \cdots, n$. When estimating $m(\cdot)$

at the observation $\mathbf{X}_k$, a weight

$$K\{h_k^{-1}(\mathbf{X}_i - \mathbf{X}_k)\} = W[h_k^{-1}\{\sum_{j=1}^{d}(X_{ij} - X_{kj})^2\}^{1/2}] \qquad (7.53)$$

is assigned to each observation $\mathbf{X}_i$. Note that (7.53) is a straightforward generalization of (2.10) but now using the Euclidean distance. We restricted the description here to local linear fitting, but mention that the LOESS procedure is formulated for local polynomial fitting in general. The LOESS fitting procedure is implemented in S-Plus software. See also Cleveland, Grosse and Shyu (1992).

Another simple way of determining a nearest neighbor type of local neighborhood is to apply, for each direction, the bandwidth (4.43). More precisely, for a given smoothing parameter k, and a point $\mathbf{x}$, construct

$$\widehat{h}_{k,j}(\mathbf{x}) = \{X^*_{\ell_j+k,j} - X^*_{\ell_j-k,j}\}/2, \qquad j = 1, \cdots, d, \qquad (7.54)$$

where $X^*_{1,j} \leq \cdots \leq X^*_{n,j}$ are the ordered X-values in the jth direction and ℓ_j is the index associated with the order statistic closest to x_j ($j = 1, \cdots, d$). The bandwidth matrix is simply

$$B = \text{diag}\{\widehat{h}_{k,1}(\mathbf{x}), \cdots, \widehat{h}_{k,d}(\mathbf{x})\}. \qquad (7.55)$$

We now illustrate the performance of a bivariate local linear regression estimator via two simulated data sets. A random sample of size $n = 400$ is simulated from the model

$$\begin{aligned} Y &= 0.7\exp[-3\{(X_1+0.8)^2 + 8(X_2-0.5)^2\}] \\ &\quad + \exp[-3\{(X_1-0.8)^2 + 8(X_2-0.5)^2\}] + \varepsilon, \end{aligned} \qquad (7.56)$$

and

$$Y = \frac{5}{\pi}\exp(-5X_1^2/8) + \varepsilon, \qquad (7.57)$$

respectively, where $\varepsilon \sim N\{0, (0.1)^2\}$, $X_1 \sim \text{Uniform}(-2, 2)$ and $X_2 \sim \text{Uniform}(0, 1)$ are independent random variables. Simulation model (7.57) was used in an illustration in Hermann, Wand, Engel and Gasser (1995). Here we consider a different noise level. We use a kernel (7.51) with $W(\cdot)$ the univariate Epanechnikov kernel. The 2×2 bandwidth matrix B is chosen to be as in (7.55). The smoothing parameter k was selected via cross-validation as described in Section 5.2.3. Figure 7.6 shows the true surfaces and the estimated surfaces, based on 51×51 grid points, for models (7.56) and (7.57). The cross-validated choice for the smoothing parameter was $k = 52$ for model (7.56) and $k = 53$ for model (7.57).

LOCAL LINEAR REGRESSION IN THE MULTIVARIATE SETTING 301

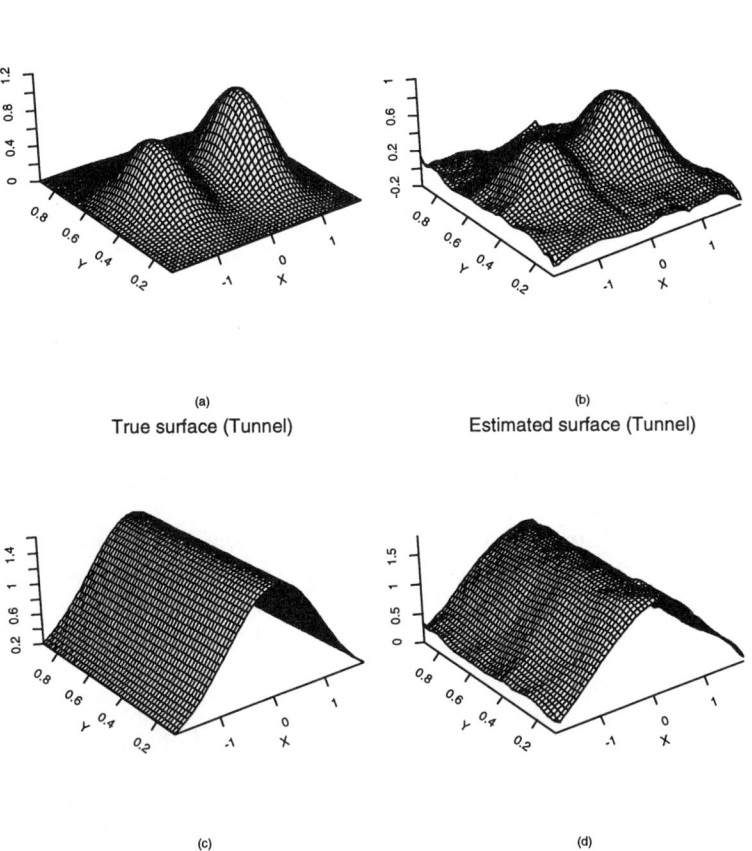

Figure 7.6. *True and estimated surfaces for models (7.56) and (7.57) using a bivariate local linear regression estimator with bandwidths as in (7.54).*

7.8.2 Bias and variance

The asymptotic analysis of estimator (7.49) can be found in Ruppert and Wand (1994). Let f denote the d-variate marginal density function of $\mathbf{X} = (X_1, \cdots, X_d)^T$. Further let $H(\mathbf{x})$ denote the Hessian matrix of m at $\mathbf{x}$. For $\mathbf{x}$ a point in the interior of the support of f, the conditional bias of the estimator $\widehat{m}(\mathbf{x})$ is given by

$$E\{\widehat{m}(\mathbf{x}) - m(\mathbf{x})|\mathbb{X}\} = \frac{1}{2}\mu_2(K)[\text{tr}\{H(\mathbf{x})BB^T\} + o_P\{\text{tr}(BB^T)\}] \quad (7.58)$$

and its conditional variance is

$$\text{Var}\{\widehat{m}(\mathbf{x})|\mathbb{X}\} = \frac{1}{n|B|}\nu_0(K)\frac{\sigma^2(\mathbf{x})}{f(\mathbf{x})}\{1 + o_P(1)\}, \quad (7.59)$$

where $\nu_0(K) = \int K^2(\mathbf{u})d\mathbf{u}$. It is easy to see that these asymptotic expressions are an extension of the univariate case established in Chapter 3.

The asymptotic expressions for the conditional bias and variance in (7.58) and (7.59) respectively lead to the following asymptotic expression for the conditional mean squared error

$$\left(\frac{1}{2}\mu_2(K)[\text{tr}\{H(\mathbf{x})BB^T\}]\right)^2 + \frac{1}{n|B|}\frac{\nu_0(K)\sigma^2(\mathbf{x})}{f(\mathbf{x})}. \quad (7.60)$$

A possible optimal choice for the bandwidth matrix B would be a matrix which minimizes the asymptotic mean squared error (7.60). Differentiating the asymptotic expression in (7.60) with respect to the matrix BB^T leads to a necessary condition for a local minimum, namely

$$\frac{1}{2}\mu_2^2(K)\text{tr}\{H(\mathbf{x})BB^T\}H(\mathbf{x}) - \frac{\nu_0(K)\sigma^2(\mathbf{x})}{2nf(\mathbf{x})|B|}(BB^T)^{-1} = 0$$

(see Rao (1973), page 72). In case the Hessian matrix $H(\mathbf{x})$ is positive or negative definite the above equation has a unique solution given by

$$BB^T = \left\{\frac{\nu_0(K)\sigma^2(\mathbf{x})|H^*|^{1/2}}{\mu_2^2(K)ndf(\mathbf{x})}\right\}^{2/(d+4)}(H^*)^{-1}, \quad (7.61)$$

with

$$H^* = \begin{cases} H & \text{for positive definite } H \\ -H & \text{for negative definite } H. \end{cases}$$

The bandwidth matrix B can then be chosen as any matrix satisfying equation (7.61).

7.8.3 Optimal weight function

We now turn to the question of how to choose the d-variate kernel function K. Substitution of the optimal choice (7.61) of the

LOCAL LINEAR REGRESSION IN THE MULTIVARIATE SETTING

bandwidth matrix into the asymptotic expression for the conditional mean squared error given in (7.60) leads to the asymptotic expression

$$\frac{d+4}{d} d^{4/(d+4)} \{\nu_0^2(K)\mu_2^d(K)\}^{2/(d+4)} \left\{\frac{\sigma^4(\mathbf{x})}{n^2 f^2(\mathbf{x})}|H^*(\mathbf{x})|\right\}^{2/(d+4)}. \tag{7.62}$$

One possibility is then to choose the weight function K such that

$$\nu_0^2(K)\mu_2^d(K) \tag{7.63}$$

is minimized subject to

$$\int K(\mathbf{u})d\mathbf{u} = 1, \quad \int \mathbf{u}K(\mathbf{u})d\mathbf{u} = 0, \quad K \geq 0,$$

and $\int u_i u_j K(\mathbf{u})d\mathbf{u} = \delta_{ij}\mu_2(K). \tag{7.64}$

The solution to the above minimization problem is the *spherical Epanechnikov kernel* defined as

$$K_0(\mathbf{u}) = \frac{d(d+2)}{2S_d}(1 - u_1^2 - \cdots - u_d^2)_+, \tag{7.65}$$

where $S_d = 2\pi^{d/2}/\Gamma(d/2)$ denotes the area of the surface of the d-dimensional unit ball. The subscript + indicates that we take the positive part. The proof can be found in Fan, Gasser, Gijbels, Brockmann and Engel (1995). Epanechnikov (1969) recommended using the product of univariate Epanechnikov kernels as a multivariate kernel function, which differs from our solution.

The moments $\mu_2(K_0)$ and $\nu_0(K_0)$, appearing in for example the optimal choice for the bandwidth matrix (see (7.61)), are easily calculated and equal

$$\mu_2(K_0) = \frac{1}{d+4} \quad \text{and} \quad \nu_0(K_0) = \frac{2d(d+2)}{(d+4)S_d}. \tag{7.66}$$

7.8.4 Efficiency

In Sections 3.4 and 3.5 we discussed the asymptotic minimax properties of the univariate local linear regression estimator. The multivariate local linear regression estimator (7.49) possesses the same minimax properties: with an appropriate choice of the bandwidth matrix and the kernel function, the multivariate local linear regression estimator achieves asymptotically the linear minimax risk, and comes asymptotically fairly close to the minimax risk. This was established in Fan, Gasser, Gijbels, Brockmann and Engel (1995).

We first need to specify the class of unknown regression functions considered in this minimax study. Let $\mathbf{x}_0$ be a point in the interior of the support of f, the marginal density of $\mathbf{X}$. Consider the class of functions

$$\mathcal{C}_2 = \{m : |m(\mathbf{z}) - (\mathbf{z} - \mathbf{x}_0)^T \nabla m(\mathbf{x}_0)| \leq \frac{1}{2}(\mathbf{z} - \mathbf{x}_0)^T C (\mathbf{z} - \mathbf{x}_0)\}, \tag{7.67}$$

where $\nabla m(\mathbf{x})$ denotes the column vector of first-order partial derivatives of $m(\mathbf{x})$, and C is a positive definite $d \times d$ matrix. Note that the class (7.67) is just the multivariate extension of the class of functions considered in (3.30) with $p = 1$. Intuitively the class $\mathcal{C}_2$ includes all regression functions whose Hessian matrix is bounded by C. We assume that the d-variate functions $\sigma^2(\cdot)$ and $f(\cdot)$ satisfy conditions (i) and (ii) of Section 3.4.

The *linear minimax risk* $R_{0,L}(n, \mathcal{C}_2)$ is defined as in (3.31) and the minimax risk $R_0(n, \mathcal{C}_2)$ is as in (3.44). Consider the multivariate local linear regression estimator with the spherical Epanechnikov kernel (7.65) and an optimal bandwidth matrix satisfying (7.61) with H^* replaced by C. Then this multivariate local linear regression estimator achieves asymptotically the linear minimax risk given by

$$R_{0,L}(n, \mathcal{C}_2) = r_d \{1 + o_P(1)\}, \tag{7.68}$$

where

$$r_d = \frac{d}{4} \left(\frac{2}{S_d}\right)^{4/(d+4)} (d+2)^{4/(d+4)} (d+4)^{-d/(d+4)}$$

$$\times \left\{ \frac{\sigma^4(\mathbf{x}_0)}{n^2 f^2(\mathbf{x}_0)} |C| \right\}^{2/(d+4)}. \tag{7.69}$$

Note that in the univariate case, the quantity r_d reduces to the factor provided in Theorem 3.5. Further, the minimax risk $R_0(n, \mathcal{C}_2)$ is asymptotically bounded by

$$r_d \geq R_0(n, \mathcal{C}_2) \geq (0.894)^2 r_d. \tag{7.70}$$

7.9 Bibliographic notes

In this chapter many approaches to nonparametric multivariate regression analysis have been discussed briefly, and giving here a complete list of references of each of these research areas is impossible. We indicate only some references, mainly those introducing the technique and the more recent publications in the area.

Generalized additive models were introduced by Hastie and Tibshirani (1986) and several applications of this class of models were presented in Hastie and Tibshirani (1987). Stone (1982, 1985) studied the rates of convergence for additive models and extended these results for generalized additive models in Stone (1986). See also Stone (1994).

Friedman and Stuetzle (1981) used the idea of backfitting in projection pursuit regression. The same idea of backfitting is the key to the alternative conditional expectation (ACE) algorithm discussed by Breiman and Friedman (1985). They consider a class of models which generalizes the class of additive models by allowing an unknown transformation of the response variable. Breiman and Friedman (1985) also investigated convergence properties of the ACE-algorithm. Linear smoothers and their use as building blocks in additive nonparametric regression were studied in detail by Buja, Hastie and Tibshirani (1989). They also prove convergence of the backfitting algorithm and related algorithms for a class of smoothers. Consistency for ACE-type methods was investigated by Koyak (1990). For certain classes of nonparametric smoothers in additive regression Härdle and Hall (1993) study convergence of the backfitting algorithm as well as uniqueness of the solution.

Härdle and Tsybakov (1990) proposed a selection procedure to determine which covariates should be included in an additive regression model. Another procedure for selecting the number of terms in an additive model, using ideas of principal components analysis, was introduced recently by Härdle and Tsybakov (1995). Turlach (1994) extended the ideas of Härdle and Tsybakov (1990) to determine the number of terms in a generalized additive model and in additive models with interaction terms.

The problem of bandwidth selection in bivariate additive modelling using local linear fitting is studied by Opsomer (1995) relying on the plug-in bandwidth ideas developed by Ruppert, Sheather and Wand (1995).

Papers studying partial spline models are Wahba (1984), Heckman (1986), Chen (1988), Speckman (1988) and Cuzick (1992) among others. Severini and Wong (1992) give an illuminating study of the partially linear model from the profile likelihood perspective. See also Severini and Staniswalis (1994) and Hunsberger (1994). Heckman, Ichimura, Smith and Todd (1995) extensively study partially linear models and their applications to the assessment of selection biases of job training programs.

Estimation and optimal smoothing in single-index models has been studied by Härdle, Hall and Ichimura (1993). Average derivative estimation was proposed by Härdle and Stoker (1989), and further explored by Stoker (1991). Härdle, Hildenbrand and Jerison (1991) and Stoker (1992) discuss the use of average derivative estimation in econometric models. Samarov (1993) used estimates of average derivative functionals in nonparametric model selection and diagnostics. Recently Chaudhuri, Doksum and Samarov (1994) dealt with nonparametric quantile regression using techniques similar to average derivative estimation.

Sliced inverse regression as a data-analytic tool for dimension reduction was introduced by Li (1991). The asymptotic behavior of the estimated effective dimension-reduction directions was studied by Duan and Li (1991). Further asymptotic results for sliced inverse regression are provided in Hsing and Carroll (1992). Recently Hall and Li (1993) investigated the shapes of low-dimensional projections from high-dimensional data. Schott (1994) focuses attention on the determination of the dimensionality in sliced inverse regression.

Local polynomial fitting in the multivariate regression setup was studied by Stone (1977, 1980, 1982) and developped and illustrated substantially via a variety of examples in Cleveland and Devlin (1988). See that paper for an introduction to the LOESS procedure. See also Cleveland, Grosse and Shyu (1992). The asymptotic bias and variance of multivariate local polynomial regression estimators have been derived by Ruppert and Wand (1994). A nice discussion on the possible smoothing parametrizations in bivariate density estimation can be found in Wand and Jones (1993).

For a detailed treatment of classifications and regression trees see Breiman, Friedman, Olshen and Stone (1993). Breiman (1991) introduced the concept of hinging hyperplanes, i.e. two hyperplanes continuously joined together at a hinge. He discusses the use of such hinging hyperplanes for regression, classification and function approximation. Recently Hastie, Tibshirani and Buja (1994) study nonparametric versions of discriminant analysis with multiple linear regression replaced by any nonparametric regression technique. They describe methods for multigroup classification that can rely on multivariate regression techniques such as multivariate adaptive regression splines, hinging hyperplanes and additive models among others.

References

Aalen, O.O. (1976). Nonparametric inference in connection with multiple decrement models. *Scand. J. Statist.*, **3**, 15–27.

Aalen, O.O. (1978). Nonparametric inference for a family of counting processes. *Ann. Statist.*, **6**, 701–726.

Abramovich, F. and Steinberg, D.M. (1996). Improved inference in nonparametric regression using L_k-smoothing splines. *J. Statist. Plann. Infer.*, to appear.

Abramson, I.S. (1982). On bandwidth variation in kernel estimates–a square root law. *Ann. Statist.*, **10**, 1217–1223.

Aerts, M., Janssen, P. and Veraverbeke, N. (1994). Bootstrapping regression quantiles. *J. Nonpar. Statist.*, **4**, 1–20.

Aigner, D., Amemiya, T. and Poirier, D. (1976). On the estimation of production frontiers: maximum likelihood estimation of the parameters of a discontinuous density function. *Internat. Econom. Rev.*, **17**, 372–396.

Akaike, H. (1970). Statistical predictor identification. *Ann. Inst. Statist. Math.*, **22**, 203–217.

Akaike, H. (1974). A new look at the statistical model identification. *IEEE Trans. on Automatic Control AC*, **19**, 716–723.

Allen, D.M. (1974). The relationship between variable and data augmentation and a method of prediction. *Technometrics*, **16**, 125–127.

An, H.Z. and Huang, F.C. (1996). Geometrical ergodicity of nonlinear autoregressive models. *Statistica Sinica*, **6**, to appear.

Andersen, P.K., Borgan, Ø., Gill, R.D. and Keiding, N. (1993). *Statistical Models Based on Counting Processes*. Springer-Verlag, New York.

Andersen, P.K. and Gill, R.D. (1982). Cox's regression model for counting processes: A large sample study. *Ann. Statist.*, **10**, 1100–1120.

Anscombe, F.J. (1948). The transformation of poisson, binomial and negative-binomial data. *Biometrika*, **35**, 246–254.

Azzalini, A. (1984). Estimation and hypothesis testing of autoregressive time series. *Biometrika*, **71**, 85–90.

Bartlett, M.S. (1948). Smoothing periodograms from time series with continuous spectra. *Nature*, **161**, 686–687.

Bartlett, M.S. (1950). Periodogram analysis and continuous spectra.

Biometrika, **37**, 1–16.

Bartlett, M.S. (1954). A note on some multiplying factors for various χ^2 approximations. *J. Royal Statist. Soc. B*, **16**, 296–298.

Becker, R.A., Chambers, J.M. and Wilks, A.R. (1988). *The New S Language: A Programming Environment for Data Analysis and Graphics*. Wadsworth & Brooks, Pacific Grove, California.

Bellman, R.E. (1961). *Adaptive Control Processes*. Princeton University Press, Princeton.

Beltrão, K.I. and Bloomfield, P. (1987). Determining the bandwidth of a kernel spectrum estimate. *J. Time Ser. Anal.*, **8**, 21–36.

Beran, R. (1981). Nonparametric regression with randomly censored survival data. *Technical report*, Department of Statistics, University of California, Berkeley.

Bhattacharya, P.K. and Gangopadhyay, A.K. (1990). Kernel and nearest-neighbor estimation of a conditional quantile. *Ann. Statist.*, **18**, 1400–1415.

Bickel, P.J. (1973). On some analogues to linear combinations of order statistics in linear model. *Ann. Statist.*, **1**, 597–616.

Bickel, P.J. (1983). Minimax estimation of a normal mean subject to doing well at a point. In *Recent Advances in Statistics* (M.H. Rizvi, J.S. Rustagi, and D. Siegmund, eds), 511–528. Academic Press, New York.

Bickel, P.J. and Doksum, K.A. (1977). *Mathematical Statistics: Basic Ideas and Selected Topics*. Holden-Day, San Francisco.

Bickel, P.J., Klaassen, A.J., Ritov, Y. and Wellner, J.A. (1993). *Efficient and Adaptive Inference in Semi-parametric Models*. Johns Hopkins University Press, Baltimore.

Bickel, P.J. and Ritov, Y. (1988). Estimating integrated squared density derivatives: Sharp order of convergence estimates. *Sankhyā Ser. A*, **50**, 381–393.

Bickel, P.J. and Rosenblatt, M. (1973). On some global measures of the deviation of density function estimates. *Ann. Statist.*, **1**, 1071–1095.

Birgé, L. (1987a). Estimating a density under order restrictions: Non-asymptotic minimax risk. *Ann. Statist.*, **15**, 995–1012.

Birgé, L. (1987b). On the risk of histograms for estimating decreasing densities. *Ann. Statist.*, **15**, 1013–1022.

de Boor, C. (1978). *A Practical Guide to Splines*. Springer-Verlag, New York.

Bowman, A.W. (1984). An alternative method of cross-validation for the smoothing of density estimates. *Biometrika*, **71**, 353–360.

Breiman, L. (1991). Hinging hyperplanes for regression, classification and function approximation. *Technical report 324*, Department of Statistics, University of California, Berkeley.

Breiman, L. and Friedman, J.H. (1985). Estimating optimal transformations for multiple regression and correlation (with discussion). *J.*

Amer. Statist. Assoc., **80**, 580–619.

Breiman, L., Friedman, J.H., Olshen, R.A., and Stone, C.J. (1993). *CART: Classification and Regression Trees* (first edition, 1984). Wadsworth, Belmont.

Breiman, L., Meisel, W. and Purcell, E. (1977). Variable kernel estimates of multivariate densities. *Technometrics*, **19**, 135–144.

Breiman, L., Stone, C.J. and Kooperberg, C. (1990). Robust confidence bounds for extreme upper quantiles. *J. Comp. Simul.*, **37**, 127–149.

Breslow, N.E. (1972). Contribution to the discussion on the paper by D.R. Cox, 'Regression and life tables'. *J. Royal Statist. Soc. B*, **34**, 216–217.

Breslow, N.E. (1974). Covariance analysis of censored survival data. *Biometrics*, **30**, 89–99.

Brillinger, D.R. (1981). *Time Series Analysis: Data Analysis and Theory*. Holt, Rinehart & Winston, New York.

Brockwell, P.J. and Davis, R.A. (1989). *Time Series: Theory and Methods*, second edition. Springer-Verlag, New York.

Brown, L.D. and Low, M.G. (1995). A constrained risk inequality with applications to nonparametric functional estimation. *Ann. Statist.*, to appear.

Bruce, A.G., Gao, H.Y. and Ragozin, D. (1993). S+ Wavelets: An object-oriented wavelet toolkit. *Technical report*, Statistical Sciences Division of MathSoft, Inc.

Buckley, J. and James, I.R. (1979). Linear regression with censored data. *Biometrika*, **66**, 429–436.

Buja, A., Hastie, T.J. and Tibshirani, R. (1989). Linear smoothers and additive models (with discussion). *Ann. Statist.*, **17**, 453–555.

Cao, R., Cuevas, A. and González-Manteiga, W. (1994). A comparative study of several smoothing methods in density estimation. *Comp. Statist. Data Anal.*, **17**, 153–176.

Carroll, R.J., Fan, J., Gijbels, I, and Wand, M.P. (1995). Generalized partially linear single-index models. *Discussion Paper #9506*, Institute of Statistics, Catholic University of Louvain, Louvain-la-Neuve, Belgium.

Carroll, R.J. and Ruppert, D. (1982). Robust estimation in heteroscedastic linear models. *Ann. Statist.*, **10**, 429–441.

Carroll, R.J. and Ruppert, D. (1988). *Transformation and Weighting in Regression*. Chapman and Hall, New York.

Čencov, N.N. (1962). Evaluation of an unknown distribution density from observations. *Soviet Math.*, **3**, 1559–1562.

Chan, K.S. and Tong, H. (1994). A note on noisy chaos. *J. Royal Statist. Soc. B*, **56**, 301–311.

Chaudhuri, P. (1991). Nonparametric estimates of regression quantiles and their local Bahadur representation. *Ann. Statist.*, **19**, 760–777.

Chaudhuri, P., Doksum, K.A. and Samarov, A.M. (1994). Nonparametric estimation of global functionals based on quantile regression. *Manuscript*.

Chen, H. (1988). Convergence rates for parametric components in a partly linear model. *Ann. Statist.*, **16**, 136–146.

Chen, R. and Tsay, R.J. (1993). Functional-coefficient autoregressive models. *J. Amer. Statist. Assoc.*, **88**, 298–308.

Cheng, B. and Robinson, P.M. (1991). Density estimation in strongly dependent non-linear time series. *Statistica Sinica*, **1**, 335–360.

Cheng, K.F. and Chu, C.K. (1993). A new kernel regression method. *Manuscript*.

Cheng, K.F. and Lin, P.E. (1981). Nonparametric estimation of a regression function. *Z. Wahr. verw. Geb.*, **57**, 223–223.

Cheng, M.Y., Fan, J. and Marron, J.S. (1993). Minimax efficiency of local polynomial fit estimators at boundaries. *Institute of Statistics Mimeo Series #2098*, University of North Carolina at Chapel Hill.

Cheng, P.E. (1995). A note on strong convergence rates in nonparametric regression. *Statist. Prob. Lett.*, **24**, 357–364.

Cheng, P.E. and Bai, Z.D. (1995). Optimal strong convergence rates in nonparametric regression. *Math. Methods in Statist.*, **4**, 405–420.

Chiu, S.T. (1991). Bandwidth selection for kernel density estimation. *Ann. Statist.*, **19**, 1528–1546.

Chow, Y.S. and Teicher, H. (1988). *Probability Theory: Independence, Interchangeability, Martingales* (first edition 1978). Springer-Verlag, New York.

Chu, C.K., and Marron, J.S. (1991). Choosing a kernel regression estimator (with discussions). *Statist. Sci.*, **6**, 404–436.

Chui, C.K. (1992). *An Introduction to Wavelets*. Academic Press, Boston.

Cleveland, W.S. (1979). Robust locally weighted regression and smoothing scatterplots. *J. Amer. Statist. Assoc.*, **74**, 829–836.

Cleveland, W.S. and Devlin, S.J. (1988). Locally-weighted regression: an approach to regression analysis by local fitting. *J. Amer. Statist. Assoc.*, **83**, 597–610.

Cleveland, W.S., Grosse, E. and Shyu, W.M. (1992). Local regression models. In *Statistical Models in S* (Chambers, J.M. and Hastie, T.J., eds), 309–376. Wadsworth & Brooks, California.

Collomb, G. (1985). Nonparametric time series analysis and prediction: uniform almost sure convergence of the window and k-NN autoregression estimates. *Statistics*, **16**, 309–324.

Collomb, G. and Härdle, W. (1986). Strong uniform convergence rates in robust nonparametric time series analysis and prediction: Kernel regression estimation from dependent observations. *Stoch. Proc. Appl.*, **23**, 77–89.

Cox, D.D. (1984). Multivariate smoothing spline functions. *SIAM J. Numer. Anal.*, **21**, 789–813.

Cox, D.D. and O'Sullivan, F. (1990). Asymptotic analysis of penalized likelihood and related estimators. *Ann. Statist.*, **18**, 1676–1695.
Cox, D.R. (1972). Regression models and life-tables (with discussion). *J. Royal Statist. Soc. B*, **34**, 187–220.
Cox, D.R. (1975). Partial likelihood. *Biometrika*, **62**, 269–276.
Cox, D.R. (1983). Some remarks on over-dispersion. *Biometrika*, **70**, 269–274.
Cox, D.R. and Oakes, D. (1984). *Analysis of Survival Data*. Chapman and Hall, London.
Craven, P. and Wahba, G. (1979). Smoothing noisy data with spline functions: estimating the correct degree of smoothing by the method of generalized cross-validation. *Numer. Math.*, **31**, 377–403.
Cutler, C.D. (1993). A review of the theory and estimation of fractal dimension. In *Dimension Estimation and Models* (H. Tong, ed.), 1–107. World Scientific, Singapore.
Cuzick, J. (1992). Semiparametric additive regression. *J. Royal Statist. Soc. B*, **54**, 831–843.
Dabrowska, D.M. (1987). Non-parametric regression with censored data. *Scand. J. Statist.*, **14**, 181–197.
Dabrowska, D.M., Doksum, K.A., Feduska, N.J., Husing, R. and Neville, P.(1992). Methods for comparing cumulative hazard functions in a semi-proportional hazard model. *Statist. Medicine*, **11**, 1465–1476.
Daniels, H.E. (1946). Discussion of paper by M.S. Bartlett. *J. Royal Statist. Soc. B*, **24**, 185–198.
Daubechies, I. (1992). *Ten Lectures on Wavelets*. SIAM, Philadelphia.
Davis, H.T. and Jones, R.H. (1968). Estimation of the innovation variance of a stationary time series. *J. Amer. Statist. Assoc.*, **63**, 141–149.
Deheuvels, P. (1977). Estimation nonparamétrique de la densité par histogrammes généralisés. *Rev. Statist. Appl.*, **35**, 5–42.
Devroye, L.P. (1981). On the almost everywhere convergence of nonparametric regression estimates. *Ann. Statist.*, **9**, 1310–1319.
Devroye, L.P. and Györfi, L. (1985). *Nonparametric Density Estimation: The L_1 View*. Wiley, New York.
Devroye, L.P. and Wagner, T.J. (1980). Distribution-free consistency results in nonparametric discrimination and regression function estimation. *Ann. Statist.*, **8**, 231–239.
Djojosugito, R.A. and Speckman, P.L. (1992). Boundary bias correction in nonparametric density estimation. *Commun. Statist. Theory Meth.*, **21**, 69–88.
Doksum, K.A., Blyth, S., Bradlow, E., Meng, X.-L. and Zhao, H. (1994). Correlation curves as local measures of variance explained by regression. *J. Amer. Statist. Assoc.*, **89**, 571–582.
Doksum, K.A. and Yandell, B.S. (1982). Properties of regression estimates based on censored survival data. In *A Festschrift for Erich Lehmann* (P.J. Bickel, K.A. Doksum and J.L. Hodges, Jr., eds), 140–

156. Wadsworth & Brooks/Cole, California.

Donoho, D.L. (1994). Statistical estimation and optimal recovery. *Ann. Statist.*, **22**, 238–270.

Donoho, D.L. (1995). Nonlinear solution of linear inverse problems by wavelet-vaguelette decomposition. *Appl. Comput. Harm. Anal.*, **2**, 101–126.

Donoho, D.L. and Johnstone, I.M. (1994). Ideal spatial adaptation by wavelet shrinkage. *Biometrika*, **81**, 425–455.

Donoho, D.L. and Johnstone, I.M. (1995a). Minimax estimation via wavelet shrinkage. *Ann. Statist.*, to appear.

Donoho, D.L. and Johnstone, I.M. (1995b). Adapting to unknown smoothness via wavelet shrinkage. *J. Amer. Statist. Assoc.*, **90**, to appear.

Donoho, D.L. and Johnstone, I.M. (1996). Neo-classical minimax problems, thresholding, and adaptive function estimation. *Bernoulli*, to appear.

Donoho, D.L., Johnstone, I.M., Kerkyacharian, G. and Picard, D. (1993). Density estimation by wavelet thresholding. *Technical report 426*, Department of Statistics, Stanford University, Stanford.

Donoho, D.L., Johnstone, I.M., Kerkyacharian, G. and Picard, D. (1995). Wavelet shrinkage: asymptopia? *J. Royal Statist. Soc. B*, **57**, 301–369.

Donoho, D.L. and Liu, R.C. (1991a). Geometrizing rate of convergence II. *Ann. Statist.*, **19**, 633–667.

Donoho, D.L. and Liu, R.C. (1991b). Geometrizing rate of convergence III. *Ann. Statist.*, **19**, 668–701.

Donoho, D.L., Liu, R.C. and MacGibbon, B. (1990). Minimax risk for hyperrectangles, and implications. *Ann. Statist.*, **18**, 1416–1437.

Donoho, D.L. and Low, M. (1992). Renormalization exponents and optimal pointwise rates of convergence. *Ann. Statist.*, **20**, 944–970.

Donoho, D.L. and Nussbaum, M. (1990). Minimax quadratic estimation of a quadratic functional. *J. Complexity*, **6**, 290–323.

Drazin, P.G. and King, G.P. (1992). *Interpretation of Time Series from Nonlinear Systems*. North Holland, Amsterdam.

Duan, N. and Li, K.-C. (1991). Slicing regression: a link-free regression method. *Ann. Statist.*, **19**, 505–530.

Duin, R.P.W. (1976). On the choice of smoothing parameters for Parzen estimators of probability density functions. *IEEE Trans. Comput.*, **C-25**, 1175–1179.

Eckmann, J.P. and Ruelle, D. (1985). Ergodic theory of chaos and strange attractors. *Rev. Mod. Phys.*, **57**, 617–656.

Eddy, W.F. (1980). Optimum kernel estimators of the mode. *Ann. Statist.*, **8**, 870–882.

Efromovich, S. (1985). Nonparametric estimation of a density with unknown smoothness. *Theory Prob. Appl.*, **30**, 557–568.

Efromovich, S. (1996). On nonparametric regression for i.i.d. observations in general setting. *Ann. Statist.*, **24**, to appear.

Efromovich, S.Y. and Pinsker, M.S. (1982). Estimation of square-integrable probability density of a random variable. *Problems of Information Transmission*, **18**, 175–189.

Efron, B. (1991). Regression percentiles using asymmetric squared error loss. *Statistica Sinica*, **1**, 93–125.

Epanechnikov, V.A. (1969). Nonparametric estimation of a multidimensional probability density. *Theor. Prob. Appl.*, **13**, 153–158.

Eubank, R.L. (1984). The hat matrix for smoothing splines. *Statist. Prob. Lett.*, **2**, 9–14.

Eubank, R.L. (1985). Diagnostics for smoothing splines. *J. Royal Statist. Soc B*, **47**, 332–341.

Eubank, R.L. (1988). *Spline Smoothing and Nonparametric Regression*. Marcel Dekker, New York.

Eubank, R.L. and Speckman, P.L. (1991). A bias reduction theorem with applications in nonparametric regression. *Scand. J. Statist.*, **18**, 211–222.

Eubank, R.L. and Speckman, P.L. (1993). Confidence bands in nonparametric regression. *J. Amer. Statist. Assoc.*, **88**, 1287–1301.

Fan, J. (1991). On the estimation of quadratic functionals. *Ann. Statist.*, **19**, 1273–1294.

Fan, J. (1992). Design-adaptive nonparametric regression. *J. Amer. Statist. Assoc.*, **87**, 998–1004.

Fan, J. (1993a). Local linear regression smoothers and their minimax efficiency. *Ann. Statist.*, **21**, 196–216.

Fan, J. (1993b). Adaptively local one-dimensional subproblems with application to a deconvolution problem. *Ann. Statist.*, **21**, 600–610.

Fan, J. (1996). Test of significance based on wavelet thresholding and Neyman's truncation. *J. Amer. Statist. Assoc.*, to appear.

Fan, J., Gasser, T., Gijbels, I., Brockmann, M. and Engel, J. (1995). On nonparametric estimation via local polynomial regression. *Discussion Paper #9511*, Institute of Statistics, Catholic University of Louvain, Louvain-la-Neuve, Belgium.

Fan, J. and Gijbels, I. (1992). Variable bandwidth and local linear regression smoothers. *Ann. Statist.*, **20**, 2008–2036.

Fan, J. and Gijbels, I. (1994). Censored regression: local linear approximations and their applications. *J. Amer. Statist. Assoc.*, **89**, 560–570.

Fan, J. and Gijbels, I. (1995a). Data-driven bandwidth selection in local polynomial fitting: variable bandwidth and spatial adaptation. *J. Royal Statist. Soc. B*, **57**, 371–394.

Fan, J. and Gijbels, I. (1995b). Adaptive order polynomial fitting: bandwidth robustification and bias reduction. *J. Comp. Graph. Statist.*, **4**, 213–227.

Fan, J., Gijbels, I., Hu. T.-C. and Huang, L.-S. (1996). An asymptotic study of variable bandwidth selection for local polynomial regression with application to density estimation. *Statistica Sinica*, **6**, 1, to appear.

Fan, J., Gijbels, I. and King, M. (1995). Local likelihood and local partial likelihood in hazard regression. *Technical report 95.3*, Department of Statistics, Chinese University of Hong Kong.

Fan, J. and Hall, P. (1994). On curve estimation by minimizing mean absolute deviation. *Ann. Statist.*, **22**, 867–885.

Fan, J., Hall, P., Martin, M. and Patil, P. (1993). Adaptation to high spatial inhomogeneity based on wavelets and on local linear smoothing. *Research report SMS-59-93*, Centre for Applied Mathematics, Australian National University.

Fan, J., Hall, P., Martin, M. and Patil, P. (1996). On the local smoothing of nonparametric curve estimators. *J. Amer. Statist. Assoc.*, to appear.

Fan, J., Heckman, N.E. and Wand, M.P. (1995). Local polynomial kernel regression for generalized linear models and quasi-likelihood functions. *J. Amer. Statist. Assoc.*, **90**, 141–150.

Fan, J., Hu, T.-C. and Truong, Y.K. (1994). Robust nonparametric function estimation. *Scand. J. Statist.*, **21**, 433–446.

Fan, J. and Kreutzberger, E. (1995). Automatic local smoothing for spectral density estimation. *Institute of Statistics Mimeo Series #2322*, University of North Carolina at Chapel Hill.

Fan, J. and Marron, J.S. (1992). Best possible constant for bandwidth selection. *Ann. Statist.*, **20**, 2057–2070.

Fan, J. and Marron, J.S. (1993). Comments on 'Local regression: automatic kernel carpentry' by Hastie and Loader. *Statist. Sci.*, **8**, 129–134.

Fan, J. and Marron, J.S. (1994). Fast implementations of nonparametric curve estimators. *J. Comp. Graph. Statist.*, **3**, 35–56.

Fan, J., Yao, Q. and Tong, H. (1996). Estimation of conditional densities and sensitivity measures in nonlinear dynamical systems. *Biometrika*, to appear.

Fang, K.T., Kotz, S. and Ng, K.W. (1990). *Symmetric Multivariate and Related Distributions*. Chapman and Hall, London.

Farrell, R.H. (1972). On the best obtainable asymptotic rates of convergence in estimation of a density function at a point. *Ann. Math. Statist.*, **43**, 170–180.

Fleming, T.R. and Harrington, D.P. (1991). *Counting Processes and Survival Analysis*. Wiley, New York.

Franke, J. and Härdle, W. (1992). On bootstrapping kernel spectral estimates. *Ann. Statist.*, **20**, 121–145.

Friedman, J.H. (1984). A variable span smoother. *LCS Technical report 5*, Stanford University.

Friedman, J.H. (1991). Multivariate adaptive regression splines (with discussion). *Ann. Statist.*, **19**, 1–141.

Friedman, J.H. and Silverman, B.W. (1989). Flexible parsimonious smoothing and additive modelling (with discussion). *Technometrics*, **31**, 3–39.

Friedman, J.H. and Stuetzle, W. (1981). Projection pursuit regression. *J. Amer. Statist. Assoc.*, **76**, 817–823.

Gasser, T. and Kneip, A. (1989). Discussion on 'Linear smoothers and Additive Models' by Hastie and Tibshirani. *Ann. Statist.*, **17**, 453–555.

Gasser, T., Kneip, A. and Köhler, W. (1991). A flexible and fast method for automatic smoothing. *J. Amer. Statist. Assoc.*, **86**, 643–652.

Gasser, T. and Müller, H.-G. (1979). Kernel estimation of regression functions. In *Smoothing Techniques for Curve Estimation*, Lecture Notes in Mathematics, **757**, 23–68. Springer-Verlag, New York.

Gasser, T. and Müller, H.-G. (1984). Estimating regression functions and their derivatives by the kernel method. *Scand. J. of Statist.*, **11**, 171–185.

Gasser, T., Müller, H.-G. and Mammitzsch, V. (1985). Kernels for nonparametric curve estimation. *J. Royal Statist. Soc. B*, **47**, 238–252.

Gijbels, I. and Veraverbeke, N. (1991). Almost sure asymptotic representations for a class of functionals of the Kaplan-Meier estimator. *Ann. Statist.*, **19**, 1457–1470.

Gill, R.D. (1980). *Censoring and Stochastic Integrals*. Mathematical Centre Tracts, **124**. Mathematisch Centrum, Amsterdam.

Godambe, V.P. and Heyde, C.C. (1987). Quasi-likelihood and optimal estimation. *Inter. Statist. Rev.*, **55**, 231–244.

Good, I.J. and Gaskins, R.A. (1971). Nonparametric roughness penalties for probability densities. *Biometrika*, **58**, 255–277.

Granovsky, B.L. and Müller, H.-G. (1991), Optimizing kernel methods: a unifying variational principle. *Inter. Statist. Rev.*, **59**, 373–388.

Green, P.J. and Silverman, B.W. (1994). *Nonparametric Regression and Generalized Linear Models: a Roughness Penalty Approach*. Chapman and Hall, London.

Green, P.J. and Yandell, B. (1985). Semiparametric generalized linear models. In *Proceedings of the 2nd International GLIM Conference*. Lecture Notes in Statistics, **32**. Springer-Verlag, Berlin.

Grenander, U. (1956). On the theory of mortality measurement, Part II. *Skand. Akt.*, **39**, 125–153.

Groeneboom, P. (1985). Estimating a monotone density. In *Proceedings of the Berkeley Conference in Honor of Jerzy Neyman and Jack Kiefer* **Vol II** (L. M. Le Cam and R. A. Olshen, eds), 539–555.

Gu, C. (1990). Adaptive spline smoothing in non-Gaussian regression models. *J. Amer. Statist. Assoc.*, **85**, 801–807.

Gu, C. (1993). Smoothing spline density estimation: A dimensionless automatic algorithm. *J. Amer. Statist. Assoc.*, **87**, 1051–1058.

Gu, C. and Qiu, C. (1993). Smoothing spline density estimation: Theory. *Ann. Statist.*, **21**, 217–234.

Györfi, L., Härdle, W., Sarda, P. and Vieu, P. (1989). *Nonparametric Curve Estimation from Time Series*. Lecture Notes in Statistics, **60**. Springer-Verlag, Berlin.

Habbema, J.D.F., Hermans, J. and van der Broeck, K. (1974). A stepwise discrimination program using density estimation. In Bruckman, G. (ed.), *Compstat 1974*, 100–110. Physica Verlag, Vienna.

Hall, P. (1987). Cross-validation and the smoothing of orthogonal series density estimators. *J. Mult. Anal.*, **21**, 181–206.

Hall, P. (1990). On the bias of variable bandwidth curve estimators. *Biometrika*, **77**, 529–535.

Hall, P. and Heyde, C.C. (1980). *Martingale Limit Theory and its Applications*. Academic Press, New York.

Hall, P. and Johnstone, I. (1992). Empirical functionals and efficient smoothing parameter selection (with discussion). *J. Royal. Statist. Soc. B*, **54**, 475–530.

Hall, P. and Jones, M.C. (1990). Adaptive M-estimation in nonparametric regression. *Ann. Statist.*, **18**, 1712–1728.

Hall, P. and Li, K.-C. (1993). On almost linearity of low dimensional projections from high dimensional data. *Ann. Statist.*, **21**, 867–889.

Hall, P. and Marron, J.S. (1988). Variable window width kernel estimates of probability densities. *Prob. Theory Rel. Fields*, **80**, 37–49.

Hall, P. and Marron, J.S. (1991). Lower bounds for bandwidth selection in density estimation. *Prob. Theory Rel. Fields*, **90**, 149–173.

Hall, P. and Marron, J.S. (1995). On the role of the ridge parameter in local linear smoothing. *Manuscript*.

Hall, P., Marron, J.S. and Park, B.U. (1992). Smoothed cross-validation. *Prob. Theory Rel. Fields*, **92**, 1–20.

Hall, P. and Patil, P. (1995a). On wavelet methods for estimating smooth functions. *Bernoulli*, **1**, 41–58.

Hall, P. and Patil, P. (1995b). Formulae for mean integrated squared error of nonlinear wavelet-based density estimators. *Ann. Statist.*, **23**, 905–928.

Hall, P., Sheather, S.J., Jones, M.C. and Marron, J.S. (1991). On optimal data-based bandwidth selection in kernel density estimation. *Biometrika*, **78**, 263–271.

Hall, P, and Wand, M.P. (1996). On the accuracy of binned kernel density estimates. *J. Mult. Anal.*, to appear.

Hall, P. and Wehrly, T.E. (1991). A geometrical method for removing edge effects from kernel-type nonparametric regression estimators. *J. Amer. Statist. Assoc.*, **86**, 665–672.

Hall, P. and Wolff, R.C.L. (1995). Properties of invariant distributions and Lyapunov exponents for chaotic logistic maps. *J. Royal Statist. Soc. B*, **57**, 439–452.

Hampel, F.R., Ronchetti, E.M., Rousseeuw, P.J. and Stahel, W.A. (1986). *Robust Statistics: the Approach Based on Influence Functions.* Wiley, New York.

Härdle, W. (1984). Robust regression function estimation. *J. Mult. Anal.*, **14**, 169–180.

Härdle, W. (1990). *Applied Nonparametric Regression.* Cambridge University Press, Boston.

Härdle, W. and Gasser, T. (1984). Robust non-parametric function fitting. *J. Royal Statist. Soc. B*, **46**, 42–51.

Härdle, W. and Gasser, T. (1985). On robust kernel estimation of derivatives of regression functions. *Scand. J. Statist.*, **12**, 233–240.

Härdle, W. and Hall, P. (1993). On the backfitting algorithm for additive regression models. *Statistica Neerlandica*, **47**, 43–57.

Härdle, W., Hall, P. and Ichimura, H. (1993). Optimal smoothing in single-index models. *Ann. Statist.*, **21**, 157–178.

Härdle, W., Hildenbrand, W. and Jerison, M. (1991). Empirical evidence on the law of demand. *Econometrica*, **59**, 1525–1549.

Härdle, W. and Marron, J.S. (1995). Fast and simple scatterplot smoothing. *Comp. Statist. Data Anal.*, **20**, 1–17.

Härdle, W. and Scott, D.W. (1992). Smoothing by weighted averaging of rounded points. *Comp. Statist.*, **7**, 97–128.

Härdle, W. and Stoker, T.M. (1989). Investigating smooth multiple regression by the method of average derivatives. *J. Amer. Statist. Assoc.*, **84**, 986–995.

Härdle, W. and Tsybakov, A.B. (1990). How many terms should be added into an additive model? *CORE discussion paper #9068*, Catholic University of Louvain, Louvain-la-Neuve, Belgium.

Härdle, W. and Tsybakov, A.B. (1995). Additive nonparametric regression on principal components. *J. Nonpar. Statist.*, **5**, 157–184.

Hart, J.D. and Vieu, P. (1990). Data-driven bandwidth choice for density estimation based on dependent data. *Ann. Statist.*, **18**, 873–890.

Hart, J.D. and Wehrly, T.E. (1986). Kernel regression estimation using repeated measurements data. *J. Amer. Statist. Assoc.*, **81**, 1080–1088.

Hasminskii, R.Z. (1978). A lower bound on the risks of nonparametric estimates densities in the uniform metric. *Theory Prob. Appl.*, **23**, 794–798.

Hasminskii, R.Z. (1979). Lower bound for the risks of nonparametric estimate of the mode. In *Contributions to Statistics: Jaroslaw Hájek Memorial Volumes* (J. Jureckova, ed.), 91–97. Academia Prague.

Hastie, T.J. (1992). Generalized additive models. In *Statistical Models in S* (Chambers, J.M. and Hastie, T., eds), 195–247. Wadsworth & Brooks, California.

Hastie, T.J. and Loader, C. (1993). Local regression: automatic kernel carpentry (with discussion). *Statist. Sci.*, **8**, 120–143.

Hastie, T.J. and Tibshirani, R. (1986). Generalized additive models (with discussion). *Statist. Sci.*, **1**, 297–318.

Hastie, T.J. and Tibshirani, R. (1987). Generalized additive models: some applications. *J. Amer. Statist. Assoc.*, **82**, 371–386.

Hastie, T.J. and Tibshirani, R. (1990). *Generalized Additive Models.* Chapman and Hall, London.

Hastie, T.J., Tibshirani, R. and Buja, A. (1994). Flexible discriminant analysis by optimal scoring. *J. Amer. Statist. Assoc.*, **89**, 1255–1270.

Heckman, J., Ichimura, H., Smith, J. and Todd, P. (1995). Nonparametric characterization of selection bias using experimental data: a study of adult males in JTPA. *Manuscript.*

Heckman, N. (1986). Spline smoothing in a partly linear model. *J. Royal Statist. Soc. A*, **48**, 244–248.

Hermann, E., Wand, M.P., Engel, J. and Gasser, T. (1995). A bandwidth selector for bivariate kernel regression. *J. Royal Statist. Soc. B*, **57**, 171–180.

Hjort, N.L. (1995). Dynamic likelihood hazard rate estimation. *Biometrika*, to appear.

Hjort, N.L. and Glad, I.K. (1995). Nonparametric density estimation with a parametric start. *Ann. Statist.*, **23**, 882–904.

Hjort, N.L. and Jones, M.C. (1996a). Locally parametric nonparametric density estimation. *Ann. Statist.*, **24**, to appear.

Hjort, N.L. and Jones, M.C. (1996b). Better rules of thumb for choosing bandwidth in density estimation. *Statistical Research Report*, Department of Mathematics, University of Oslo, Norway.

Hjort, N.L. and Pollard, D. (1996). Asymptotics for minimisers of convex processes. *Ann. Statist.*, to appear.

Hsing, T. and Carroll, R.J. (1992). An asymptotic theory for sliced inverse regression. *Ann. Statist.*, **20**, 1040–1061.

Huang, L.S. (1995). On nonparametric estimation and goodness-of-fit. *Ph.D. Dissertation.* Department of Statistics. The University of North Carolina, Chapel Hill.

Huber, P.J. (1964). Robust estimation of a location parameter. *Ann. Math. Statist.*, **35**, 73–101.

Huber, P.J. (1981). *Robust Estimation.* Wiley, New York.

Hunsberger, S. (1994). Semiparametric regression in likelihood-based models. *J. Amer. Statist. Assoc.*, **89**, 1354–1365.

Hurvich, C.M. and Beltrão, K.I. (1990). Cross-validatory choice of a spectral estimate and its connections with AIC. *J. Time Ser. Anal.*, **11**, 121–137.

Ibragimov, I.A. (1962). Some limit theorems for stationary processes. *Theor. Prob. Appl.*, **7**, 349–382.

Ibragimov, I.A. and Hasminksii, R.Z. (1984). On nonparametric estimation of a linear functional in a Gaussian white noise. *Theory Prob. Appl.*, **29**, 19–32.

Ioffe, M.O. and Katkovnik, V.Ya. (1989). Pointwise and uniform convergence with probability 1 of nonparametric regression estimators. *Automat. and Remote Control*, No. 12, 1659–1667.

Ioffe, M.O. and Katkovnik, V.Ya. (1990). Rate of pointwise and uniform convergence with probability 1 of nonparametric estimates of a regression function. *Automat. and Remote Control*, **51**, 23–30.

James, I.R. and Smith, P.J. (1984). Consistency results for linear regression with censored data. *Ann. Statist.*, **12**, 590–600.

Jones, M.C. (1989). Discretized and interpolated kernel density estimates. *J. Amer. Statist. Assoc.*, **84**, 733–741.

Jones, M.C. (1990). Variable kernel density estimates and variable kernel density estimates. *Aust. J. Statist.*, **32**, 361–371.

Jones, M.C. (1994). A variation on local linear regression. *Manuscript*.

Jones, M.C. and Lotwick, H.W. (1984). A remark on Algorithm AS 176. Kernel density estimation using the fast Fourier transform. Remark AS R50. *Appl. Statist.*, **33**, 120–122.

Jones, M.C., Marron, J.S. and Sheather, S.J. (1992). Progress in data-based bandwidth selection for kernel density estimation. *Manuscript*.

Jones, M.C., Marron, J.S. and Sheather, S.J. (1995). A brief survey of bandwidth selection for density estimation. *J. Amer. Statist. Assoc.*, **90**, to appear.

Jones, M.C. and Sheather, J.S. (1991). Using non-stochastic terms to advantage in estimating integrated squared density derivatives. *Statist. Prob. Lett.*, **11**, 511–514.

Kalbfleisch, J.D. and Prentice, R.L. (1980). *The Statistical Analysis of Failure Time Data*. Wiley, New York.

Kaplan, E.L. and Meier, P. (1958). Nonparametric estimator from incomplete observations. *J. Amer. Statist. Assoc.*, **53**, 457–481.

Karn, M.N. and Penrose, L.S. (1951). Birth weight and gestations in relation to maternal age, parity and infant survival. *Ann. Eugen.*, **16**, 147–164.

Katkovnik, V.Ya. (1979). Linear and nonlinear methods of nonparametric regression analysis. *Avtomatika*, No. 5, 35–46.

Kerkyacharian, G. and Picard, D. (1993a). Density estimation by kernel and wavelet methods, optimality in Besov Spaces. *Manuscript*.

Kerkyacharian, G. and Picard, D. (1993b). Linear wavelet methods and other periodic kernel methods. *Manuscript*.

Kimeldorf, G.S. and Wahba, G. (1970). A correspondence between Bayesian estimation on stochastic processes and smoothing by splines. *Ann. Math. Statist.*, **41**, 495–502.

Kimeldorf, G.S. and Wahba, G. (1971). Some results on Tchebysheffian spline functions. *J. Math. Anal. Appl.*, **33**, 82–95.

Koenker, R. and Bassett, G. (1978). Regression quantiles. *Econometrica*. **46**, 33–50.

Koenker, R., Ng, P. and Portnoy, S. (1994). Quantile smoothing splines. *Biometrika*, **81**, 673–680.

Koenker, R., Portnoy, S. and Ng, P. (1992). Nonparametric estimation of conditional quantile function. *Proceedings of the conference of L_1-Statistical Analysis and Related Methods* (Y. Dodge, ed.), 217–229. Elsevier.

Kolmogorov, A.N. and Rozanov, Yu.A. (1960). On strong mixing conditions for stationary Gaussian processes. *Theory Prob. Appl.*, **52**, 204–207.

Kooperberg, C. and Stone, C.J. (1991). A study of logspline density estimation. *Comp. Statist. & Data Anal.*, **12**, 327–347.

Kooperberg, C., Stone, C.J. and Truong, Y.K. (1995a). Logspline estimation of a possibly mixed spectral distribution. *J. Time Series Anal.*, **16**, 359–388.

Kooperberg, C., Stone, C.J. and Truong, Y.K. (1995b). Rate of convergence for logspline spectral density estimation. *J. Time Series Anal.*, **16**, 389–401.

Korostelev, A.P. and Tsybakov, A.B. (1993). *Minimax Theory of Image Reconstruction.* Lecture Notes in Statistics, **82**. Springer-Verlag, New York.

Koul, H., Susarla, V. and Van Ryzin, J. (1981). Regression analysis with randomly right-censored data. *Ann. Statist.*, **9**, 1276–1288.

Koyak, R. (1990). Consistency for ACE-type methods. *Ann. Statist.*, **18**, 742–757.

Kushner, R.F. (1992). Bioelectrical impedance analysis: A review of principles and applications. *J. Amer. College Nutrition*, **11**, 199–209.

Lai, T.L, and Ying, Z. (1991). Large sample theory of a modified Buckley-James estimator for regression analysis with censored data. *Ann. Statist.*, **19**, 1370–1402.

Lai, T.L., Ying, Z. and Zheng, Z. (1995). Asymptotic normality of a class of adaptive statistics with applications to synthetic data methods for censored regression. *J. Mult. Anal.*, **52**, 259–279.

Le Cam, L. (1985). *Asymptotic Methods in Statistical Decision Theory.* Springer-Verlag, New York.

Lejeune, M. (1985). Estimation non-paramétrique par noyaux: régression polynomiale mobile. *Revue de Statist. Appliq.*, **33**, 43–68.

Lepskii, O. (1992). On problems of adaptive estimation in white Gaussian noise. *Topics in Nonpar. Est.*, **12**, 87–106.

Leurgans, S. (1987). Linear models, random censoring and synthetic data. *Biometrika*, **74**, 301–309.

Li, G. and Doss, H. (1995). An approach to nonparametric regression for life history data using local linear fitting. *Ann. Statist.*, **23**, 787–823.

Li, K.-C. (1982). Minimaxity of the method of regularization on stochastic processes. *Ann. Statist.*, **10**, 937–942.

Li, K.-C. (1985). From Stein's unbiased risk estimates to the method of generalized cross validation. *Ann. Statist.*, **13**, 1352–1377.

Li, K.-C. (1986). Asymptotic optimality for C_L, and generalized cross-validation in ridge regression with application to spline smoothing. *Ann. Statist.*, **14**, 1101–1112.

Li, K.-C. (1991). Sliced inverse regression for dimension reduction (with discussion). *J. Amer. Statist. Assoc.*, **86**, 316–342.

Loader, C.R. (1995). Local likelihood density estimation. *Ann. Statist.*, to appear.

Low, M.G. (1993). Lower bounds for the integrated risk in nonparametric density and regression estimation. *Ann. Statist.*, **21**, 577–589.

Macauley, R.R. (1931). *The Smoothing of Time Series*. National Bureau of Economic Research, New York.

Mack, Y.P. and Müller, H.-G. (1989). Convolution type estimators for nonparametric regression estimation. *Statist. Prob. Lett.* **7**, 229–239.

Mallat, S. (1989). A theory for multiresolution signal decomposition: The wavelet representation. *IEEE Trans. Pattern Anal. Machine Intell.*, **11**, 674–693.

Mallows, C.L. (1973). Some comments on C_p. *Technometrics*, **15**, 661–675.

Mammen, E. (1991a). Estimating a smooth monotone regression function. *Ann. Statist.*, **19**, 724–740.

Mammen, E. (1991b). Nonparametric regression under qualitative smoothness assumptions. *Ann Statist.*, **19**, 741–759.

Marron, J.S. (1989). Automatic smoothing parameter selection: a survey. *Empirical Econ.*, **13**, 187–208.

Marron, J.S. and Nolan, D. (1988). Canonical kernels for density estimation. *Statist. Prob. Lett.*, **7**, 195–199.

Marron, J.S. and Padgett, W.J. (1987). Asymptotically optimal bandwidth selection from randomly right-censored samples. *Ann. Statist.*, **15**, 1520–1535.

Marron, J.S., Park, B.U. and Kim, W.C. (1994). Asymptotically best bandwidth selectors in kernel density estimation. *Statist. Prob. Lett.*, **19**, 119–127.

Masry, E. and Fan, J. (1993). Local polynomial estimation of regression functions for mixing processes. *Institute of Statistics Mimeo Series #2311*, University of North Carolina at Chapel Hill.

McCullagh, P. and Nelder, J.A. (1989). *Generalized Linear Models*. Chapman and Hall, London.

Messer, K. (1991). A comparison of a spline estimate to equivalent kernel estimate. *Ann. Statist.*, **19**, 817–829.

Meyer, Y. (1990). *Ondelettes*. Hermann, Paris.

Meyn, S.P. and Tweedie, R.L. (1993). State-dependent criteria for convergence of Markov chains. *Ann. Appl. Prob.*, **4**, 149–168.

Meyn, S.P. and Tweedie, R.L. (1994). *Markov Chains and Stochastic Stability*. Springer-Verlag, London.

Miller, R.G. (1976). Least squares regression with censored data. *Biometrika*, **63**, 449–464.

Miller, R.G. (1981). *Survival Analysis.* Wiley, New York.
Miller, R.G. and Halpern, J. (1982). Regression with censored data. *Biometrika*, **69**, 521–531.
Müller, H.-G. (1987). Weighted local regression and kernel methods for nonparametric curve fitting. *J. Amer. Statist. Assoc.*, **82**, 231–238.
Müller, H.-G. (1988). *Nonparametric Regression Analysis of Longitudinal Data.* Lecture Notes in Statistics, **46**. Springer-Verlag, Berlin.
Müller, H.-G. (1991). Smooth optimum kernel estimators near endpoints. *Biometrika*, **78**, 521–530.
Müller, H.-G. (1993). On the boundary kernel method for nonparametric curve estimation near endpoints. *Scand. J. Statist.*, **20**, 313–328.
Müller, H.-G. and Stadtmüller, U. (1987). Variable bandwidth kernel estimators of regression curves. *Ann. Statist.*, **15**, 182–201.
Müller, H.-G., Stadtmüller, U. and Schmitt, T. (1987). Bandwidth choice and confidence intervals for derivatives of noisy data. *Biometrika*, **74**, 743–749.
Müller, H.-G. and Wang, J.-L. (1990). Analyzing changes in hazard functions: An alternative to change-point models. *Biometrika*, **77**, 610–625.
Müller, H.-G. and Wang, J.-L. (1994). Hazard rate estimation under random censoring with varying kernels and bandwidths. *Biometrics*, **50**, 61–76.
Nadaraya, E.A. (1964). On estimating regression. *Theory Prob. Appl.*, **9**, 141–142.
Nadaraya, E.A. (1989). *Nonparametric Estimation of Probability Densities and Regression Curves* (Translated by S. Kotz). Kluwer Academic Publishers, Boston.
Nason, G.P. and Silverman, B.W. (1994). The discrete wavelet transform in S. *J. Comp. Graph. Statist.*, **3**, 163–191.
Nelder, J.A. and Wedderburn, R.W.M. (1972). Generalized linear models. *J. Royal Statist. Soc. A*, **135**, 370–384.
Newey, W.K. and Powell, J.L. (1987). Asymmetric least squares estimation and testing. *Econometrica*, **50**, 43–61.
Newey, W.K. and Stoker, T.M. (1993). Efficiency of weighted average derivative estimators and index models. *Econometrica*, **61**, 1199–1223.
Nussbaum, M. (1985). Spline smoothing in regression models and asymptotic efficiency in L_2. *Ann. Statist.*, **13**, 984–997.
Nychka, D. (1988). Bayesian 'Confidence' intervals for smoothing splines. *J. Amer. Statist. Assoc.*, **83**, 1134–1143.
Nychka, D. (1995). Splines as local smoothers. *Ann. Statist.*, **23**, 1175–1197.
Nychka, D., Ellner, S., Gallant, A.R. and McCaffrey, D. (1992). Finding chaos in noisy systems. *J. Royal Statist. Soc. B*, **54**, 399–426.
Opsomer, J. (1995). Optimal bandwidth selection for fitting an addi-

tive model by local polynomial regression. *Ph.D. Dissertation.* Cornell University.

O'Sullivan, F. (1986). A statistical perspective on ill-posed inverse problems. *Statist. Sci.*, **1**, 502–527.

O'Sullivan, F. (1988). Nonparametric estimation of relative risk using splines and cross-validation. *SIAM J. Sci. Statist. Comput.*, **9**, 531–542.

O'Sullivan, F., Yandell, B. and Raynor, W. (1986). Automatic smoothing of regression functions in generalized linear models. *J. Amer. Statist. Assoc.*, **81**, 96–103.

Padgett, W.J. (1986). A kernel-type estimation of a quantile function from right-censored data. *J. Amer. Statist. Assoc.*, **81**, 215–222.

Parzen, E. (1962). On estimation of a probability density function and mode. *Ann. Math. Statist.*, **33**, 1065–1076.

Pawitan, Y. and O'Sullivan, F. (1994). Nonparametric spectral density estimation using penalized Whittle likelihood. *J. Amer. Statist. Assoc.*, **89**, 600–610.

Pinsker, M.S. (1980). Optimal filtering of square integrable signals in Gaussian white noise. *Problems Inform. Transmission*, **16**, 52–68.

Pollard, D. (1991). Asymptotics for least absolute deviation regression estimators. *Econometrics Theory*, **7**, 186–199.

Prakasa Rao, B.L.S. (1983). *Nonparametric Functional Estimation.* Academic Press, New York.

Priestley, M.B. (1981). *Spectral Analysis and Time Series.* Academic Press, London.

Priestley, M.B. and Chao, M.T. (1972). Nonparametric function fitting. *J. Royal Statist. Soc. B*, **34**, 384–392.

Pyke, R. (1965). Spacings. *J. Royal Statist. Soc. B*, **27**, 395–436.

Rao, C.R. (1973). *Linear Statistical Inference and Its Applications.* Wiley, New York.

Reinsch, C. (1967). Smoothing by spline functions. *Numer. Math.*, **10**, 177–183.

Rice, J.A. (1984). Boundary modification for nonparametric regression. *Commun. Statist. Theory Meth.*, **13**, 893–900.

Rice, J.A. and Rosenblatt, M. (1981). Integrated mean squared error of a smoothing spline. *J. Approx. Theory*, **33**, 353–365.

Rice, J.A. and Rosenblatt, M. (1983). Smoothing splines, regression, derivatives and convolution. *Ann. Statist.*, **11**, 141–156.

Ritov, Y. (1990). Estimation in a linear regression model with censored data. *Ann. Statist.*, **18**, 303–328.

Robinson, P.M. (1983). Nonparametric estimators for time series. *J. Time Series Anal.*, **4**, 185–297.

Robinson, P.M. (1986). On the consistency and finite sample properties of nonparametric kernel time series regression, auto regression, and density estimators. *Ann. Inst. Statist. Math.*, **38**, 539–549.

Rose, S.J. (1992). *Social Stratification in the United States.* The New Press, New York.

Rosenblatt, M. (1956a). Remarks on some nonparametric estimates of a density function. *Ann. Math. Statist.*, **27**, 832–837.

Rosenblatt, M. (1956b). A central limit theorem and strong mixing conditions. *Proc. Nat. Acad. Sci.*, **4**, 43–47.

Rosenblatt, M. (1969). Conditional probability density and regression estimates. In *Multivariate Analysis II* (Krishnaiah, ed.), 25–31. Academic Press, New York.

Rosenblatt, M. (1991). *Stochastic Curve Estimation.* NSF-CBMS Regional Conference Series in Probability and Statistics, **3**. Institute of Mathematical Statistics, California.

Roussas, G.G. (1990). Nonparametric regression estimation under mixing conditions. *Stoch. Proc. Appl.*, **36**, 107–116.

Roussas, G.G. and Tran, L.T. (1992). Asymptotic normality of the recursive kernel regression estimate under dependence conditions. *Ann. Statist.*, **20**, 98–120.

Rousseauw, J., du Plessis, J., Benade, A., Jordaan, P., Kotze, J., Jooste, P. and Ferreira, J. (1983). Coronary risk factory screening in three rural communities. *South African Medical J.*, **64**, 430–436.

Rudemo, M. (1982). Empirical choice of histograms and kernel density estimators. *Scand. J. Statist.*, **9**, 65–78.

Ruppert, D., Sheather, S.J. and Wand, M.P. (1995). An effective bandwidth selector for local least squares regression. *J. Amer. Statist. Assoc.*, **90**, 1257–1270.

Ruppert, D. and Wand, M.P. (1994). Multivariate weighted least squares regression. *Ann. Statist.*, **22**, 1346–1370.

Rutkowski, L. (1982). Orthogonal series estimates of a regression function with applications in system identification. In: *Probability and Statistical Inference* (W. Grossmann et al., eds.), 343–347. North Holland, Amsterdam.

Sacks, J. and Ylvisaker, D. (1978). Linear estimation for approximately linear models. *Ann. Statist.*, **6**, 1122–1137.

Sacks, J. and Ylvisaker, D. (1981a). Asymptotically optimum kernels for density estimation at a point. *Ann. Statist.*, **9**, 334–346.

Sacks, J. and Ylvisaker, D. (1981b). Variance estimation for approximately linear models. *Math. Operationsforch. Statist. Ser. Statist.*, **12**, 147–162.

Samarov, A.M. (1993). Exploring regression structure using nonparametric functional estimation. *J. Amer. Statist. Assoc.*, **88**, 836–847.

Schick, A., Susarla, V. and Koul, H. (1988). Efficient estimation of functionals with censored data. *Statist. Decis.*, **6**, 349–360.

Schmidt, G., Mattern, R. and Schüler, F. (1981). Biomechanical investigation to determine physical and traumatological differentiation criteria for the maximum load capacity of head and vertebral column with and without protective helmet under effects of impact. EEC Research Program on Biomechanics of Impacts. Final Report Phase III, Project 65, Institut für Rechtsmedizin, Universität Heidelberg, Germany.

Schoenberg, I.J. (1964). Spline functions and the problem of graduation. *Proc. Nat. Acad. Sci.* USA, **52**, 947–950.

Schott, J.R. (1994). Determining the dimensionality in sliced inverse regression. *J. Amer. Statist. Assoc.*, **89**, 141–148.

Schucany, W.R. and Sommers, J.P. (1977). Improvement of kernel type density estimators. *J. Amer. Statist. Assoc.*, **72**, 420–423.

Schuster, E.F. (1985). Incorporating support constraints into nonparametric estimates of densities. *Commun. Statist. Theory Meth.*, **14**, 1123–1126.

Schwarz, G. (1978). Estimating the dimension of a model. *Ann. Statist.*, **6**, 461–464.

Scott, D.W. (1992). *Multivariate Density Estimation: Theory, Practice, and Visualization*. Wiley, New York.

Scott, D.W., Tapia, R.A. and Thompson, J.R. (1977). Kernel density estimation revisited. *Nonlinear Analysis*, **1**, 339–372.

Scott, D.W. and Terrell, G.R. (1987). Biased and unbiased cross-validation in density estimation. *J. Amer. Statist. Assoc.*, **82**, 1131–1146.

Seifert, B., Brockmann, M., Engel, J. and Gasser, T. (1994). Fast algorithms for nonparametric curve estimation. *J. Comp. Graph. Statist.*, **3**, 192–213.

Seifert, B. and Gasser, T. (1995). Finite sample variance of local polynomials: analysis and solutions. *J. Amer. Statist. Assoc.*, to appear.

Severini, T.A. and Staniswalis, J.G. (1994). Quasi-likelihood estimation in semiparametric models. *J. Amer. Statist. Assoc.*, **89**, 501–511.

Severini, T.A. and Wong, W.H. (1992). Generalized profile likelihood and conditional parametric models. *Ann. Statist.*, **20**, 1768–1802.

Sheather, S.J. and Jones, M.C. (1991). A reliable data-based bandwidth selection method for kernel density estimation. *J. Royal Statist. Soc. B*, **53**, 683–690.

Shibata, R. (1976). Selection of the order of an autogressive model by Akaike's information criterion. *Biometrika*, **63**, 117–126.

Silverman, B.W. (1982). Kernel density estimation using the fast Fourier transform method. *Appl. Statist.*, **31**, 93–99.

Silverman, B.W. (1984). Spline smoothing: the equivalent variable kernel method. *Ann. Statist.*, **12**, 898–916.

Silverman, B.W. (1985). Some aspects of the spline smoothing approach to nonparametric regression curve fitting (with discussion). *J. Royal Statist. Soc. B*, **47**, 1–52.

Silverman, B.W. (1986). *Density Estimation for Statistics and Data Analysis*. Chapman and Hall, London.

Smith, P.L. (1982). Curve fitting and modeling with splines using statistical variable selection methods. NASA, Langley Research Center, Hampla, VA, NASA Report 166034.

Smith, R. (1992). Estimating dimension in noisy chaotic time series. *J. Royal Statist. Soc. B*, **54**, 329–351.

Speckman, P. (1981). The asymptotic integrated mean square error for smoothing noisy data by splines. *Unpublished manuscript.*

Speckman, P. (1988). Kernel smoothing in partial linear models. *J. Royal Statist. Soc. B*, **50**, 413–436.

Staniswalis, J.G. (1989). The kernel estimate of a regression function in likelihood-based models. *J. Amer. Statist. Assoc.*, **84**, 276–283.

Stein, C. (1956). Inadmissibility of the usual estimator for the mean of a multivariate normal distribution. In *Proceedings on the Third Berkeley Symposium on Mathematical Statistics and Probability*, Vol. 1, 197–206. University of California Press, Berkeley.

Stoker, T.M. (1991). Equivalence of direct, indirect and slope estimators of average derivatives. In *Nonparametric and Semiparametric Methods in Econometrics and Statistics*, Proceedings of the Fifth International Symposium in Economic Theory and Econometrics; (Barnett, W.A., Powell, J.L. and Tauchen, G., eds), 99–118. Cambridge University Press, Cambridge.

Stoker, T.M. (1992). *Lectures on Semiparametric Econometrics.* Center for Operational Research and Econometrics, Université Catholique de Louvain, Louvain-la-Neuve, Belgium.

Stone, C.J. (1977). Consistent nonparametric regression. *Ann. Statist.*, **5**, 595–645.

Stone, C.J. (1980). Optimal rates of convergence for nonparametric estimators. *Ann. Statist.*, **8**, 1348–1360.

Stone, C.J. (1982). Optimal global rates of convergence for nonparametric regression. *Ann. Statist.*, **10**, 1040–1053.

Stone, C.J. (1985). Additive regression and other nonparametric models. *Ann. Statist.*, **13**, 689–705.

Stone, C.J. (1986). The dimensionality reduction principle for generalized additive models. *Ann. Statist.*, **14**, 590–606.

Stone, C.J. (1990a). Large sample inference for logspline model, *Ann. Statist.*, **18**, 717–714.

Stone, C.J. (1990b). L_2 rate of convergence for interaction spline regression. *Technical report 268*, Dept. of Statist., Univ. of California, Berkeley.

Stone, C.J. (1991). Multivariate logspline conditional models. *Technical report. 320*, Dept. of Statist., Univ. of California, Berkeley.

Stone, C.J. (1994). The use of polynomial splines and their tensor products in multivariate function estimation (with discussion). *Ann. Statist.*, **22**, 118–184.

Stone, C.J. and Koo, C.Y. (1986). Logspline density estimation. *Contemporary Mathematics*, **59**, 1–15.

Stone, M. (1974). Cross-validatory choice and assessment of statistical predictions (with discussion). *J. Royal Statist. Soc. B*, **36**, 111–147.

Strang, G. (1989). Wavelets and dilation equations: a brief introduction. *SIAM Review*, **31**, 614–627.

Strang, G. (1993). Wavelet transforms versus fourier transforms. *Bulletin Amer. Math. Soc.*, **28**, 288–305.

Stute, W. and Wang, J.-L. (1993). A strong law under random censorship. *Ann. Statist.*, **21**, 1591–1607.

Swanepoel, J.W. and van Wyk, J.W.J. (1986). The bootstrap applied to spectral density function estimation. *Biometrika*, **73**, 135–142.

Tibshirani, R. and Hastie, T.J. (1987). Local likelihood estimation. *J. Amer. Statist. Assoc.*, **82**, 559–567.

Tjøstheim, D. (1990). Nonlinear time series and Markov chains. *Adv. Appl. Prob.*, **22**, 587–611.

Tong, H. (1990). *Non-Linear Time Series: A Dynamical System Approach*. Oxford University Press, Oxford.

Truong, Y.K. (1989). Asymptotic properties of kernel estimators based on local medians. *Ann. Statist.*, **17**, 606–617.

Truong, Y.K. (1991). Nonparametric curve estimation with time series errors. *J. Statist. Plann. Infer.*, **28**, 167–183.

Truong, Y.K. (1992). Robust nonparametric time series regression. *J. Mult. Anal.*, **41**, 163–177.

Truong, Y.K. and Stone, C.J. (1992). Nonparametric function estimation involving time series. *Ann. Statist.*, **20**, 77–97.

Tsiatis, A.A. (1990). Estimating regression parameters using linear rank tests for censored data. *Ann. Statist.*, **18**, 354–372.

Tsybakov, A.B. (1986). Robust reconstruction of functions by the local-approximation method. *Problems of Information Transmission*, **22**, 133–146.

Turlach, B.A. (1994). Computer-aided additive modeling. *Ph.D. Dissertation*. Institut de Statistique, Université Catholique de Louvain, Louvain-la-Neuve, Belgium.

Utreras, F.D. (1980). Sur le choix du parametre d'adjustement dans le lissage par fonctions spline. *Numer. Math.*, **34**, 15–28.

Uzunoğullari, Ü. and Wang, J.-L. (1992). A comparison of the hazard rate estimators for left truncated and right censored data. *Biometrika*, **79**, 297–310.

Vieu, P. (1991) Nonparametric regression: optimal local bandwidth choice. *J. Royal Statist. Soc. B*, **53**, 453–464.

Volkonskii, V.A. and Rozanov, Yu.A. (1959). Some limit theorems for random functions. *Theory Prob. Appl.*, **4**, 178–197.

Wahba, G. (1975). Smoothing noisy data with spline functions. *Numer. Math.*, **24**, 383–393.

Wahba, G. (1977). A survey of some smoothing problems and the method of generalized cross-validation for solving them. In *Appli-*

cations of Statistics (P.R. Krisnaiah, ed.), 507–523. North Holland, Amsterdam.

Wahba, G. (1978). Improper priors, spline smoothing and the problem of guarding against model errors in regression. *J. Royal Statist. Soc. B*, **40**, 364–372.

Wahba, G. (1980). Automatic smoothing of the log periodogram. *J. Amer. Statist. Assoc.*, **75**, 122–132.

Wahba, G. (1984). Partial spline models for semiparametric estimation of functions of several variables. In *Statistical Analysis of Time Series*, Proceedings of the Japan U.S. Joint Seminar, Tokyo, 319–329. Institute of Statistical Mathematics, Tokyo.

Wahba, G. (1990). *Spline Models for Observational Data.* SIAM, Philadelphia.

Wahba, G. and Wang, Y. (1990). When is the optimal regularization parameter insensitive to the choice of the loss function? *Commun. Statist.*, **5**, 1685–1700.

Waldmeir, M. (1961). *The Sunspot Activity in the Years 1610–1960.* Schulthess, Zurich.

Wand, M.P. (1994). Fast computation of multivariate kernel estimators. *J. Comp. Graph. Statist.*, **3**, 433–445.

Wand, M.P. and Jones, M.C. (1993). Comparison of smoothing parametrizations in bivariate kernel density estimation. *J. Amer. Statist. Assoc.*, **88**, 520–528.

Wand, M.P. and Jones, M.C. (1995). *Kernel Smoothing.* Chapman and Hall, London.

Watson, G.S. (1964). Smooth regression analysis. *Sankhyā Ser. A*, **26**, 359–372.

Wedderburn, R.W.M. (1974). Quasilikelihood functions, generalized linear models and the Gauss-Newton method. *Biometrika*, **61**, 439–447.

Wegman, E.J. (1969). A note on estimating a unimodal density. *Ann. Math. Statist.*, **40**, 1661–1667.

Wegman, E.J. (1970). Maximum likelihood estimation of a unimodal function. *Ann. Math. Statist.*, **41**, 457–471.

Wei, C.Z. and Chu, J.K. (1994). A regression point of view toward density estimation. *J. Nonpar. Statist.*, **4**, 191–201.

Weisberg, S. and Welsh, A.H. (1994). Adapting for the missing link. *Ann. Statist.*, **22**, 1674–1700.

Welsh, A.H. (1994). Robust estimation of smooth regression and spread functions and their derivatives. *Manuscript.*

Whittle, P. (1962). Gaussian estimation in stationary time series. *Bulletin of the International Statistical Institute*, **39**, 105–129.

Wittaker, E.T. (1923). On a new method of graduation. *Proc. Edinburgh Math. Soc.*, **41**, 63–75.

Wolff, R.C.L. (1992). Local Lyapunov exponents: looking closely at chaos. *J. Royal Statist. Soc. B*, **54**, 353–371.

Wong, W.H. (1984). On constrained multivariate splines and their approximations. *Numer. Math.*, **43**, 141–152.

Wong, W.H. (1986). Theory of partial likelihood. *Ann. Statist.*, **14**, 88–123.

Wong, W.H. and Shen, X. (1995). Dimension reduction in regression. *Manuscript*.

Woodroofe, M. (1970). On choosing a delta-sequence. *Ann. Math. Statist.*, **41**, 1665–1671.

Woodroofe, M. (1982). On model selection and the arc sine laws. *Ann. Statist.*, **10**, 1182–1194.

Yao, Q. and Tong, H. (1994a). Quantifying the inference of initial values on nonlinear prediction. *J. Royal Statist. Soc. B*, **56**, 701–725.

Yao, Q. and Tong, H. (1994b). On prediction and chaos in stochastic systems. *Phil. Tran. Roy. Soc. Lond. A*, **348**, 357–369.

Yao, Q. and Tong, H. (1995). Asymmetric least squares regression estimation: a nonparametric approach. *J. Nonpar. Statist.*, to appear.

Zhang, P. (1992). On the distributional properties of model selection criteria. *J. Amer. Statist. Assoc.*, **87**, 732–737.

Zheng, Z. (1987). A class of estimators of the parameters in linear regression with censored data. *Acta Mathem. Applic. Sinica*, **3**, 231–241.

Zheng, Z. (1988). Strong consistency of nonparametric regression estimates with censored data. *J. Math. Res. Exposit.*, **8**, 307–313.

Zhou, M. (1992). Asymptotic normality of the 'synthetic data' regression estimator for censored survival data. *Ann. Statist.*, **20**, 1002–1021.

Author index

Aalen, O.O. 214
Abramovich, F. 44
Abramson, I.S. 107
Aerts, M. 216
Aigner, D. 231
Akaike, H. 42
Allen, D.M. 44
Amemiya, T. 231
An, H.Z. 261
Andersen, P.K. 160, 175, 179, 214
Anscombe, F.J. 51
Azzalini, A. 260

Bai, Z.D. 106
Bartlett, M.S. 190, 260
Bassett, G. 216
Becker, R.A. 11
Bellman, R.E. 264
Beltrão, K.I. 261
Beran, R. 168
Bhattacharya, P.K. 216
Bickel, P.J. 31, 47, 55, 107, 149, 171, 216
Birgé, L. 56
de Boor, C. 40, 56
Bloomfield, P. 261
Blyth, S. 113
Borgan, Ø. 160, 175, 179, 214
Bowman, A. 157
Bradlow, E. 113
Breiman, L. 40, 54, 107, 265–267, 283, 305, 306
Breslow, N.E. 184

Brillinger, D.R. 234, 235
Brockmann, M. 75, 76, 86, 88, 99, 100, 106, 303, 304
Brockwell, P.J. 234, 257
van der Broeck, K. 157
Brown, L.D. 107
Bruce, A.G. 39
Buckley, J. 166, 214
Buja, A. 305, 306

Cao, R. 157
Carroll, R.J. 26, 202, 206, 279–281, 306
Čencov, N.N. 55
Chambers, J.M. 11
Chan, K.S. 225, 261
Chao, M.T. 16
Chaudhuri, P. 216, 306
Chen, H. 305
Chen, R. 261
Cheng, B. 260
Cheng, K.F. 16, 106
Cheng, M.Y. 18, 51, 76, 89–91, 106
Cheng, P.E. 106
Chiu, S.T. 158
Chow, Y.S. 218
Chu, C.K. 17, 18, 68, 106
Chui, C.K. 28
Cleveland, W.S. 18, 24, 26, 99, 105, 106, 117, 172, 200, 299, 300, 306
Collomb, G. 260
Cox, D.D. 44, 215

AUTHOR INDEX

Cox, D.R. 160, 175, 214, 215
Craven, P. 45
Cuevas, A. 157
Cutler, J. 261
Cuzick, J. 305

Dabrowska, D.M. 163, 215
Daniels, P.J. 260
Daubechies, I. 28, 31–37
Davis, H.T. 235
Davis, R.A. 234, 257
Deheuvels, P. 149
Devlin, S.J. 106, 299, 306
Devroye, L.P. 55
Djojosugito, R.A. 106
Doksum, K.A. 47, 113, 149, 160, 163, 215, 306
Donoho, D.L. 28–31, 35, 38, 39, 52, 53, 92, 107, 129, 137
Doss, H. 215
Drazin, P.G. 261
Duan, N. 290, 306
Duin, R.P.W. 157

Eckmann, J.P. 261
Eddy, W.F. 55
Efromovich, S. 29, 107
Efron, B. 231, 232
Ellner, S. 261
Engel, J. 75, 76, 86, 88, 99, 100, 106, 300, 303, 304
Epanechnikov, V.A. 303
Eubank, R.L. 12, 40, 45, 56, 106, 157

Fan, J. 17, 18, 23, 28, 50–53, 60, 62, 67–81, 85–100, 104–107, 112, 113, 118, 121, 125–133, 151, 157, 167–174, 183, 196–201, 210, 215, 216, 220, 224–227, 236–239, 243–246, 279–281, 303, 304
Fang, K.T. 292
Farrell, R.H. 55, 107

Feduska, N.J. 163
Fleming, T.R. 160, 175, 179, 185, 187, 214
Franke, J. 261

Gallant, A.R. 261
Gangopadhyay, A.K. 216
Gao, H.Y. 39
Gaskins, R.A. 44
Gasser, T. 15, 59, 64, 65, 69, 70, 75, 76, 86, 88, 99, 100, 106, 153, 201, 216, 300, 303, 304
Gijbels, I. 18, 62, 67, 70–76, 106, 81, 86, 88, 112, 113, 118, 121, 125–127, 132, 133, 151, 167–174, 183, 215, 243, 279–281, 303, 304
Gill, R.D. 160, 175, 179, 214
Glad, I.K. 20
Godambe, V.P. 215
González-Manteiga, W. 157
Good, I.J. 44
Granovsky, B.L. 75, 106
Green, P.J. 12, 40, 171, 215, 264, 273
Grenander, U. 56
Groeneboom, P. 56
Grosse, E. 24, 299, 300, 306
Gu, C. 56, 215
Györfi, L.P. 12, 55

Habbema, J.D.F. 157
Hall, P. 28, 52, 53, 56, 106, 107, 114, 128, 129, 157, 158, 216, 252, 261, 275, 305, 306
Halpern, J. 160
Hampel, F.R. 199
Härdle, W. 12, 15, 44, 107, 150, 201, 216, 221, 260, 261, 264, 275, 276, 305, 306
Harrington, D.P. 160, 175, 179, 185, 187, 214
Hart, J.D. 157, 260
Hasminskii, R.Z. 55, 107

Hastie, T.J. 12, 18, 21, 70, 106, 157, 215, 264, 267–271, 306
Heckman, J. 305
Heckman, N.E. 157, 196–199, 210, 215, 305
Hermann, E. 300
Hermans, J. 157
Heyde, C.C. 215, 252
Hildenbrand, W. 306
Hjort, N.L. 20, 47, 52, 80, 157, 210, 215
Hsing, T. 306
Hu, T.-C. 62, 67, 106, 127, 201, 216
Huang, F.C. 261
Huang, L.-S. 62, 67, 106, 158, 114, 127
Huber, P.J. 199, 200
Hunsberger, S. 215, 305
Hurvich, C.M. 261
Husing, R. 163

Ibragimov, I.A. 107, 219
Ichimura, H. 275, 305, 306
Ioffe, M.O. 106

James, I.R. 166, 214
Janssen, P. 216
Jerison, M. 306
Johnstone, I. 28–31, 35, 38, 39, 52, 53, 107, 129, 131, 157
Jones, M.C. 12, 47, 49, 52, 80, 107, 110, 114, 148, 153–158, 216, 264, 306
Jones, R.H. 235

Kalbfleisch, J.D. 160, 175, 214
Kaplan, E.L. 164
Karn, M.N. 195
Katkovnik, V.Ya. 106
Keiding, N. 160, 175, 179, 214
Kerkyacharian, G. 28, 52, 129
Kim, W.C. 158
Kimeldorf, G.S. 56

King, G.P. 261
King, M. 183
Klaassen, C.J. 171
Kneip, A. 99, 153
Koenker, R. 216
Köhler, W. 153
Kolmogorov, A.N. 219
Koo, C.Y. 54
Kooperberg, C. 42, 54, 55, 261
Korostelev, A.P. 107
Kotz, S. 292
Koul, H. 168, 215
Koyak, R. 305
Kreutzberger, E. 236, 238, 239
Kushner, R.F. 135

Lai, T.L. 214, 215
Le Cam, L. 107
Lejeune, M. 64, 106
Lepskii, O. 107
Leurgans, S. 168, 215
Li, G. 215
Li, K.-C. 56, 107, 290–293, 306
Lin, P.E. 16
Liu, R.C. 92, 107
Loader, C.R. 18, 21, 52, 106, 157
Lotwick, H.W. 107
Low, M. 107

MacGibbon, B. 107
Macauley, R.R. 105
Mack, M.P. 17, 68
Mallat, S. 35
Mallows, C.L. 42
Mammen, E. 56
Mammitzsch, V. 59, 64, 65, 70, 75, 106
Marron, J.S. 15–18, 23, 51, 60, 68, 76, 89–91, 95–100, 106, 107, 114, 148, 153, 157, 158, 215
Martin, M. 28, 52, 53, 128, 129, 157
Masry, E. 220

AUTHOR INDEX

Mattern, R. 2
McCaffrey, D. 261
McCullagh, P. 189, 215
Meier, P. 164
Meisel, W. 107
Meng, X.-L. 113
Messer, K. 44
Meyer, Y. 28, 31
Meyn, S.P. 261
Miller, R.G. 160, 214
Müller, H.-G. 12, 15, 17, 59, 64–70, 75, 76, 106, 107, 119, 215

Nadaraya, E.A. 12, 15
Nason, G.P. 39, 131
Nelder, J.A. 189, 215
Neville, P. 163
Newey, W.K. 231, 232, 276
Ng, P. 216
Ng, W. 292
Nolan, D. 15
Nussbaum, M.S. 29, 107
Nychka, D. 44, 56, 261

O'Sullivan, F. 56, 215, 261
Oakes, D. 160, 175, 214
Olshen, R.A. 40, 265, 283, 306
Opsomer, J. 305

Padgett, W.J. 215
Park, B.U. 157, 158
Parzen, E. 55
Patil, P. 28, 52, 53, 128, 129, 157
Pawitan, Y. 261
Penrose, L.S. 195
Picard, D. 28, 52, 129
Pinsker, M.S. 29, 107
Poirier, D. 231
Pollard, D. 209, 210
Portnoy, S. 216
Powell, J.L. 231, 232
Prakasa Rao, B.L.S. 55

Prentice, R.L. 160, 175, 214
Priestley, M.B. 16, 234
Purcell, E. 107
Pyke, R. 156

Qiu, C. 56

Ragozin, D. 39
Rao, C.R. 302
Raynor, W. 215
Reinsch, C. 56
Rice, J.A. 56, 106
Ritov, Y. 107, 171, 214
Robinson, P.M. 260
Ronchetti, E.M. 199
Rose, S.J. 229
Rosenblatt, M. 12, 55, 56, 106, 218, 219, 260
Roussas, G.G. 260
Rousseauw, J. 270
Rousseeuw, P.J. 199
Rozanov, Yu.A. 219
Rudemo, M. 157
Ruelle, D. 261
Ruppert, D. 18, 26, 61, 63, 70, 78, 106, 153, 202, 206, 301, 305, 306
Rutkowski, L. 55

Sacks, J. 91, 107
Samarov, A.M. 276, 295, 306
Sarda, P. 12
Schick, A. 215
Schmidt, G. 2
Schmitt, T. 119
Schoenberg, I.J. 56
Schott, J.R. 306
Schucany, W.R. 106
Schüler, F. 2
Schuster, E.F. 106
Schwarz, G. 43
Scott, D.W. 46, 107, 153, 157, 264
Seifert, B. 99, 100, 106

Severini, T.A. 215, 305
Sheather, S.J. 49, 110, 114, 148, 153, 154, 158, 305
Shen, X. 276, 295
Shibata, R. 43
Shyu, W.M. 24, 299, 300, 306
Silverman, B.W. 12, 39, 40, 44–47, 60, 107, 121, 131, 149, 152, 171, 264, 273, 290
Smith, J. 305
Smith, P.J. 214
Smith, P.L. 40
Smith, R. 261
Sommers, J.P. 44, 106, 157, 305
Stadtmüller, U. 107, 119
Stahel, W.A. 199
Staniswalis, J.G. 215, 305
Stein, C. 29
Steinberg, D.M. 44
Stoker, T.M. 276, 306
Stone, C.J. 18, 40, 42, 54, 55, 105, 107, 260, 261, 265, 267, 283, 305, 306
Stone, M. 45
Strang, G. 17, 28, 31, 32
Stuetzle, W. 265, 266, 305
Stute, W. 215
Susarla, V. 168, 215
Swanepoel, J.W. 261

Tapia, R.A. 153
Teicher, H. 218
Terrell, T.R. 157
Thompson, J.R. 153
Tibshirani, R. 12, 70, 106, 157, 215, 264, 267–271, 305, 306
Tjøstheim, D. 261
Todd, P. 305
Tong, H. 218, 224–227, 230–233, 244–248, 261
Tran, L.T. 260
Truong, Y.K. 42, 201, 216, 260, 261
Tsay, R.J. 261
Tsiatis, A.A. 214

Tsybakov, A.B. 106, 107, 201, 216, 305
Turlach, B.A. 305
Tweedie, R.L. 261

Utreras, F.D. 56
Uzunoğullari, Ü. 215

Van Ryzin, J. 168
Veraverbeke, N. 215, 216
Vieu, P. 12, 157
Volkonskii, V.A. 219

Wagner, T.J. 55
Wahba, G. 12, 40, 44, 45, 56, 240, 260, 305
Waldmeir, M. 222
Wand, M.P. 12, 18, 61, 63, 70, 78, 106, 107, 153, 157, 196–199, 210, 215, 264, 279–281, 300, 301, 305, 306
Wang, J.-L. 215
Wang, Y. 45
Watson, G.S. 15
Wedderburn, R.W.M. 215
Wegman, E.J. 56
Wehrly, T.E. 106, 260
Wei, C.Z. 106
Weisberg, S. 277
Wellner, J.A. 171
Welsh, A.H. 201, 204, 206, 208, 216, 277
Whittle, P. 235, 237
Wilks, A.R. 11
Wittaker, E.T. 40
Wolff, R.C.L. 261
Wong, W.H. 56, 177, 276, 295, 305
Woodroofe, M. 43, 152
van Wyk, J.W.J. 261

Yandell, B.S. 160, 215
Yao, Q. 224–227, 230–233, 244–248

Ying, Z. 214, 215
Ylvisaker, D. 91, 107

Zhang, P. 43
Zhao, H. 113
Zheng, Z. 167, 215
Zhou, M. 214

Subject index

Accelerated lifetime model 162–163
Adaptive order 80–83
Additive models 264, 265–267
Additive regression spline model 289
Adjusted dependent variable 268–269
Adjusting constants 119
Akaike's information criterion 42
Alternating conditional expectation 267
ARMA model 240
Autocovariance function 234
Automatic boundary correction, *see* Boundary
Average derivative 275

B-spline basis, *see* Splines
Backfitting algorithm 266
Bandwidth 4, 15
    Constant bandwidth 66, 120–125, 281
    Global bandwidth 66
    Matrix 298–304
    Variable bandwidth 66, 122–128, 281
Bandwidth selection
    Cross-validation 44–45, 149–150, 172
    Extended residual squares criterion 147
    Generalized cross-validation 45–46

Least squares cross-validation 149–150
Likelihood cross-validation 151
Nearest neighbor 151, 167, 172
Normal reference rule 47–49, 133, 149, 225
Plug-in 49, 152–154
Refined bandwidth selector 123–125, 132, 238–239
Residual squares criterion 117–122, 225
Rule of thumb 110–113, 149, 184, 198, 202
Variable bandwidth 122–123, 124–132
Baseline hazard function 162–163, 173, 184, 187
Bernoulli model 7, 191, 194
    See also GLIM
Bias 8, 14, 17–22, 60–83, 113–116, 143–145, 196–197, 301–302
Binning 50–51, 96–99
    Bin averages 96
    Bin counts 50–51, 96
    Binwidth 97
    Linear binning 97–99
    Simple binning 97–99
Biweight kernel, *see* Kernel
Boundary effects 69, 106
    Automatic correction 69–74
    Boundary points 69
    Kernel methods 18, 69
    Reflection methods 18, 69
Breakdown points 199

SUBJECT INDEX

Breslow estimator 184, 189

Canonical exponential family 190, 276
  See also GLIM
Canonical link function 190, 268
  See also GLIM
Canonical parameter 190
  See also GLIM
Canonical regression model 28
Censoring 160–189
  Distribution-based unbiased transformation 168
  Independent 162, 187–188
  Local average unbiased transformation 167
  Local linear unbiased transformation 167
  Mechanism 160
  Noninformative 162, 187–188
  Unbiased transformations 165, 167
Change point model 173
Classification and regression trees 265
Coefficient of variation 192
Conditional densities 218, 224–228
Conditional inferences 175
Conditional interquartile range 206
Confidence intervals 116–118
Convexity lemma 209–210
Convolution 99
Cross-validation, *see* Bandwidth selection
Cumulative hazard function 184
Curse of dimensionality 264

Data mining 30
Data reduction 290
Design
  densities 17, 57
  Equispaced designs 16, 28, 96
  Non-equispaced designs 16

Density estimation 46–55, 120–121, 125, 134–141, 224–228
Deviance, *see* GLIM
Discrete wavelet transforms, *see* Wavelets
Dispersion parameter, *see* GLIM
Distribution-based unbiased transformations, *see* Censoring
Downweighted 24

Edge effects, *see* Boundary effects
Effective dimension-reduction
  direction, *see* SIR
  space, *see* SIR
Embedding dimension 244
Epanechnikov kernel, *see* Kernel
Equispaced designs, *see* Design
ERSC bandwidth selector 147
Euclidean distance 244, 299
Expectiles 228, 230–233
Exponential family 189, 190, 276
  See also GLIM
Extended residual squares criterion, *see* Bandwidth selection
Extremal phase family, *see* Wavelets

Fast Fourier transform 99
Father wavelet, *see* Wavelets
Fisher information 245, 268
Fisher scoring 195, 238, 268
Forward regression 291, 295
Fourier frequency 234

Generalized additive models 265–272
  See also GLIM
  See also Additive models
Generalized cross-validation, *see* Bandwidth selection
Generalized linear models 6, 189–199, 263, 267
  Deviance 192

Deviance residual 193
Dispersion parameter 190
Link function 190–191
Quasi-likelihood 193–195, 278–280
Generalized partially linear single-index models 272, 276–277
  Fully iterated algorithm 279–280
  One-step algorithm 279
GLIM, abbreviation for Generalized Linear Models

Hard-thresholding, *see* Thresholding
Hazard
  rate (risk) 162–164, 173, 175–189
  regression 159, 162, 179–183
Heteroscedasticity 50–52, 135, 204–206, 236–243
High-pass filter 36–39
Homoscedasticity 40, 51, 115, 191, 192, 230, 236
Huber's
  bisquare function 26, 200
  proposal 208
  ψ-estimator 200

Interaction terms 264, 271, 277, 283–290
Inverse regression 291
Iterative procedure 152, 266, 268–269

Jittered 13

Kaplan-Meier estimator, *see* Product-limit estimator
Kernel 5, 15
  Biweight 15
  Boundary 18
  Epanechnikov 15
  Equivalent 64
  Minimum variance 75
  Optimal 75, 302–303
  Spherical Epanechnikov 303
  Tricube kernel 24
  Triweight kernel 15
  Uniform 15
Knots 4, 40
  deletion 40, 42
Kullback-Leibler discriminant information 151, 244

Least asymmetric family, *see* Wavelets
Least squares cross-validation, *see* Bandwidth selection
Leave-one-out 150, 275
Level of primary resolution, *see* Wavelets
Likelihood cross-validation, *see* Bandwidth selection
Linear binning, *see* Binning
Linear minimax risk, *see* Minimax
Linear regression 1
Linear smoother 84
Local adaptability 129
Local average unbiased transformation, *see* Censoring
Local likelihood smoothed log-periodogram, *see* Spectral density
Local linear fit 20
Local linear unbiased transformation, *see* Censoring
Local log-likelihood 142–143, 147, 180
Local modelling 4–7, 142
  See also Local polynomial regression
Local partial likelihood 179–181
Local polynomial regression 19, 58, 170, 295
Local quasi-likelihood 194–195, 278

SUBJECT INDEX

Local variable bandwidth, *see* Bandwidth
Location-scale model 57
LOESS 24, 271, 287, 299–300
Log-linear model 192
Logit transformation 191
Logspline, *see* Splines
LOWESS 24–27, 187, 200
Low-pass filter 36–39
Lyapounov exponent 245

Main effect 283, 284
Mallat's pyramid algorithm 35
Mallows' C_p criterion 42
Marginal inferences 176
Martingale residual 186–187
Mean absolute deviation error 83
Mean integrated squared error 14
Mean squared error 14
Minimax
   Efficiency 85–88, 91, 303
   Linear risk 85, 303–304
   Risk 92, 94
Minimum variance kernel, *see* Kernel
Missing link 277
Mixing conditions 218–219
   ρ-mixing 219
   Strongly mixing 218
   Uniformly mixing 219
Model complexity 8, 20
Modulus of continuity 92–93
Mother wavelet, *see* Wavelets
Multiple linear regression model 264
Multiresolution analysis 31
Multivariate adaptive regression splines 289–290
Multivariate local linear regression estimator 297–299
Multivariate regression problems 263
Multivariate tensor-product splines, *see* Splines

Nearest neighbor 151, 167, 172
Nearly universal principle 110, 148
Newton-Raphson 195, 238, 268
Noise amplification 218, 243, 247
Noise to signal ratio 113
Non-equispaced designs, *see* Design
Nonlinear prediction 217, 218
Nonlinear time series 217
Nonparametric regression 13
Normal reference bandwidth selector, *see* Bandwidth selection

Optimal kernels, *see* Kernel
Oracle estimator 30
Orthogonal series method 4
   See also Wavelets
Outliers 22
Over-dispersion 193, 206
Over-parametrization 19

Partial likelihood 175–179
Partial residuals 266
Partial spline models 273
Partially linear models 264, 272–274
Percentiles, *see* Quantiles
Periodogram 233, 234–237
Pilot bandwidth 110
Pilot estimates 111
Plug-in technique, *see* Bandwidth selection
Polynomial regression 3
Power basis, *see* Splines
Power spectrum 222
Predictive intervals 201, 203–204
Primary information 29, 31
Product-limit estimator 164
Projection pursuit 265
Proportional hazards model 6, 162

Quadratic approximation lemma 210
Quantiles
    Regression 200–203
    Regression percentiles 228–230
Quasi-likelihood function, see GLIM

Refined bandwidth selector, see Bandwidth selection
Reflection methods, see Boundary
Regression percentiles, see Quantiles
Regular 93
Residual squares criterion, see Bandwidth selection
Residual sum of squares 42, 115, 139, 147, 192
Richness 93
Robust
    likelihood 200
    method 22, 199–200, 296
    weights 25
RSC bandwidth selector 120
Rule of thumb, see Bandwidth selection

Scale function 31
Scaling equation 32
Seasonal effects 222
Sensitivity analysis 173
Sensitivity measure 218, 243–244
Shrinkage 30, 34, 35
Sieve method 31
Simple binning, see Binning
Single-index model 273, 274–276
    See also Generalized partially linear single-index models
    See also Projection pursuit
SIR, abbreviation for Sliced Inverse Regression
Skeleton 225, 226
Sliced inverse regression 264, 290
    Effective dimension-reduction direction 291

Effective dimension-reduction space 291
Standardized e.d.r. direction 292
Smoothed log-periodogram, see Spectral density
Smoothed periodogram, see Spectral density
Smoothing parameter 15
    See also Bandwidth
Smoothing splines 4, 40, 43–45
Soft-thresholding, see Thresholding
Spatial adaptation 20, 28, 128–132
Spectral density 217, 233–234
    Local-likelihood smoothed log-periodogram 234
    Smoothed log-periodogram, 234, 235
    Smoothed periodogram 235, 242
Splines 4, 39–41
    B-spline basis 40
    Function 42
    Logspline 42, 54
    Power basis 40
    Tensor-product 264, 290
    Thin plate 264
    See also Smoothing splines
Statistical efficiency 199
Stochastic system 244
Strongly mixing, see Mixing
Surface smoothing 264
Survivor function 163

Thin plate splines, see Splines
Thresholding
    Hard 30, 53
    Parameter 30
    Soft 31, 53
    Universal 39
    Variable thresholding 53
Thresholding model 173, 182
Tricube kernel, see Kernel

SUBJECT INDEX 341

Triweight kernel, *see* Kernel
Two-sample problem 49
Two-term interactions 284

Uniform kernel, *see* Kernel
Uniformly mixing, *see* Mixing conditions
Universal optimal weighting scheme 59, 74–76
  See also Kernel
Universal thresholding, *see* Thresholding
Updating algorithm 99–100

Variable bandwidth, *see* Bandwidth
Variable order 80–83
Variable thresholding, *see* Thresholding
Variance 8, 14, 60–83, 113–116, 145–146, 196–197, 301–302
Variance-stabilizing transformation 51
Visual shrinkage 35

Wavelets
  Coefficients 35
  Discrete transforms 35–39
  Extremal phase family 32, 33, 37
  Father wavelet 31
  Least asymmetric family 32, 33
  Level of primary resolution 34
  Mother wavelet 34
Whittle likelihood 234–235